Farmyard and car buried in fan from nearby gully, Peterson Minnesota, 12 Sept. 1939. See Figure 4.19, page 108.

# Historical Agriculture and Soil Erosion in the Upper Mississippi Valley Hill Country

# Historical Agriculture and Soil Erosion in the Upper Mississippi Valley Hill Country

Stanley W. Trimble

CRC Press
Taylor & Francis Group
Boca Raton London New York

CRC Press is an imprint of the
Taylor & Francis Group, an **informa** business

CRC Press
Taylor & Francis Group
6000 Broken Sound Parkway NW, Suite 300
Boca Raton, FL 33487-2742

First issued in paperback 2017

© 2013 by Taylor & Francis Group, LLC
CRC Press is an imprint of Taylor & Francis Group, an Informa business

No claim to original U.S. Government works

ISBN-13: 978-1-4665-5574-7 (hbk)
ISBN-13: 978-1-138-07161-2 (pbk)

This book contains information obtained from authentic and highly regarded sources. Reasonable efforts have been made to publish reliable data and information, but the author and publisher cannot assume responsibility for the validity of all materials or the consequences of their use. The authors and publishers have attempted to trace the copyright holders of all material reproduced in this publication and apologize to copyright holders if permission to publish in this form has not been obtained. If any copyright material has not been acknowledged please write and let us know so we may rectify in any future reprint.

Except as permitted under U.S. Copyright Law, no part of this book may be reprinted, reproduced, transmitted, or utilized in any form by any electronic, mechanical, or other means, now known or hereafter invented, including photocopying, microfilming, and recording, or in any information storage or retrieval system, without written permission from the publishers.

For permission to photocopy or use material electronically from this work, please access www.copyright.com (http://www.copyright.com/) or contact the Copyright Clearance Center, Inc. (CCC), 222 Rosewood Drive, Danvers, MA 01923, 978-750-8400. CCC is a not-for-profit organization that provides licenses and registration for a variety of users. For organizations that have been granted a photocopy license by the CCC, a separate system of payment has been arranged.

**Trademark Notice:** Product or corporate names may be trademarks or registered trademarks, and are used only for identification and explanation without intent to infringe.

**Visit the Taylor & Francis Web site at**
**http://www.taylorandfrancis.com**

**and the CRC Press Web site at**
**http://www.crcpress.com**

# *Dedication*

*First, I wish to dedicate this book to the late Dr. Stafford C.Happ and his colleagues, whose data and assistance were vital to this study. They were ahead of their time, and their contributions to soil conservation and to fluvial geomorphology are just now being more fully recognized.*

*Equally, I dedicate this book to my family, who so willingly supported me in the long endeavor described herein.*

*Thirdly, I dedicate it to the people of the Hill Country, who persevered in the adversity of a grim period, helped solve the problem, and continued to improve their countryside.*

This thought-provoking book demonstrates how processes of landscape transformation, usually illustrated only in simplified or idealized form, play out over time in real, complex landscapes. Trimble illustrates how a simple landscape disturbance, generated in this case by colonizing farmers, can spread an astonishing variety of altered hydrologic and sedimentation processes throughout a drainage basin. And the changes have spatial and temporal patterns forced on them by the distinctive topographic structure of drainage basins. Through painstaking field surveys, comparative photographic records, careful dating, a skilful eye for subtle landscape features, and a geographer's interdisciplinary understanding of landscape processes, the author leads the reader through the arc of an instructive and encouraging story. Settlers, whose unfamiliarity with new environmental conditions led initially to landscape destruction, impoverishment, and instability, eventually adapted their land use and settlement practices and, supported by government institutions, recovered and enriched the same working landscape.

For the natural scientist, the book illustrates how an initially simple alteration of land cover can set off a train of unanticipated changes to runoff, erosion, and sedimentation processes that spread through a landscape over decades, impoverishing downstream landscapes and communities. Distinct zones of the landscape respond differently and in sequence. The effects take a surprisingly long time to spread through a landscape because sediment moves short distances during storms and can persist for decades or centuries in relatively stable forms where it resists further movement because of consolidation, plant reinforcement, and low gradients. For the social scientist, the book raises questions of whether and how people can be alerted early to their potential for environmental disturbance, but also for learning and adopting restorative practices. Trimble's commitment to all aspects of this problem should energize both groups.

**Thomas Dunne**
*University of California - Santa Barbara*

# Contents

Figures ..................................................................................................................................... xi
Foreword .............................................................................................................................. xxix
Preface .................................................................................................................................. xxxi
Biography ........................................................................................................................... xxxix
Acknowledgments ................................................................................................................ xli
Introduction ........................................................................................................................ xliii

**Chapter 1   The physical region and primeval landscape** ............................................ 1
Physiography ........................................................................................................................... 5
Climate ...................................................................................................................................... 8
Vegetation and soils .............................................................................................................. 14
Streams ................................................................................................................................... 19
Conclusion .............................................................................................................................. 22

**Chapter 2   European settlement and changes of land use** ........................................ 23
Early settlement ..................................................................................................................... 24
Overview of historical agriculture in the region ............................................................... 26
Historical crops and other agricultural land use ............................................................... 33
    Crops .................................................................................................................................. 33
Grazing and animal husbandry ........................................................................................... 37
Land use management from the time of settlement to the 1930s ................................... 41
A revolution in agricultural land management ................................................................. 44
Conservation agencies .......................................................................................................... 47
The continuing soil conservation revolution ..................................................................... 51
A composite of erosive land use over the historical period ............................................ 52

**Chapter 3   The systematic effects of historical agriculture on the physical
            landscape** ........................................................................................................... 55
The role of modified hydrology (in particular, the role of rills and gullies) ................. 55
Increasing hydrologic change and soil erosion in the Hill Country ............................... 57
Effects of soil conservation on the physical landscape .................................................... 65
Sediment budgets over the historical period ..................................................................... 68
Zones of physical processes within stream basins of the Hill Country ......................... 70
    Tributaries ......................................................................................................................... 70
    Upper main valley (UMV) .............................................................................................. 74
    Lower main valley (LMV) .............................................................................................. 75
Stream erosion of high terraces and high banks ............................................................... 80

**Chapter 4 Upland gully erosion and its effects** ......................................................... 85
Hillside gullies and their fans .................................................................................................. 86
   The Appleby farm ................................................................................................................. 86
   The Zink farm ........................................................................................................................ 86
High terrace gullies and their fans .......................................................................................... 98
   The Buffalo and Black River terraces ................................................................................ 98
   Proksch Coulee ...................................................................................................................... 98
   Ratz gully ............................................................................................................................. 105
   The Peterson event ............................................................................................................. 105

**Chapter 5 The tributaries: Zone of early, complex changes of process and form** ..... 109
Villages ..................................................................................................................................... 109
   Fairwater, Minnesota ........................................................................................................ 109
   Similar villages ................................................................................................................... 112
Other locations and functions ............................................................................................... 113
   Farms and farmsteads ....................................................................................................... 113
   Roads, railroads, bridges, and communication lines ................................................... 118
   Mills, reservoirs, and water power .................................................................................. 122
   Fish habitat .......................................................................................................................... 125

**Chapter 6 The upper main valleys: Zone of later complex changes of process and form** ........................................................................................................... 129
Villages ..................................................................................................................................... 129
   Elba, Minnesota ................................................................................................................. 129
   Coon Valley, Wisconsin .................................................................................................... 131
   Freeburg, Minnesota ......................................................................................................... 134
Farms and farmsteads ............................................................................................................. 135

**Chapter 7 The lower main valleys: Zone of perennial sedimentation** ....................... 141
Villages and towns .................................................................................................................. 141
   Chaseburg, Wisconsin ...................................................................................................... 142
   Beaver and Whitewater Falls, Minnesota ...................................................................... 147
   Soldiers Grove and Gays Mills, Wisconsin ................................................................... 149
   Galena, Illinois and Potosi, Wisconsin .......................................................................... 151
   Elkport and Garber, Iowa ................................................................................................. 153
   Village Creek, Iowa ........................................................................................................... 157
   Ion, Iowa .............................................................................................................................. 157
   Rushford, Minnesota ......................................................................................................... 160
   Arcadia, Wisconsin ........................................................................................................... 161
Farms and farmsteads of the LMV ....................................................................................... 162
Roads, bridges, and communication lines .......................................................................... 169
Mills and reservoirs ................................................................................................................ 175

**Chapter 8 The great flood of August 2007 and its implications** ................................. 177
The storm .................................................................................................................................. 177
Mass movements ..................................................................................................................... 179
Upland slopes .......................................................................................................................... 181
Gullies ....................................................................................................................................... 182
Tributaries ................................................................................................................................ 186

Upper main valley ........................................................................................... 188
Lower main valley .......................................................................................... 188
Sediment yield or efflux ................................................................................ 195
Conclusions ..................................................................................................... 197

Conclusions ..................................................................................................... 207

References ....................................................................................................... 211

Glossary ........................................................................................................... 221

Appendix: Unit Conversion Factors ........................................................... 233

Index ................................................................................................................. 237

# Figures

**Figure 0.1** The Upper Mississippi River Hill Country. The delineations shown were used because of previous historical work done on the region. This is also quite similar to the delineation of the region into a Major Land Resource Area (MLRA) by the USDA-NRCS (Argabright et al., 1996). Termed by the USDA as the Northern Mississippi Valley Loess Hills, it would exclude most of Dunn County, Wisconsin, and Carroll County, Illinois, but include Clinton County, Iowa. ...................................... xxxiii

**Figure 0.2** The Great Flood of 1907 in lower Chaseburg, Wisconsin, the highest on record up to that time. The photo was made looking east on Mill Street near the front of the present Hideaway Tavern. Since that time, the floodplain has aggraded to a level about 4 or 5 ft (1.2–1.4 m) above the flood level of 1907. The remains of lower Chaseburg were purposefully demolished in 2009 because of the continued flooding. ........................................................................................................ xxxiv

**Figure 0.3** Natural and human effects in geomorphology. (From A. Ward and S. Trimble, *Environmental Hydrology*, 2nd ed., CRC, Boca Raton, 2004. With permission.) ........................................................................................................... xliii

**Figure 1.1** Top: A sketch of the northeastern Iowa landscape by the famed naturalist David Dale Owen, 1852. A similar sketch appeared in Owen (1847) that had slightly more detail but the Native Americans were not included. Bottom: A sketch of the Kickapoo River Valley close to the junction with the Wisconsin River. Note the settler's house and farmstead with rail fencing (Owen, 1847). In all his sketches, Owens portrays the stream bottoms as mostly treeless, but this might have been for the sake of clarity. .................................................................................... 2

**Figure 1.2** Geologic cross section of Houston County, MN, drawn in a west to east direction. (USDA, NRCS, *Soil Survey of Houston County, Minnesota*, 1984.) ............................ 4

**Figure 1.3** Local relief in the Hill Country. (Modified from Trewartha, G. and Smith, G.-H, *Annals, Assoc. Am. Geogs*, 31, 26, 1941.) ................................................... 6

**Figure 1.4** Physiographic diagram of the Hill Country. (Modified from Trewartha, G. and Smith, G.-H., *Annals, Assoc. Am. Geogs*, 31. foldout map 1941.) Refer to Figure 0.1 for more detailed place names. ................................................................ 7

**Figure 1.5** Frequency-magnitude relationships for precipitation events, Hill Country or Driftless Area (humid continental) and Puget Sound and United Kingdom (marine west coast). (From Trimble, S., in Bennett, S., and Simon, A. (eds.), *Riparian Vegetation and Fluvial Geomorphology*. American Geophysical Union,

Washington, 2004.) Note that the 24-hour, 100-year event is about 2 times greater in the Hill Country while the 1-hour, 100-year event is 3 times greater. ...................................... 9

**Figure 1.6** Infiltration capacities of soils versus average rainfall rates, United States and United Kingdom. The difference is excess rainfall that flows off the surface as overland flow. Note that it is much greater for the United States than for the United Kingdom. (Modified from Trimble, S., in Brunn, S. (ed.), *Engineering Earth*. Springer, Berlin.) ........................................................................................................................................... 11

**Figure 1.7** Historical precipitation trends in the Hill Country (from Trimble, S., and Lund, S., *US Geol. Surv. Prof. Paper*. 1243, 1982). The data show a nadir of precipitation in the early 20th century and increasing wetness and storms since 1930. See also the annual flood series for the Upper Mississippi River for the period c. 1850–1975, which shows a similar pattern (Knox et al., 1975). .............................................. 12

**Figure 1.8** Sketch by David Dale Owen (1852) showing what were later called "goat prairies," where the sunnier southern and western hill slopes were often bare of trees. While he terms these "northern and southern," they are more likely northern (forested) and western (treeless) in this sketch. ............................................................................ 16

**Figure 2.1** The US Rectangular Survey System (redrawn from Johnson, H., *Order upon the Land*, Oxford, 1976). In the Hill Country, tracts as small as 40 acres (16 ha) could be purchased, and tracts were not always contiguous. See page 24 for addition to caption. ............................................................................................................................................ 24

**Figure 2.2** Sketch of the Lead Region (from *Harper's New Monthly Magazine*, 1853). Note the barren, almost lunar-like quality of the landscape and the scattered mine pits, or "diggings." .............................................................................................................................. 25

**Figure 2.3** Regionalization of the Hill Country based on historical trends of agricultural land use. ......................................................................................................................... 26

**Figure 2.4** Improved land, 1860, 1880, 1900, and 1980 (US Census of Agriculture) ............. 30

**Figure 2.5** Agricultural land use, 1860–1982, Winneshiek, Co., IO (Region I), and Buffalo Co., WI (Region II). US Census of Agriculture. ........................................................... 32

**Figure 2.6** The decline of wheat cultivation. Proportion of land in wheat, 1880 and 1900 (US Census of Agriculture). .......................................................................................................... 34

**Figure 2.7** Corn and soybean cultivation. Proportion of land in corn and soybeans, 1900, 1940, 1960, and 1980 (US Census of Agriculture). ............................................................. 36

**Figure 2.8** An eroded hillside from overgrazing, probably 1950s. (Photo credit: R. Sartz.) ...................................................................................................................................................... 38

**Figure 2.9** Density of grazing animals (cows, sheep, horses), 1880, 1900, 1920, 1940 (US Census of Agriculture). ...................................................................................................................... 40

**Figure 2.10** Grazing animals, 1850–1975, and grazed forest, 1925–1970, Vernon Co., WI (Trimble, S. and Lund, S., *US Geol. Surv. Prof. Paper* 1234, 1982). .......................................... 41

**Figure 2.11** Grazed woodland, percentage of county area, 1925 and 1969 (US Census of Agriculture). ........................................................................................................................... 42

*Figures*　　　　　　　　　　　　　　　　　　　　　　　　　　　　　　　　　　　　*xiii*

**Figure 2.12** Air photos of Coon Creek landscape, 1934 and 1967, just north of Coon Valley (SE1/4,T15N, R5W, Vernon Co.). 1934: Note rectangular fields and gully systems extending into upland cultivated fields. 1967: Note contoured and strip cropped fields with no rills or gullies. (From Trimble, S. and Lund, S., *US Geol. Surv. Prof. Paper* 1234, 1982.) ..................................................................................................................43

**Figure 2.13** Changes of land use and management in 10 sample basins, Coon Creek, WI, 1934 and 1974. (From Trimble, S. and Lund, S., *US Geol. Surv. Prof Paper* 1234, 1982.) ..................................................................................................................46

**Figure 2.14** Partial farm plan for John Haugen for his 160-acre farm south of Coon Valley, WI. 1939. (SE1/2 T14 N R5W, Vernon Co.) (Courtesy of Ernest and Joseph Haugen, Coon Valley, WI, Photo credit: Nadine Kleinhenz, July 2006.) .......................48

**Figure 2.15** Spread of contour-strip farming from the Coon Creek Conservation Demonstration Area, WI., 1939–1967. (From Trimble, S. and Lund, S., *US Geol. Surv. Prof. Paper* 1234, 1982.) ..................................................................................................49

**Figure 2.16** CCC project to stabilize high Pleistocene terrace undercut by stream. Exact location unknown but probably Trempealeau Co., WI., late 1930s. SCS Photo. .........50

**Figure 2.17** Erosive land use, Hill Country, 1880–1980. See text for definitions and explanation. ..............................................................................................................................53

**Figure 3.1** Schematic of response of hillsides to agriculturally induced overland flow from upland areas. Top: Early; with increasing crops on the upland and deteriorating soil hydrologic conditions, runoff increasingly becomes surface (Horton Model) rather than subsurface (Hewlett Model), but hillside woodlands are able to reinfiltrate much of this. Bottom: With increasing overland flow from above and decreased infiltration on grazed hillsides, gullies form and create fans at the toe of the slope. ....................................................................................................................56

**Figure 3.2** Decrease of gullies in the upper Coon Creek basin as the result of improved land use and management, 1938–1978. This decrease of drainage density is strong evidence of decreased overland flow and, in turn, decreases the rapidity of runoff. (Redrawn from Fraczek, W., Unpublished MS thesis, University of Wisconsin, Madison, 1987.) See also Trimble and Lund, 1982, p. 6. ........................................58

**Figure 3.3** Estimated change of stream discharge from a small tributary of Coon Creek for a moderately large storm, 1853–1995. The vertical axis is discharge in cubic meters per second ($1\ m^3 = 35\ ft^3$). Calculations suggest the significance of (a) land use and management and (b) the formation of gullies for increasing storm flow peaks. The short dashed line suggests the role of response lag discussed in the text. (Redrawn from Trimble, S., *Geomorphology* 108, 8–23, 2009.) .........................................59

**Figure 3.4** Alarming erosion and sedimentation in the Hill Country, 1929–1931. Top: Severe gullying in the Black River terrace, 3 mi (5 km) east of North Bend, Jackson Co., WI, June 15, 1929. See also book cover. Bottom: a 40-acre (16 ha) fan, deposited mainly 1923–1929, covers formerly good hay land on the Buffalo River floodplain, 10/24/31. (Photos credit: C.G. Bates.)..........................................................62

**Figure 3.5** Early private erosion control. Sheet metal flume used to convey water around the head of gullies, Proksch Coulee gully, Vernon Co., WI (see Chapter 4). (Photo credit: Nadine Kleinhenz.) ........................................................................................63

**Figure 3.6** Generalized upland soil erosion from water, 1933, Hill Country. Compiled from state maps of WI, IO, MN, and IL. USDA-SES *Reconnaissance Erosion Survey*, 1933, 1:500,000. ............................................................................................................. 64

**Figure 3.7** Nonlinear and hysteretic relationship of erosion and sedimentation to erosive land use, 1853–1975, Coon Creek. Sedimentation rates were measured but erosion was modeled with the Universal Soil Loss Equation. The lags are largely functions of (a) Soil condition and (b) connectivity between uplands and bottoms by rills and gullies. See text for more explanation. (From Trimble, S. and Lund, S., *US Geol. Surv., Prof. Paper* 1234, 1982.) ...................................................................................... 66

**Figure 3.8** Sediment budgets for Coon Creek, 1853–1993, showing the complexity of sources, sinks, fluxes, and efflux over time. Numbers are annual averages for the periods in tonnes/yr (1 tonne = 1.1 short tons). All values are direct measurements except net upland sheet and rill erosion, which is a tare: the sum of all sinks and the efflux minus the measured sources. The lower main valley and tributaries are sediment sinks, whereas the upper main valley has been a sediment source. Note that sediment yield to the Mississippi River has held relatively constant over historical time. (From Trimble, S, *Science* 285, 1244–46, 1999. With permission.) ................... 69

**Figure 3.9** Differential stream and valley sediment budgets, Coon Creek, WI, 1853–1993 (1 Mg = 1 tonne = 1.1 short tons). (Redrawn from Trimble, S. *Geomorphology* 108, 8–23, 2009.) ................................................................................................................................. 70

**Figure 3.10** (A) Schematic model of changes of historic stream and valley morphology for Coon Creek and other Driftless Area tributaries, 1860 to 1974 (from page 16 of a mimeographed pamphlet by S.W. Trimble for a field trip to the Driftless Area, April 1975, sponsored by the Association of American Geographers and led by G. Dury, J.C. Knox, W.C. Johnson and S.W. Trimble [Trimble, 1975b]). Lateral migration of the stream is not shown in this model. (B) Changes of stream power and the transformation of stream and valley morphology. This model assumes a constant discharge for each stage. With the small stream channel of stage 1, floods spread out over the floodplain, keeping depth, velocity, and stream power low. With accretion of the floodplain and stream banks with historic sediment in stage 2 (c. 1900), greater flows were restricted to the channel, thus increasing depth, velocity, and stream power so that the channel erosion shown in the 1900 stage (left) must have been very rapid. In stages 3 and 4, the channel erodes laterally, so that floods are spread, with decreases of depth, velocity, and stream power. By the latter stage, fine sediment covered the old gravel meander plains and new floodplains are formed as shown to the left (from presentation to the Association of American Geographers, San Diego, CA., April 20, 1992. Ron Shreve of UCLA made important suggestions for preparation of this diagram in 1991, from Ward, A. and Trimble, S., *Environmental Hydrology*, CRC, Boca Raton, 2004. .................................................................... 72

**Figure 3.11** Changes of bankfull discharge capacity (BFQ) for tributaries, 1930s vs. 1970s, Coon Creek (1 $m^3$ = 35 $ft^3$, 1 $km^2$ = 0.37 $mi^2$). Alluvial reaches downstream of tributaries (upper and lower main valley) were not calculated for the 1970s, but this analysis suggests a different set of channel forming conditions from the tributaries, at least for the 1930s. The overall relationship between BFQ and drainage area (DA) in the 1930s is a third-degree polynomial curve (Gatwood, 1989). For selected streams elsewhere in the Hill Country, Knox and others found a log-linear relationship for bank widths and DA (e.g., Knox, 1987). The relationship of stream size in the Hill Country, however measured, is a function of many factors, including land use and basin morphology, particularly drainage network. ............................................. 73

**Figure 3.12** Evolution of tributary channels as explained in the text and in Figures 3.10 and 3.11. North Branch, Whitewater River, MN, 1 mi (1.6 km) west of Elba, MN, looking upstream (west) (NE1/4, Sec9, T107NR10W, Winona Co.). For reference, note barn, bridge, and road intersection. Top: c. 1905, Note narrow, apparently deep stream lined with trees and dense riparian vegetation. Middle: By 1940, by lateral erosion, the stream has widened 2 to 5 times in response to increased storm flow and the abundant supply of bedload from hillside gullies and the eroded stream itself. Indeed, a hillside gully and rock fan is barely visible in the top right quadrant. 1976: The stream has narrowed again to similar dimensions as seen in 1905 with dense riparian vegetation. Top 2nd page: 1905, close-up photo of stream perhaps near the barn shown in photo. Bottom, 2nd page: 1940, close-up of the reach upstream of the barn looking east (downstream). (Photo credit: 1905 and 1940 photos by Soil Conservation Service [SCS].) ................................................................. 74

**Figure 3.13** Lewiston Fork of Garvin Branch, 2 mi (3 km) SW of Stockton, MN, on US Hwy 14 looking upstream. Top: 11/2/39. This is the typical "gravel road" tributary of that period, which Stafford Happ described here as a "bouldery torrent plain formed by bank erosion." Middle: July 1977; rocks have been covered with silt, and the floodplain is a fine pasture. Note the eroded silt banks where cows have trampled them. Bottom: July 2009; With cessation of grazing, floodplain has now grown up with woody plants. ......................................................................................................... 78

**Figure 3.14** Downstream end of the Upper Main Valley of Coon Creek, as defined by the process of a stream eroding a high bank and depositing a lower bank on the other side of the stream adjusted to a lower flood regime. Top: 1940; looking downstream from bridge on Lietke Lane, just off WI Hwy 162, stream is fairly straight at this point. (Photo credit: S. C. Happ.) Bottom: Late 1980s. Note that the stream has cut away the bank to the left foreground and is depositing a slightly lower bank in the right foreground. The same pattern exists for two more meander loops in the distance (looking downstream from bridge over Coon Creek, NW1/4, Sec 24,T14N, R6W). ........................................................................................................................... 80

**Figure 3.15** Deposition rates at selected Lower Main Valley sites in Coon Creek, 1853–1977. These are derived from cross-sectional reconstructions of deposition rates based on surveys, excavations of soils, bridges, roadways, and other dating methods. (From Trimble, S. and Lund, S., *US Geol. Surv. Prof. Paper* 1234, 1982). ................ 82

**Figure 3.16** Longitudinal profile of Coon Creek stream banks showing depth of historical sediment accretion for the entire main valley, 1850 and 1975. Dashed lines in the longitudinal profile are straight lines for reference. Profile 16 is 2 mi (3 km) upstream from Coon Valley and Range 36 is less than a mile (1.6 km) from the Mississippi River. Note that the longitudinal profile in 1850 was concave upward, typical for a humid area stream in equilibrium. By 1975, however, the profile has become almost straight, indicating a stream overloaded with sediment. Indeed, from Range 32 to 35, where erosion from high Pleistocene terraces in Wing Hollow was especially severe, the profile became convex upward. Note also that the raising of the base level of about 8 ft (2.5 m) by navigation pool 8 in 1938 seems to have had negligible effects on sediment depths in the lower 5 mi of the stream. In fact, the accretion rate near the mouth 1938–75 was less than in mid-valley. ...................................... 83

**Figure 3.17** Destabilization of stream reach by collapse of undercut high terrace, showing natural and human factors. This reach may be observed by looking upstream (south) from WI. State Highway 162 bridge over Coon Creek (see Fig. 8.14). ...... 84

**Figure 4.1** Stabilization of hillside and gullies, North Fork Whitewater River, looking north just east of Fairwater, MN (SE1/4, Sec 5, T107N, R10W, pan of two photos made 10/23/40). Top: 1940, note treeless "goat prairie" and two hillside gullies with fans at base of slope. Bottom: July 2009, note the transformation from "goat prairie" to mostly deciduous forest. Gullies and fans are relatively inactive. Compare to Chapter 5, Figure 5.1 of the same slopes. (See also Trimble, 2009a, p. 18.) ...... 87

**Figure 4.2** Gully heads (overfalls) threatening economic activity in the late 1930s. Top: A threatened farm possibly in Trempealeau Co. WI. Bottom: A threatened roadway in Wabasha Co, MN, 2 mi (3 km) SW of Weaver. ........................................................ 88

**Figure 4.3** The Appleby farm, Whitewater Valley looking north (NW1/4, Sec 10, T108N, R10W, Winona County, MN). Top: The farm seen across the floodplain, grown too wet for grazing. Thus, Appleby was forced to turn his animals loose on the hillside behind the house, and gullies are visible. Bottom: Close-up of gullied hillside behind house with partially buried stone wall visible. Corner of house visible to left. Both photos 8/27/37. A present remake of the top photo would show the lake formed in the foreground as the main channel of the Whitewater River aggraded to a level higher than the distal floodplain here (see Chapter 7). ......................... 89

**Figure 4.4** Changes at the Zink farm about 1 mi (1.6 km) southeast of Stoddard, WI, c. 1900–1940 (NE1/4, Sec33, T14N, R7W). In 1900, Coon Creek still ran on the east side of the valley close to the Zink farmstead. By 1924, the small tributary from the southeast had formed an alluvial fan that has pushed the creek across to the west side of the valley. This process clearly demonstrates the rapid onset of disastrous erosion in the early 20th century. ............................................................................................ 91

**Figure 4.5** The Zink farm, Coon Creek basin. Top: Looking southeast upstream from bridge on County Road O. Note dike in middle background to protect farm from flooding (arrow). The former gully and fan, now a stream, flows toward viewer on the bridge. The side channel flows in from the left (north) between the barn and the house. (Photo credit: Nadine Kleinhenz, 2006.) Bottom: Looking upstream (northeast). Mostly dormant rock fan between barn and house reshaped into flume with dikes to either side. The fan formed in the early 20th century and carried water and rock with most rainfall events in the early 20th century. Material from the fan was used to construct the dikes. Flume is shown after 4 in. (10 cm) rainfall event, c. 1990. ........................ 92

**Figure 4.6** The Zink farm. Standing on fan looking east (upstream) at bridge on County Road O. Top: 1975. Wingwalls of buried earlier bridge are still visible. An excavation 5 ft (1.5 m) deep here in 1975 and 1976 did not reach the base of the wingwalls nor did it fill with water, indicating the lack of permanent flow and significant underflow at that time. Bottom: 1999. The wingwalls have been buried in the aggrading fan. Most notable to observe is the permanent streamflow, which began in 1988. For scale, the person to the right is 77 in. (193 cm) tall. .................................. 93

Figures                                                                                                                                   xvii

**Figure 4.7** Examples of gully fans covering roads and railroads. Top: Sand fan over MN highway 74, 2 mi (3 km) west of Weaver, 9/7/38, looking south. (Photo credit: Vince McKelvey). Farm in background was already abandoned. Bottom: sand fan over railroad, location unknown. (Photo credit: C.G.Bates, 8/2/30.) ........................................ 95

**Figure 4.8** Top: cars halted and partially buried by sand fan on US Highway 61 about 2.5 mi (4 km) north of Weaver, Wabasha Co., MN, 6/7/40. The fan covered 0.2 acre (0.08 ha), was 7 ft (2.1 m) thick at the apex, and contained 4.6 acre-feet (5,700 m3). Bottom: The gully that created the fan, 6/9/40. Note the almost vertical walls. (Photo credit Soil Conservation Service [SCS].) .......................................................... 96

**Figure 4.9** Examples of elaborate drop structures to control hillside gullies creating rock fans on transportation routes. Top: Near Genoa, WI, to protect railroad, 6/27/30. Bottom: Between Wabasha and Lake City, MN, 1932. (Photos credit: C.G. Bates.) .................................................................................................................................. 97

**Figure 4.10** Untrenched drainageways in Pleistocene terraces showing how such valleys appeared before trenching. Such untrenched valleys are now rare. Historical sediment usually covers the valley floor to a depth of several feet. Top: Legue Coulee, 6 mi (10 km) NE of Ettrick, Trempealeau Co., WI. (Photo credit: Stafford Happ, 9/26/40.) Bottom: Small valley leading into Coon Creek. (Photo made from Cedar Valley Drive, looking downstream (south) about 1 mi [1.6 km] southeast of Proksch Coulee gully [SE1/4, Sec 26, T14N, R 7W], c. 1977.) ....................................... 99

**Figure 4.11** Explanatory diagram of Proksch Coulee Trench (Gully) development based on intensive field research and precise surveys done 1974–1977. ............................. 100

**Figure 4.12** Schematic explanation of the trenching of Proksch Coulee. Concentrating the flow and increasing the depth increased the power of the stream to erode (see Chapter 3, Figure 3.10). This diagram may apply to many if not most of the trenched Pleistocene terraces of the Hill Country. ............................................................ 101

**Figure 4.13** Top: Elmer Miller in 1977 points to a portion of the ditch his grandfather dug in 1905 to drain the bottom of the channel. Subsequent trenching bypassed this reach of the ditch and cut around the left side (to the viewer here looking upstream) leaving this part isolated and elevated. Bottom, 1977: Looking downstream at the terrace floor remnant on which Miller stands in top photo. Dr. Stafford Happ stands in the gully remnant. For some reason, perhaps a blocked channel, the gully cut around to the right and bypassed this site. The two white flags on slope to right (arrow) mark the buried dark soil of the original (pre-trench) valley bottom, covered with about 6 ft (2 m) of historical sediment. The gully floor in the foreground to the right is now covered with about 6 ft (2 m) of alluviation adjusted to the aggraded floodplain of Coon Creek 1000 ft (300 m) downstream. ............................................................ 102

**Figure 4.14** Proksch Coulee Trench. Top: air photo, Feb 1934. North is to the left and the photo covers about 4000 ft of the trench. Bottom: photo made in the Clement farmyard (arrow in top photo) looking up the valley, late 1930s. Note the bare, almost lunar-like, grazed landscape and the recent, vertical quality of the gully walls in the background. ............................................................................................ 103

**Figure 4.15** Alluvial fan from Proksch Trench on Coon Creek floodplain looking northeast up the channel. (Photo credit: S.C. Happ, 11/11/38 (NW1/4, Sec27, T14N, R7W, Vernon Co.) The extremely straight channel appears to have been dug and is completely occluded (filled up) in the foreground. Note sandy wash on floodplain and obviously wet conditions. However, this huge influx of sand has created an elevated fan out across the floodplain quite visible on the most recent USGS *Stoddard* sheet (1:24,000, 1983). ............................................................................................... 104

**Figure 4.16** Ratz gully. Top left: 11/27/39, looking south (between Sec 3 and 4, T108N, R10W, Winona County, MN). Line fence was rebuilt 1933–34, but just 6 years later was undercut and suspended about 10 ft (3 m) above the valley floor. Distance to far bank was about 50–60 ft (16–19 m). Top right: repeat of photo, made July 1979. Distance to far bank was about 275 ft (84 m). Bottom: c. 1940, looking downstream (east) toward remains of Ratz bridge, constructed in the 1930s after the gully became too deep to cross. Gully has cut around right side, isolating the Ratz farm to the left (north). ................................................................................................................................ 105

**Figure 4.17** Cross-sectional reconstruction, growth, and change of Ratz gully based on photos, surveys, and soils. (Redrawn from Trimble, S., *Catena* 32, 1998, 283–304.) ....... 107

**Figure 4.18** Mouth of Glendale hollow and fan of Ratz gully on Whitewater River floodplain looking northwest, 1939 (NW1/4, Sec 10, T108N, R10W). Note sandy deposits in the foreground and partially buried fence in middle background. In far left background is the face of the Pleistocene terrace with a new gully actively incising. ........................................................................................................................................... 107

**Figure 4.19** Fan from gully near Peterson, MN, 9/12/38. Top: standing on Pleistocene terrace face looking south toward the village of Peterson on the Root River. Methodist church in village in left background for reference. Gully to rear of viewer. Bottom: detail of car and house. (SCS Photos.) ......................................................... 108

**Figure 5.1** Fairwater, MN. Top: c. 1905 photo made from bluff west of town looking northeast (downstream). Note tree-lined stream (North fork, Whitewater River) and millrace (arrow). Bottom: 1940. Stream has widened by 2–5 times. Bridge in left foreground has been destroyed and a new one placed about 500 ft (150 m) upstream. Millrace is filled with sediment. Note bare "goat prairies" in background. 1975: stream channel is again lined with trees and is no longer visible from this vantage point. Goat prairies are becoming revegetated. ...................................................... 110

**Figure 5.2** Fairwater, MN. Photos made from bluff on N. side of town (visible in Figure 5.1) looking upstream (west). c. 1905. Stream and millrace (arrow) lined with trees. Note small sand fan on right from hillside gully from valley to the right (north). Bottom: October 22, 1940. Stream has widened by 2–5 times, millrace (arrow) has filled with sediment, and gully from the right (north) has enlarged and deepened, and has created boulder fan into the creek. The coarse bedload from this and other sources in this reach probably have played an important role in the severe bank erosion and widening of the stream. 1975: channel is again narrow, lined with trees and all is stable. 2001: old gully from the north now stable (arrow on 1940 and 1975). Note trees growing on old gully bottom. This is presently typical of most old gullies in the Hill Country. ................................................................................................ 112

**Figure 5.3** Fairwater, MN. looking downstream (east) Top: c. 1905. Pleistocene terrace in fore and middle background. Stream in middle background is hidden by riparian trees. Goat prairies on hillsides in background to left on south side of hills. 1940: Stream has widened and eroded deeply into terrace. As the result of floods and the coarse sediment, the stream is now braided. Goat prairies in background now have more trees (see Figure 4.1 for a close-up of the same hillsides). The same farmhouse is marked by an arrow in all photos for reference. 1975: stream had cut further into very high Pleistocene terrace. The arrow to the right points to the same spot on all three photos and indicates the extent of lateral bank erosion by 1940 and 1975. Thus, it can be seen that the stream had cut several hundred feet into the high terrace. Note that the high cut bank of the terrace in 1975 is partially obscured by a tree branch. See Figure 5.4. ................................................................................................... 115

**Figure 5.4** Fairwater, destruction of the high Pleistocene terrace, looking north/northwest (upstream) from hill on southeast side of the village. c. 1905: stream is narrow and tree-lined. House (arrow) is the same as marked in 5.3. Large, dark building just upstream from house is the mill. 1940: same view but taken farther back. Using the house as a marker, it will be seen that the stream had eroded laterally perhaps 150–200 ft to the left (west) and a comparable distance toward the viewer (south, see Figure 5.3). Not only is the high terrace eroding from stream action but is also eroding from agricultural runoff as evidenced by the lateral gullies. As a result of flooding and coarse sediment, the stream is now braided. The visible bridge is a replacement for a bridge similar to the one seen in Figure 5.5. 2001: view now shifted to north/northeast because old camera position was eroded away. For reference, note different perspective on house. Stream has cut several hundred additional feet to the west and south and is now cutting deeply into the high terrace in the immediate foreground and on which the viewer is standing. A comparison of this area c. 1940 (Brown and Nygard, 1940) with the 1972 USGS topographic sheet Elba (1:24,000) suggests that about 4 acres (1.6 ha) of terrace were eroded away in those 32 years. Conservatively, assuming an average net removal depth of 10 ft (3 m), about 40 acre-feet (49, 400 m$^3$) of material were lost here with much being moved to the main valley of the Whitewater River. Note that goat prairie in background now covered with cedar (same as seen in Figure 5.1). .................... 117

**Figure 5.5** Fairwater, bridge crossing visible in 5.1 and 5.2 looking upstream. c. 1905: the structure is probably slightly later than the one visible in 5.1 and 5.2 but is at the same crossing. Note stability of stream and banks. 1940: stream widened by about three times here so that right pier is near the center of the widened stream. The left pier is still visible (2006) in the village, but the recovering (narrowing) stream has buried the right pier in the new bank. ................................................................. 119

**Figure 5.6** Fairwater, road from Elba to Fairwater about 1 mi east of Fairwater, 9/38, looking west. Several hundred feet of roadway eroded away. ............................................. 120

**Figure 5.7** Newton, WI, on the north branch of Bad Axe River (Sec 23, T13N, R6W, Vernon Co.), 1896 (from Hood, E., *Plat Book of Vernon County*, 1896). The mill and dam are now buried and only one original building remains. The river flows westward (right to left). ........................................................................................................... 121

**Figure 5.8** Millville, Clayton Co., IA, on Little Turkey River near the confluence with Turkey River. In 1886, it was already an established village built around a millsite. In 1989 (top), only these two buildings were left, huddled inside a flood dike. The highway in the background is on the far dike. The nearest, a once-beautiful Federal-style building suggests the one time wealth of Millville and how long it has existed. ...................................................................................................................... 122

**Figure 5.9** Hass house, Deadman's Creek (NE 1/4 Sec 19, T10N, R6W, Vernon County, WI). 1915: Creek in the foreground flowing to the left (east). House sits behind flood dike made of boulders brought down by creek. 1977: stable, grassy floodplain, stream out of view in foreground. House foundation is buried in floodplain in middle ground. ................................................................................................ 123

**Figure 5.10** Bridges over tributaries destroyed by flooding and sediment. Top: Undermined bridge in the Coon Creek basin, exact location unknown, 1938 (from McKelvey, V.E. 1939. Stream and valley sedimentation in the Coon Creek drainage basin, Wisconsin: M.A. thesis, University of Wisconsin-Madison.. Bottom: Abutment of destroyed bridge over Trout Brook near Dumfries, MN. Zumbro River Valley, 1939 (Sec 10, T110N, R11W, Wabasha County) (SCS Photos). ..................... 124

**Figure 5.11** Oium Mill dam site, Timber Coulee, Coon Creek basin (N1/2, Sec8, T14N, R4W, Vernon County, WI), looking upstream from bridge on Lars Hill Road. 1919. Note typical "gravel road" appearance of stream and the almost lunar, bare quality of the heavily grazed landscape behind it. The expendable rock dam, rebuilt after every flood, diverted water into a race (to the right, not visible) that led to a reservoir excavated into a terrace downstream. This was above most floods thus saving it from destruction. The mill was about ¼ mi (400 m) downstream at the edge of the terrace (Files of Wisconsin Railroad Commission). ............................. 125

**Figure 5.12** Lake Como, Hokah, MN, looking south 1926: lake full and 20 ft. deep. Note barriers in water (left middle ground) to prevent swimmers from venturing too far into deep lake. 1936: lake completely filled with sediment. Old lake bed now used by villages for recreation and gardens. (Photo credit Soil Conservation Service [SCS].) ................................................................................................................................ 126

**Figure 5.13** Lanesboro Lake, Lanesboro, MN, on the upper Root River. C. 1890: large expanse of water. 1938: completely filled with sediment. Arrow marks top of dam in both photos. (Photo credit Soil Conservation Service [SCS].) ................................. 127

**Figure 6.1** Nick Siebenaler's house just upstream from Elba, MN, on a terrace above the Whitewater River. Flood 9/8/38. Top: Floodwater was up halfway on screen door. The men are pumping out the basement. Bottom: Siebenaler's chicken house was moved off the foundation by the flood. Note sediment deposits in foreground. (Photo credit Soil Conservation Service [SCS].) ....................................................... 130

**Figure 6.2** Bank erosion and channel enlargement of Whitewater River at Elba, MN, 1940–1990. In 1940, the river flowed through a flume-like reach with high banks (the aggraded historical floodplain) to either side. Between 1940 and 1990, the floodplain aggraded an additional 1 ft (30 cm) or so and the channel was later greatly enlarged by bank erosion. This process, creating a greater floodway but sending copious amounts of sediment downstream, has been typical of the Upper Main Valley Zone since soil conservation measures took full effect. (From Ward, A., and Trimble, S., *Environmental Hydrology*, CRC, Boca Raton, 2004. With permission.) ..... 131

**Figure 6.3** Coon Valley, Coon Creek, looking upstream Vernon Co., WI. Top, 1912: photo by Lloyd Thrune. Streambanks are about 6 ft (2 m) high, and terrace face in background with cheese factory (white building) is about 10 ft (3 m) high. 1976: bank to right is about 12 to 13 ft (4 m) high and is eroding, but new bank (new floodplain) accreting to left is only about 6 ft (2 m) high, diagnostic of the

*Figures* xxi

newer, milder flood regime of the stream. Mid-1980s: photo made farther back and standing about 12 ft (4 m) higher. From this view, one can more clearly see the meandering stream with the retreating high bank to the right and the advancing low bank to the left. In the background, it will also be seen that the floodplain has aggraded to the level of the old cheese factory. As at Elba, a large floodway was being created here that would make Coon Valley safer from flooding, but that process has been artificially curtailed at Coon Valley (Figure 6.4)....................................... 133

**Figure 6.4** Aggrading "new" floodplain in Coon Valley Village Park. Floodplains were cut down to facilitate handicapped access ramps (arrow) to Coon Creek (seen on far right). Subsequently, the stream has attempted to rebuild its floodplains to the height dictated by the present flood regime, that height being about 6 ft (2 m). The elevation of the lowered floodplain is shown by the paved walk on which the person stands, and the recent deposits are up to 2 ft (60 cm) thick, However, the formerly retreating high banks have been stabilized with rock so that the previously expanding floodway is now stable and is decreasing in size. Photo made July 2008.............................................................................................................................. 135

**Figure 6.5** Freeburg, Crooked Creek, Houston Co., MN. Crooked Creek Township Hall. Top: Turn of century, photo made from road. Viewer's eyes are at a level about 4 ft (1.2 m) below the bottom of wooden siding on the building. Bottom: Again standing on the road, 1980s. Viewer's eyes are level with bottom of first board below windows. This suggests that road is now about 8 ft (2.5 m) higher. ..................................... 137

**Figure 6.6** Daffinrud house, near confluence of Spring Creek and Coon Creek about 2 mi west of Coon Valley (SE1/4, SE1/4, Sec 5, T14N, R5W, Vernon Co.). House was originally built on a terrace about 7 to 8 ft (2.5 m) above the floodplain. Top: Wedding at Daffinrud house in 1890. The crowd in front of the house makes it difficult to perform oblique photogrammetry, but the eye of the viewer appears to be below the base of the house. Bottom: Same view, 2010. The eye level is now slightly above the bottom of the windows. Note that there is a dike about 2½ ft (75 cm) high around the house on which Carl Daffinrud stands and which hides the bottom of the house. The floodplain is now almost level with the terrace on which the house sits............................................................................................................................... 138

**Figure 6.7** An active agricultural reach of Upper Main Valley showing typical UMV processes and morphology, virtually untouched and unhindered by artificial modification. In 1940, the stream channel was like a flume with high banks on either side. Since then, it has cut away the high banks and left low ones as seen in Range C (inset) and the process continues. Just this short reach has furnished perhaps 20–30 acre-feet (25,000–38,000 m$^3$) of sediment to downstream locations since c. 1940. This morphological transformation has both benefits and drawbacks to the landowner. The new, low floodplains, being sandy, are less fertile than the old floodplain, and they flood much more frequently. However, the increased size of the floodway between the high banks means that flooding on the historical floodplain is less probable. The old floodplain is now a new fluvial terrace and is now more valuable. Stream flows from left to right (westward). ............................................................ 139

**Figure 6.8** Diagram showing how UMV reaches have developed since c. 1940 and why they might not be a long-term net sediment source: the expanding "new" floodplains may eventually accumulate more sediment than the high-cut banks lose. The inset shows how these processes move downstream. (From Ward, A., and Trimble, S., *Environmental Hydrology*, CRC, Boca Raton, 2004. With permission.) Stream flows right to left (northwestward). ................................................................................ 140

**Figure 7.1** Chaseburg, WI, on Coon Creek in 1896 (Hood, E. *Plat book of Vernon County*, 1896). Note that all buildings except two are in lower Chaseburg (on or near the island). By 2010, only one building was left in the lower part of the village. Stream flows right to left (northwestward). ................................................................................ 143

**Figure 7.2** Cross section of Chaseburg from field research and excavations, 1976–1977. The limestone foundation to the left is that of the Martiny house, built 1903. A new surface survey in 1993 indicated about 3–6 in. (7–15 cm) of additional sediment accumulation since 1977. ................................................................................................................ 144

**Figure 7.3** Chaseburg, refer to Figures 7.1 and 7.2. Top: View southward across the village, c. 1905. The Martiny house west of Main Street is marked with an arrow. Middle: View to the northeast. Martiny house marked with arrow. Dark building with white trim to far left is the mill. The white Italianate building to the right of, and beyond, the mill is the village hotel. In the background is the site of the "new" town. Photo date unknown but after 1903. Bottom: View across the millpond and village looking northwest. Martiny house west of Main Street marked with arrow. Note recent sediment deposits on banks of millpond. Date: between 1903 and 1915. ....... 145

**Figure 7.4** Bridge over Coon Creek on Mill Street in Chaseburg just below auxiliary dam (seen in background), looking southwest (upstream), June 1914. Note rock shoals below dam. The sills of the bridge were about 13 ft (4 m) above the water. For locational reference, note Lutheran church in background. 1975: Replacement bridge at the same site as above, looking southeast (upstream). The newer bridge sat about one foot (30 cm) higher than the old bridge but had only about 4 ft (1.2 m) of clearance above the water in 1977. The 2011 water level here is 10–12 ft (3–4 m) higher than in 1914. 148

**Figure 7.5** The main dam at Chaseburg looking upstream; refer to Figure 7.2. Top: wooden dam in 1914 at head of shoals. The race on the left extends 140 ft (44 m) to the mill, which originally gave 3 ft (1 m) more head or fall. This much fall meant that the dam was built on bedrock and that the bedrock extended at least to the mill. Middle: Concrete dam built downstream at mill in 1932. Note sluice gate (center). Bottom: 1946. Dam has been demolished and is rapidly being buried. Note rectangular concrete pier on dam to hold flash boards, probably a delaying tactic to keep the mill going a few more months. 1977: All has been buried in floodplain. The person is 6 ft (2 m) tall. Last: rectangular pier to hold flashboards (seen in the 1946 photo), excavated 1977. An elevation placed on the dam in 1932 tied into earlier surveys by the WI Railroad Commission and allowed reconstruction of village elevations back to 1914 (Figure 7.2). ............................................................................................. 150

**Figure 7.6** Martiny house, west of Main Street, Chaseburg, refer to Figure 7.2. Top: Excavation of foundation, August 1975. The vertical pipe marks the spot. Bottom: Base of excavated foundation, August 1975. Flag on left marks base of foundation. Flag on right marks contact of old, dark Mollisol and the lighter, highly stratified historical sediment deposited since 1903. ................................................................................. 154

**Figure 7.7** Early harbingers of a dark future for lower Chaseburg. Looking northwest at the cheese factory on south side of town, early 1900s. Floodplain of tributary from south in foreground with buried fence posts by road (now County Road K). Chaseburg cheese factory to left. For reference, Lutheran cemetery in right background. Bottom, raising the Main Street (now County Road K) bridge over the millpond on south side of village, 1915 presumably, to place it above floods. This channel is now completely occluded. ................................................................................... 155

**Figure 7.8** Top: Gardner's garage (later Vernon County Highway Dept. garage) under construction on top of terrace, 1915. This was directly across Main Street from the Martiny House by the bridge. Viewer stands on road (now WI Highway 162), but eye level is well below the floor of the garage. The house in the background sits on a terrace about 3 ft (1 m) higher than the base of the garage. Middle: County highway garage, 1977. Façade had been changed, building remains the same. Viewer stands on road, eye level far up on the building. More than 2 ft (60 cm) of building are buried in fill. Approximately 10 ft (3 m) of stream and floodplain aggradation have occurred here, and the road has been raised accordingly (see Figure 7.2). Note that the floodplain had built up the level of the terrace on which the house in background sits. Bottom, July 2008. Two large floods had passed though in the preceding 10 months and the building was abandoned and derelict. ........ 156

**Figure 7.9** Beaver, MN, on Beaver Creek, Whitewater River valley, Winona County, looking upstream Beaver Creek flows right to left (eastward). Top, Beaver in 1898. The downstream millpond is visible in the right foreground. Bottom: Repeat of previous shot, 10/15/40. Only two buildings remain. Presently, only a forested valley floor can be seen from this photo site. ................................................................................ 158

**Figure 7.10** A partially buried house in Beaver, 11/28/39. See H. Johnson (1976, pp. 212–213) for another example). ........................................................................................... 159

**Figure 7.11** Galena, Ill. Overlay of modern topographic map on 1828 map showing how the Galena River has narrowed from the influx of sediment. As late as 1850, large steamboats could turn around in the river. Early map from the Wisconsin State Library and Archives. ................................................................................................... 159

**Figure 7.12** Elkport, Clayton County, IO, 1886. At confluence of Elk Creek and the Turkey River. Arrow marks Linn Street and photo site in Figure 7.13. Note Lutheran church farther down the street on right and Catholic Church and other buildings off to the left. (*Plat Book of Clayton Co, Iowa*, 1886). Stream flows left to right (eastward). 160

**Figure 7.13** View down Linn Street in Elkport in 2009 (see arrow in Figure 7.12). The Lutheran church on higher ground to the right is the only building left. The remains of the other buildings lie beneath the sediment on the field to the left now growing corn. The whole village is now protected by a flood dike from which this photo was made. ......... 161

**Figure 7.14** Village Creek, Iowa, on Village Creek about 10 mi (16 km) southwest of Lansing, Iowa, 1886. Note the number of water-powered developments there (*Plat Book of Alamakee County, Iowa*, 1886). There is presently no trace of any of these dams, factories, and canals, all presumably buried by sediment. Stream flows eastward (left to right). .................................................................................................... 162

**Figure 7.15** Village Creek. The schoolhouse, built in 1860, is the only old building left. Although in a higher part of the village, it has been buried 4 ft (1.2 m) so that downstairs windows are at ground level. (Photo credit: Nadine Kleinhenz, 2006.) ......... 163

**Figure 7.16** Rushford, MN, Root River. Top: 1874. Stream flows eastward (left to right). Note the large "industrial park" of canals, dams, mills, and houses south of the Root River (*Illustrated Historical Atlas of the State of Minnesota*, 1874). 1980s (USGS 1:24,000 topographic maps): Note the straighter configuration of both the Root River and Rush Creek coming from the north and running through North Rushford. There is now little trace of the old industrial area, it being buried, but there are the remains of a hydroelectric power station about 500 ft (150 m) east of the south end of the present causeway. Styled in Art Deco, the power station was probably built in the early 1920s. ............................................................................................................................ 164

**Figure 7.17** Arcadia, WI. Turton Creek at an old mill site, looking downstream towards the Trempeleau River, 2011. The stream flows along the north side of the town. Note that the water surface of the creek at normal flow is presently almost as high as the land on the other side of the flood dike. The dike to the left is about 4 ft (1.2 m) high. The fact that most of the houses in town had a basement suggests that the original water level was at least 10 ft (3 m) lower. Even the level of the millpond here would have been considerably lower than the present level of the stream. Otherwise, it would have flooded the areas to either side and would have raised local groundwater levels. The higher present stream levels, including the Trempealeau River, just 2 mi (3 km) downstream from this site, have caused groundwater to rise and flood basements in the town. ......................................................... 165

**Figure 7.18** Top; drawing of the residence of Jacob Roth, Crooked Creek (NW ¼, Sec 27, R4W, T13N, Houston County, MN, from Illustrated Historical Atlas of the State of *Minnesota*, 1874). Note the stone wall in front of the house which, in comparison to the house and the figures, was about 9–10 ft (3 m) high. Note further that the stream in the foreground is at least 5 ft (1.5 m) lower than the base of the wall. For this water to flow out, the floodplain (to the left or south) would have had to be even lower. Thus the house sat at least 14–15 ft (4.5 m) above the floodplain. Also note the train in the distance going up the valley. The tracks to Caledonia were completed in 1871. Bottom: View of the Roth house from the public road across the floodplain, 2009. The level of the floodplain appears to be close to the base of the house, suggesting aggradation of more than 10 ft (3 m) since 1874 ................................................... 166

**Figure 7.19** Floodplain covered with sandy sediment with buried and damaged fences, Whitewater River Valley, 1940. Note three generations of fences in lower photo. (From Trimble, S. and Lund, S., *U.S. Geol.Surv. Prof. Paper* 1234. 1982. ..................... 168

**Figure 7.20** La Crosse and Southeastern railroad, Coon Creek Valley, built 1904. Top: A railroad bridge seen near Chaseburg in the early 20th century, probably before 1915 based on the apparel. Note that the top of the rails is about 7 ft (2.2 m) above the channel, giving a small bridge opening. Note also woody debris from flooding partially occluding the bridge opening. Bottom: Surveyed cross section between Chaseburg and Stoddard. Railroad embankment was built here on edge of terrace about 10 ft (3 m) above the old floodplain. By 1938, the embankment

*Figures* xxv

was only 1–2 ft (30–60 cm) high. By 1974, it was completely buried. Note also the highway (left or north side of the profile), which had to be raised to stay well above the aggrading floodplain of Coon Creek (NW1/4, Sec. 29, T14N, R6W). ............. 170

**Figure 7.21** Minnesota State Highway 74 along the north and west side of Whitewater River on the floodplain. In an attempt to escape the increasing flooding and sediment accretion, it was built in 1918 on a 6 ft (2 m) high embankment with a narrow concrete roadway. By the late 1930s, the floodplain had aggraded to the level of the road or even above it. Top, 6/2/38 near Beaver. Bottom: Reach of Highway 74, probably near Beaver. The floodplain had aggraded above the highway, which had to be cleared. ........................................................................................................................ 172

**Figure 7.22** Minnesota State Aid Highway 9 on the south side of the Whitewater River, 11/6/37. Bridge in right foreground was covered by 18 in. (45 cm) of sediment by June 1938. This road, the lifeline for people on the south side of the river, was becoming impassable by the late 1930s. By 1989, it was underwater, impassable even on foot. (Photo credit Soil Conservation Service [SCS].) ........................................................ 173

**Figure 7.23** The south abutment of an old Dodson Hollow Road bridge over Coon Creek (SE1/4, Sec 22, T14N, R6W, Vernon, Co., WI). This bridge was replaced in 1926 as the creek rapidly aggraded, was later buried in the floodplain. When built, probably in the early 1900s, this abutment was at least 10 ft (3 m) above the water. A small lateral shift by the stream in 1993 exposed the old abutment. The newest bridge (c. 1990) is seen to the far left. ........................................................................... 174

**Figure 7.24** Looking southward across the lower main valley of Coon Creek along Range 28 as seen in Figure 7.20. In the left middle ground (arrow) is the trace of an old private road across the valley. The tree canopies to the right are willows marking an old house site. Both the private road and the house site are buried in the floodplain, and only a difference in vegetation marks them. In the distance, the old railroad bed runs down the valley but is totally buried (see profile in Figure 7.20) and there is little to mark it at the surface. .......................................................................... 175

**Figure 8.1** 24-hour precipitation totals for the August 18–19, 2007, storm for Coon Creek and the Whitewater River. 1 in. = 2.5 cm. Data from National Weather Service. ... 178

**Figure 8.2** Mass movements from the August 2007 storm. Top: What appears to be a debris flow about 2 mi (3 km) north of Chaseburg, WI, at junction of Hohlfeld Road and County K, looking west from north end of bridge (P > 12 in. or 30 cm NW1/4, Sec 21, T14 N. R 6W). Note that the sediment flow from the undercut road bank moved across the road, across the floodplain, and into Swain Creek. Bottom: Mass movement in Timber Coulee about 3 mi (5 km) north of Westby (P = 6–8 in., or 15–20 cm). (Photos credit: Edyta Zygmunt, 2–9 September 2007.) ........................................ 180

**Figure 8.3** No-till fields after crops had been removed, April 2008. Top two photos: Note slopes > 10% and several hundred feet long with no evident erosion, P = 12+ in. (30 cm). Bottom: Rill caused by runoff from road. All 3 scenes in this figure are on Brinkman Ridge Road, about 4 mi (7 km) north of Chaseburg, WI. Secs 4&9, T14N, R6W. ..................................................................................................................... 182

**Figure 8.4** A sediment trap after the 2007 storm. Top: A "dry" floodwater retarding dam (arrow) controlling about 8–10 acres (3–4 ha) of mostly no-till crops, 5 mi (8 km) north of Altura, MN, on Beaver-Altura Road looking west. P = 10 in. (25 cm).

Corn is to the left and soybeans are to the right. Bottom: Bottom of dry pond. Note lack of any sediment accumulation. NW1/4, Sec. 19, T18N, R9W, Winona County, MN. Photos made Sept. 2–9, 2007. ..................................................................................... 184

**Figure 8.5** Severe soil erosion with lesser levels of conservation management, storm of August 2007. Top: Newly planted field with no apparent conservation measures, Beaver-Altura Road about 4 mi (7 km) north of Altura looking west. P = 10 in. (25 cm). Sec 25, T108N, R10W, Winona Co. Middle: Gullying on animal trails, heavily grazed area 2 mi (3 km) south of Stoddard, WI, on County O, NW ¼ Sec 3, T13N, R7W. Vernon Co. P = 12 inches (30cm). Bottom: Severe rills in contour strip cropping, about 5 mi (8 km) SE of Stoddard. P = 12 in. (30 cm). (All photos made September 2–9, 2007 by Edyta Zygmunt.) ............................................................................... 185

**Figure 8.6** Damage to Zink farm from 2007 storm, 1 mi (1.6 km) SW of Stoddard (see Chapter 4, P > 12 in., or 30 cm). Top: View of fan across Coon Creek floodplain from bridge on County O looking west, September 2007. Fresh deposits of mostly coarse material, 1–5 in. (2.5–7.5 cm) thick. Cottonwood tree in right foreground is the one with buried trunk mentioned in Chapter 4. Middle: between house and barn, looking north up dormant flume formed from rock fan in 1930s to protect house and farmyard. Photo made July 2007. Bottom: The same flume September 2007, looking northeast. Accumulated debris from the August 2007 storm has been cleaned out. (Photos credit: Edyta Zygmunt.) ........................................................................... 187

**Figure 8.7** Flume formed from rock fan to protect farm, similar to that at Zink farm (see Figure 8.6), photo made from County P looking north, Timber Coulee, NW1/4, Sec 8, T14N, R4W, Vernon Co. P = 7–8 in. (20 cm). Top: July 2007, dormant flume. Bottom: September 2007. Note "gravel road" appearance. (Photo credit: Edyta Zygmunt.) .................................................................................................................................. 189

**Figure 8.8** Proksch Coulee Trench, looking downstream at the dam and drop structure built in 1936 by Soil Conservation Service to stabilize gully, 2 mi (3 km) east of Stoddard WI (P > 12 in., or 30 cm). Viewer stands on reservoir fill. Top: Dam and drop inlet, looking downstream, July 2007. Arrows point to top-right corner of drop structure. Note sediment filled to level of drop inlet (see Figure 4.10). Bottom: Breached dam, September 2007. Note drop inlet to left covered and blocked with woody debris. Gully in foreground extends about 200 ft (60 m) upstream in accumulated sediment. Bottom of gully visible in right background is about 30 ft (9 m) lower than dam. (Photos credit: Edyta Zygmunt.) ............................................................. 190

**Figure 8.9** Gully 1 mi (1.6 km) north of Chaseburg, WI, looking downstream from bridge on County K, P > 12 in. or 30 cm (SW1/4 Sec21 T14N R6W, Vernon Co.). Top: July 1, 1938. Note about 3 ft (1 m) of light-colored, historical sediment accretion on bank and apparent recentness of gully incision. A storm the next day destroyed the timber and rock revetment. Photo by S.C. Happ. Middle: 1974. Note that gully more than doubled in size after 1938 but was stable by 1974. Bottom: September 2007 after flood. Note line of rock following thalweg, perhaps rip-rap from plunge pool in foreground, but gully is otherwise stable. (Photo credit: Edyta Zygmunt.) ........................ 191

**Figure 8.10** Upper Mitchell debris basin, built in 1941 to protect WI Hwy 56 from gully debris, 4 mi (7 km) SE of Genoa (P = 10–12 in. or 30 cm, Sec 36, T13 N R6W). Top: Surveying the basin, 1977. Basin filled in 4 years (1941–1945). Dam to left, sediment basin outlined by dark vegetation, Hwy 56 to top left. Arrow marks edge

*Figures* *xxvii*

of fill against dam. Bottom, September 2007. Light material is 1–5 in. (2.5–7.5 cm) of sediment delivered by August 2007 storm. Note that level of fill by dam is hardly higher than in 1977. .................................................................................................................. 193

**Figure 8.11** North Fork of Whitewater River 1 mi (1.6 km) west of Elba, looking upstream from bridge as seen in Figure 3.12; note barn for reference. Top: 1940. Note disturbed condition of channel. Middle: 1975, condition was stable, and this has continued to the present. Bottom, September 2007, P = 9–11 in. (22–27 cm). (Photo credit: Edyta Zygmunt.) ........................................................................................................ 194

**Figure 8.12** Bohemian Creek about 3 mi (5 km) NW of Coon Valley, WI, looking upstream from County G. (SW1/4 Sec 34T15NR5W, La Crosse Co.). Top: Photo made 1940 by S.C. Happ as the "typical" tributary in the Hill Country. Middle: 1974, note stable, grassed floodplain and narrow stream. Bottom: September 2007, completely stable after storm of 4–7 in. (10–18 cm), which is greater than the 100-year storm. Bohemian Creek had only scattered bank erosion from this storm, and I found no mass movements or visible soil erosion of any kind on agricultural fields. ...................... 196

**Figure 8.13** Collapse of high stream banks 2 weeks after the flood of August 2007, looking downstream Upper Main Valley, Coon Creek, 2 mi (3 km) east of Coon Valley WI, R = 4–10 in. (10–25 cm). For exact location, see photo site in Figure 6.7. Top: July, 2007. Range C from Pleistocene terrace (hump on floodplain in distance) to distant high bank seen in Figure 6.7 is noted by arrows. Bottom: About 10–14 days after the flood. Note logs in background on old floodplain suggesting inundation of the old floodplain (now a terrace). Some erosion of the high bank in the foreground was evident. However, inundation plus the rainfall saturated the normally dry bank, increasing pore pressure and weakening the bank, thus allowing it to collapse and be more susceptible to erosion. (Photos credit: Edyta Zygmunt.) ................ 198

**Figure 8.14** Erosion of grazed Lower Main Valley streambanks by August 2007 flood, looking upstream from bridge over Coon Creek, 3 mi (5 km) west of Chaseburg, WI (NW1/4 Sec 30 T14N R6W). High bank and terrace in background is shown in Figure 3.17. Top: July 2007; note trampled banks from grazing. Bottom: September 2007. Note severe erosion in left foreground, but far bank in middle ground appears to have accretion. (Photos credit: Edyta Zygmunt.) .............................. 199

**Figure 8.15** Stable ungrazed grass banks in the Lower Main Valley after 2007 flood. Top: Lower Coon Creek looking downstream from Hwy 35 bridge 1 mi (1.6 km) south of Stoddard WI. Bottom: North Fork, Bad Axe River looking downstream from Hwy 56 bridge at Romance WI (SE1/4 Sec 36 T13N R&W. Vernon Co.). (Photos credit: Edyta Zygmunt.) ........................................................................................................ 200

**Figure 8.16** Extremely severe erosion of ungrazed grass streambanks, looking downstream from bridge over Whitewater River at Beaver MN, 6 mi (10 km) north of Elba, R = 10–14 in. (25–35 cm). Top: July 2007. Bottom: September 2007. (Photos credit: Edyta Zygmunt.) ........................................................................................................ 201

**Figure 8.17** Sediment deposition on Lower Main Valley floodplains from August 2007 flood, highly variable in distribution and thickness, Top: Splays of sand ranging up to almost 2 ft (60 cm) thick just downstream of bridge in Figure 8.14. Bottom: Splays of sediment on Coon Creek floodplain looking southwest from

Cedar Valley Drive, 2 mi (3 km) east of Stoddard, WI (E1/2 Sec 26 T14N R7W, Vernon Co.). (Photos credit: Edyta Zygmunt.) .................................................................. 202

**Figure 8.18** Repeat views of Coon Creek delta front in Pool 8 of Mississippi River, looking northeast from scenic overlook on bluff, off Hwy 35, 3 mi (5 km) north of Genoa, WI. Top: July 2007, arrow indicates westernmost growth of vegetation on delta front. Bottom: Sept 2007. There was no discernible change. A photo in July 2008 also showed no change, and there was no change from July 2006. (Photos credit: Edyta Zygmunt.) .................................................................................................................. 203

**Figure 8.19** Instrumented cross-sectional survey of north bank of the Coon Creek delta extending into navigation Pool 8 of the Mississippi River, 1975, 1992, and July 2008. The profile is located 500 ft (160 m) downstream of the railroad bridge seen in Figure 8.15 and off to the right and out of sight in Figure 8.18. The period 1975–1992 included one basin-wide, 100-year storm, and the period 1992–2008 included the August 2007 flood and another large storm in June 2008. The average rate of accretion for the earlier period was *greater* than for the later period including the extreme storms of August 2007 and June 2008. These data do not support the concept that extreme floods necessarily scour out legacy sediment, at least in humid area streams with broad floodplains. .................................................................................... 204

# Foreword

*Andrew S. Goudie*
University of Oxford

That great British Geographer, the late Sir Henry Clifford Darby (1909–1992), often asserted that the two most important thrusts of geography were historical geography and geomorphology. This was because of his concern with understanding the evolution of landscapes, and he himself wrote in detail about the evolution of some great examples of these, including the English Fenland (Darby, 1956). Since that time there has, building upon the work of G.P. Marsh (1864) in the nineteenth century, been a fine tradition of tracing the way in which humans have modified the face of the Earth by, for example, draining wetlands and clearing woodlands. Notable here was the magisterial work of the late Michael Williams (1970; 2003). Environmental history has become a fertile field. Stan Trimble has already shown this in the context of the Southern Piedmont in the USA (Trimble, 2008). This present study examines the ways in which another of the great landscapes of the USA has been transformed by human activities. It is based on almost 4 decades of dedicated research and has many of the characteristics of a gripping detective story, based on meticulous forensic investigation. It demonstrates the unfashionable virtue of long immersion in a region, of the use of archival material, of intensive fieldwork, and of the value and magnificent collection of repeat photographs. It is an exemplification of true geography, which makes a major contribution, by both methodology and substance, not only to historical geography but also to geomorphology, It will also be avidly devoured by local historians.

To gain an impression of the value of this work, one needs, paradoxically, to start by looking at the volume's conclusions. It anwers the question as to why it was that the region's farmers, unwittingly, caused landscape destruction within a few decades: a climate that was more erosive than Europe, the presence of steep, erodible slopes, and the nature of the crops grown. It also demonstrates how even with simple technology—horse-drawn implements—farmers can cause severe erosion and gully formation. It also raises the question as to whether land cover change or climate change was the main cause of the degradation that occurred and indicates very clearly that it was the former rather than the latter. However, another conclusion is that in addition to causing this spasm of degradation, the region's farmers, by instituting new land management and conservation strategies, were also responsible for 'the almost miraculous recovery of the region' at the time that there was an increase in agricultural productivity. It has been the change now that it is difficult to discern 'the environmental violence and human frustrations that once existed there.' There were aided in this by well-designed government programs. The book also demonstrates the importance of forest-regrowth—the rebirth of America's forests—and its attendant environmental improvements. The geomorphologist will learn much from the complexity of stream processes that is evident and from the linkages that occur in different parts of the river basins between erosion and deposition. They will also be fascinated by the rather limited impact that the great floods of 2007 had on the landscape. There is a lesson here, and in the book as a whole, with respect to our current preoccupation with the effects of future anthropogenic climate changes. We will do well to remember the importance of future land cover changes as well.

The book's preface, however, must not be skipped. It is there, in a handful of graphic introductory paragraphs, that the justification of this remarkable work is outlined—the story of 'the fall and recovery of paradise.'

## References

Darby, H.C. 1956. *The Draining of the Fens*. Cambridge: Cambridge University Press.
Marsh, G.P. 1864. *Man and Nature*. New York: Scribner.
Trimble, S.W. 2008. *Man-induced Soil Erosion on the Southern Piedmont* (2nd. edition). Ankeny, Iowa: Soil and Water Conservation Society.
Williams, M. 1970. *The Draining of the Somerset Levels*. Cambridge: Cambridge University Press.
Williams, M. 2003. *Deforesting the Earth. From Prehistory to Global Crisis*. Chicago: The University of Chicago Press.

# *Preface*

Imagine that as a small child in the mid-19th century, your family left New England, or upstate New York, or Norway, or any of several places of origin in the United States or Western Europe. They settle in what is now called the Hill Country of southwestern Wisconsin, where your father builds your house in a small, beautiful valley with a cold, clear, ever-flowing spring. A narrow creek filled with brook trout flows near the house. Your neighbors and dear friends who migrated with you build a large house on a low terrace overlooking the broad, fertile floodplain of a larger stream nearby.

The decades pass. The land is well watered and bountiful. Crops are good and the amenities of "civilization" are increasingly available. Roads expand, and railroads and small industries such as mills and cheese factories and even breweries are constructed. Churches are built and other civic and social organizations are formed. Villages and small towns prosper. Life is good. The Hill Country is a sort of rural paradise, an Arcadia in America.

But by about the sixth or seventh decade of your life, things begin to change drastically. The once-clear spring, your only source of potable water, is no longer as dependable and sometimes dries up. The beautiful creek flowing past your house has become turbid. It dries up on occasion, but then rages when even moderate rainstorms bring down torrents of mud, sand, and boulders, all threatening your house and farmstead. Of course, the brook trout are long gone. Even after you have used the stream-transported boulders to build a protective dike, your house and farmyard often flood, and running the farm becomes difficult. Upland fields above the house, once so fertile and productive, now have deep gullies, some gushing even more water and debris into the farmyard. All of this makes it more difficult to farm, your income declines greatly, and so does your whole quality of life. You know of other people in similar situations, some of whom have abandoned their house and farmyard sites. Your family must decide whether to stay and fight the elements, or just move the farmstead to a higher place. Since you are now the family elder, they look to you for advice.

Your neighbors on the larger stream are having their problems, too. Once above all floods of the larger stream, their terrace location is now being flooded, bringing great hardship. Even worse, the floodplain is aggrading with silt and sand, bringing floodwaters ever closer and sometimes destroying their crops and fences and making the land less fertile. Floodplain pastures are increasingly swamped, and animals must now be pastured behind the house on the steep, fragile hillside, which begins to gully, bringing water and debris onto their terrace location and into their house. Their farm income decreases and their farmstead, once a showplace, now appears derelict. They, too, feel threatened and consider moving to higher ground or leaving the region altogether.

Matters on the general landscape are hardly better. Hillside gullies bring water and debris with every storm, sometimes closing roadways and railroads. Flooding tributaries rip away roadways, demolish bridges, and undercut telephone and power lines. Farther downstream, bridges and roadways are being threatened with every storm, and some are even buried. Mills are washed away or buried so flour, cornmeal, and sawed lumber are more difficult to procure. People lose their homes and sometimes drown in floods. In some locations, entire villages are being flooded and even buried. A greatly increased proportion of township, county, and state tax money must go to road repair and for public safety in general, given the increasing hazards. With transportation and communications so tenuous, it becomes more difficult to obtain supplies, to market crops and milk, to communicate and socialize with distant friends, and even to go to church.

What sounds like the plot of a horror movie or a bizarre story by Franz Kafka was reality in much of the Hill Country along the upper Mississippi River. And, indeed, much of it happened within the lifetime of many people there today. It is a sad and often tragic story, but it is more than counterbalanced by the almost miraculous and mostly happy story of the recovery, again which took place within a lifetime. Both the fall and recovery of paradise are the stories I want to tell.

## Background and perspective

Everyone knows about the buried civilizations of the ancient world. In the Middle East, soil erosion was so severe that cities were buried and ports became landlocked. Archaeologists have been excavating the Greek and Roman ruins for centuries, and we marvel at these unearthed treasures and the environmental processes that caused these interments. But very few know that civilizations in mid-America have been entombed, not over the time span of centuries and millennia, but during just short decades.

The Upper Mississippi Valley Hill Country, reaching from Hastings, Minnesota, downriver past Dubuque, Iowa (Figure 0.1) is one of the most beautiful landscapes in the world. It has many names—the Paleozoic Plateau, Coulee Country, and Bluff Country. The U.S. Department of Agriculture (USDA) terms it the Northern Mississippi Valley Loess Hills. Often perceived as an unglaciated island in a sea of glacial drift and thus generally termed the "Driftless Area," it is better seen as an island of hilly country, largely unglaciated, surrounded by a sea of gently rolling landscape, all glaciated. Extravagantly praised by early explorers before European* settlement, it is now a serene, bucolic region of wooded hills, green vales (termed *coulees*), clear streams often filled with brook trout, and model farms. Locals, with good cause, call it "God's Country." Like good music or poetry, it is an entity that sooths the soul and reassures one of the goodness of the world.

But it was not always so. In the early 20th century, the region went through a dark and desperate period of soil erosion, stream sedimentation, and flooding, much as happened millennia earlier in the Mediterranean area. Roads, bridges, farms, and villages were literally buried, but here in a much shorter time. Aquatic habitats disappeared as once-limpid brooks became turbid and began to resemble gravel roads or huge ditches. Deep, raw gullies fingered far up valleys and hillsides. Frequent floods covered lowland fields with water and sediment, and even blocked roads or eroded them away, often cutting off communication between communities. In a real sense, the landscape was truly out of control. All this was a disaster on the scale of a huge hurricane, a tsunami, or an earthquake—but instead of one huge event, it occurred in increasingly severe increments—a disaster in slow motion. One might think that the relatively slower development of this disaster would have given people more chance to escape it. But no one knew the eventual dimensions

---

* "European" here refers to the culture involved, especially the agriculture. It may include recent European immigrants, "old Americans," Native Americans, African Americans, and others.

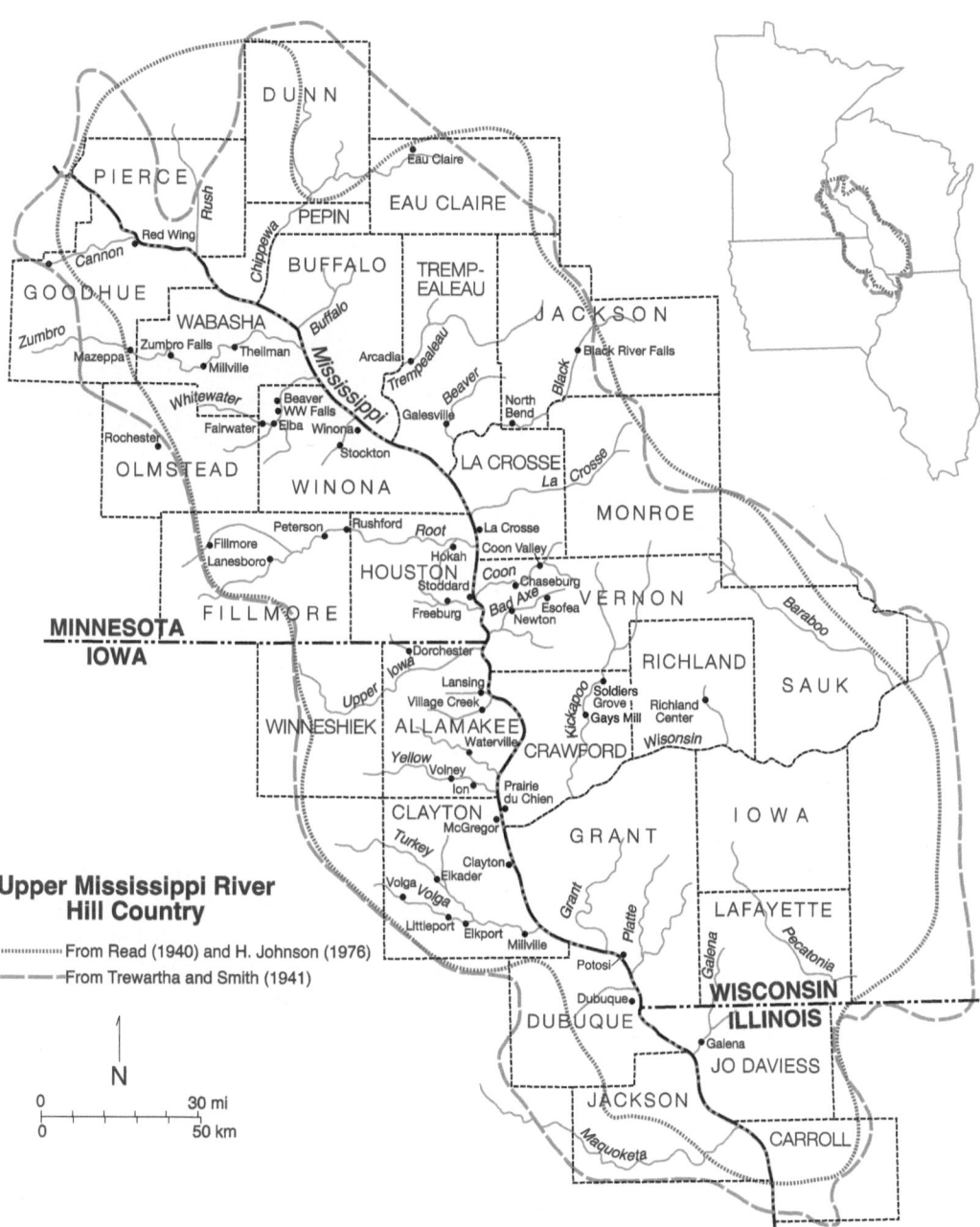

*Figure 0.1* The Upper Mississippi River Hill Country. The delineations shown were used because of previous historical work done on the region. This is also quite similar to the delineation of the region into a Major Land Resource Area (MLRA) by the USDA-NRCS (Argabright et al., 1996). Termed by the USDA as the Northern Mississippi Valley Loess Hills, it would exclude most of Dunn County, Wisconsin, and Carroll County, Illinois, but include Clinton County, Iowa.

*Figure 0.2* The Great Flood of 1907 in lower Chaseburg, Wisconsin, the highest on record up to that time. The photo was made looking east on Mill Street near the front of the present Hideaway Tavern. Since that time, the floodplain has aggraded to a level about 4 or 5 ft (1.2–1.4 m) above the flood level of 1907. The remains of lower Chaseburg were purposefully demolished in 2009 because of the continued flooding.

of the problems—the magnitude in space and time—so they had little idea about how to react. Few people had the financial and logistical ability to do much about it, but many eventually *had* to react when the water and sediment literally came in the front door.

Because all of this seems so unbelievable when related in the abstract, just consider one case, that of Ole Martiny in the once-idyllic village of Chaseburg, Wisconsin, on Coon Creek just south of La Crosse. After living in Chaseburg most of his life and knowing the village's collective observation and perception of past stream behavior, he decided in 1903 to build his house on a low alluvial terrace by the creek. On July 21, 1907, an unprecedented flood, the largest since European settlement, occurred, rising four feet (1.2 m) on his house (Figure 0.2). Believing this to be simply a quirk of nature, he promptly raised the house 4 ft (1.2 m), thinking that would put him above any future floods. But the flooding and sedimentation continued and in less than two decades, had driven him from the house, the foundation of which now lies about 10 ft (3 m) under the present floodplain (see Chapter 4).

Eventually, there was some salvation for the region. The human hand, which had brought such destruction here, was with time just as capable in restoring the same landscape. And so it happened: the second half of the 20th century, continuing even now, saw the restoration of this region, and it was done by the progeny of the original settlers, in some cases, perhaps, just barely within the lifetimes of those original settlers. Soil erosion was curtailed, scarred hillsides were healed, streams once again assumed their sylvan nature and appearance, and brook trout again bred and flourished in now-clear streams.

All is not perfect, however, because there is a painful legacy of this sediment. Once-productive agricultural fields in bottomlands are now marsh or even open water. Towns, villages, homes, and agricultural fields in some places lie huddled behind dikes to mitigate flooding caused by aggraded channels and floodplains. Sediment, eroded from slopes perhaps a century or more ago, deposited along the stream, and now termed *legacy sediment*, is presently migrating and creating more problems in some downstream locations. In several ways, the Hill Country is a region trapped in its own history.

A person visiting the region or even having grown up there may well not know this story. Those who did knew it only in part and, in any case, are already history themselves. As we now admire a wooded hillside, we may not know that gullies once cut across it; as we now gaze at a stream, we may not know that it once flowed at a level many feet lower than now, or as we look across a floodplain, we may not know that a farm or even a village lies buried there. And even if we know that, do we know how and why it happened? To reveal such landscape secrets, often literally buried, is the role of historical geography. And that is what this book is about—revealing the secrets of past landscapes and how they evolved, and why.

My love of this quintessential American landscape must be clear to the reader by now. Although I'm not a native son, this is my second home in the world. More than just telling you about it, I want to equip you with the means to understand and really appreciate it. But more than that, I hope to give you the means of going out on the landscape and visualizing *what* happened, right on the ground *where* it happened. It is a thrill to see and experience a landscape in its historical totality. Professional geographers used to say that "one learns geography through the soles of one's shoes." I still believe that. Even those who cannot physically visit the region can use topographic maps and earth images from the internet to be there virtually.

I emphasize that this study primarily concerns watersheds tributary to the Mississippi River and deals only to a small degree with the Mississippi River itself. For that, one is directed to the excellent studies by Burke (2000), Anfinson (2003), and especially by Fremling (2005).

The approaches and tools of the geographer are eclectic. This research project, over the past 39 years of my career, has employed methods and techniques ranging from hydraulic engineering to rural sociology, with much archeology thrown in. But my primary goal here is to show the human side of these profound landscape changes. Over the past 39 years, I and others, in many publishing outlets and often in technical terms, have shown how humans have effected the changes, but now the emphasis here is also on how all these landscape changes, in turn, have affected humans. Of course, I will have to discuss many of the physical aspects to make all this understandable, but I will hold that to a minimum and then make reference to the appropriate publications so that those interested persons can read further.

I hasten to point out that I am not the first to study these profound changes in this region. In particular, I mention the work of Carlos Bates and O. R. Zeasman in the 1920s, Stafford C. Happ and his associates and students in the 1930s and '40s, Richard Sartz and his colleagues in the 1960s and '70s, Hildegard Binder Johnson in the 1970s, and James C. Knox and his students starting in the early 1970s. My own work there dates from 1973. These and many other studies will be cited in due time, but this study will remain highly personal.

While I hope that professional earth scientists will find this work interesting and useful, my primary desire is to engage the interested and aware nonspecialist. Thus, there is a concerted attempt to make the work accessible to those in other fields. Professional earth scientists should consult the cited literature for more technical and comprehensive explanations and discussions. To further facilitate accessibility, I use both English (Imperial) and metric (SI) units of measure and provide a glossary of technical terms.

This book has been in the making for a long time. My earlier studies of the Southern Piedmont (e.g., Trimble, 1969, 1970, 1974) prepared me well for this work because it was there that I fully realized the power of humans to drastically modify the landscape in both negative and positive ways. The effects of modern soil conservation measures, and the dedication of those who implemented them, impressed me greatly. I also learned much about physical processes on slopes and in streams and what drove those processes. A big lesson was the role that sediment storage gain and loss played in determining the sediment yield of streams (e.g., Trimble, 1975c, 1975d, 1976b). Perhaps most important, I saw in action what we now refer to as the "complex response," that is, different stream reaches respond in different ways and different times to what seem to be the same stimuli (Schumm, 1973). But the good news is that it is somewhat systematic and predictable. Discovering this was a great education in fluvial processes, as well as the human impact, and it carried over to the present study.

While working in the Piedmont, I met Dr. Stafford C. Happ (1905–1993), the person who, as a US Department of Agriculture (USDA) scientist, had designed many stream research projects in the Piedmont, the Hill Country, and elsewhere in the United States (Trimble, 2008). When advising me on the Piedmont work, he mentioned his pioneer work in Coon Creek, Wisconsin. As luck would have it, my first university teaching job was at the University of Wisconsin at Milwaukee. As soon as possible, I reconnoitered Coon Creek in depth and obtained a research grant from the US Geological Survey (USGS). Then, from several places, including the National Archives, I was able to dig out much of the relevant data. But the best thing I did was to ask Dr. Happ to advise me on the project. We began fieldwork in earnest in late May 1974 and for the next few summers, my family, my students, and I "camped" in a rented house either in Stoddard or in Coon Valley, Wisconsin. While nearly all of the local farmers gave us permission to enter their property, some were unsure about what we were doing out on their floodplains and in their streams, but when they saw us out from 7 am to 7 pm six days per week, they were convinced that this was serious business. Although he was 69 years old at the time we began, Dr. Happ often worked right along with us in temperatures that once went to 112°F (44°C) and sometimes in water up to our necks. And we were always pestered with insects like mosquitoes, sweat bees, horseflies, deerflies, chiggers, and deer ticks. Not only did he never complain, he could work harder and longer than any of us. When we finally quit for the day, he went to his cramped motel room and laboriously plotted the surveys we had done that day. While we did not always agree on research priorities and matters of precision, I must say that he was the best field person I've ever known, and the project would have been almost impossible without his not infallible but elephant-like memory concerning the location of profile ends, benchmarks, and physical conditions of the 1930s and '40s.

Happ's research designs are given in detail in Trimble (2008), and briefly described in Chapter 3. These were highly labor intensive, being based largely on precisely cored and surveyed stream and valley cross-sectional profiles, and for the record, I wish here to briefly note what was involved in resurveying them. There were only a very few permanent benchmarks set, nearly all profiles were marked by blazed "witness" trees, and most benchmarks were merely common nails set low on trees. Such markings would be hard to recover in a dense forest, but there were very few trees in the bottoms during the late 1930s because the land was cultivated or grazed, so the markings were then fairly visible. The original intention had been to return within five or so years and resurvey, but World War II intervened and the projects were forgotten by all but Dr. Happ. By the mid-1970s, there was less grazing so that many bottomland areas had succeeded to undergrowth or young forest. This was 30 years before GPS so we had to find range ends by using

Happ's careful descriptions of trees and the surrounding landscape, and scaling from air photos on which Happ had precisely plotted the profiles. However, tree growth had covered the small nails and many marked trees had died or had been removed. Moreover, undergrowth was often so thick that even permanent markers such as steel pipes set in the ground were virtually invisible. Thus, finding some profiles was literally like looking for a needle in a haystack. On many occasions, I saw Dr. Happ spend hours with a small hand ax, carefully chipping away at the side of a tree trying to find the suspected nail, which he usually did, and I might add without disturbing the elevation on the nail.

To be useful, a resurvey had to be *precisely* on the old line so just finding the exact line was a major task. If we could find one end or any point along the line, we could use the precisely recorded azimuth to help find the line. To do this we calibrated our theodolite to Polaris and then corrected for the considerable change of magnetic declination since the surveys. Because of so many magnetic anomalies, however, we usually obtained only an approximate line by this method, so we surveyed trial lines, trying to conform to unchanged topography. This was highly problematic because many of the profiles were well over a thousand feet (300 m) long (one was over half mile, or 800 m) so that a small directional error could depart far from the original line. Moreover, all or parts of the line had to be cleared of brush and trees for a clear line of sight for our optical level. Most of my student volunteers had been through my field research class at UCLA so they knew how to survey. Thus, clearing the line usually fell to me while the students did the surveying, sometimes under Happ's direction. To do this, I used a brush hook, which is a heavy but extremely sharp, scimitar-shaped blade mounted on an ax handle. The best I can say about it was that it kept me in good shape. The end result of all these problematics was that the resurvey of some profiles was exhausting and took more than a week. At first, we were squeamish about wading through stinking, anoxic swamps and the deep turbid streams of the lower valley, but soon paid it no mind. One student, however, a recent veteran of the Vietman war, in 1974 pronounced the Hill Country swamps worse than anything he had experienced in the jungles of Vietnam. On most days, we were soaking wet by 9 am, either from wading or sweating or both. We measured the severity of the day by the amount of water we drank. Luckily, no one ever fell out from the heat but several went to the emergency room for cuts and insect bites.

Note that Happ's design and our new survey data gave us stream and valley cross-sections for c. 1853 (based on borings to the old soil), c. 1938, and c. 1975. Other than resurveying the profiles, my contribution in this respect was to determine fluvial changes between those dates, thus establishing differential rates and then trying to connect the geomorphic changes to hydrologic changes driven by land use and climate. I did that to a large degree by floodplain excavations using archaeological techniques with datable landscape features such as bridges, roads, milldams, fences, and buildings, and by historic photos and descriptions. These receive much attention in this book.

My original intention was to report the work in journal articles, preferably *Science*, and I have done that now for over 35 years. However, while I was a visiting professor at the University of Chicago in 1978, Professor Karl Butzer heard a presentation I gave on the work and invited me to write up the work as a possible book in the Geography Research Paper series sponsored by the Department of Geography there. I saw this as a heavy charge, and I started thinking in those terms, but it just seemed that there was always much more to do and to discover before writing a book. And indeed, I have found out more every year, and I believe that waiting truly has made the book immeasurably more complete and mature. Even as I write, I discover new insights and connections that had eluded me for 39 years.

At one time, I perceived the book as a technical compendium of all the fluvial geomorphology work done in the region. But eventually, I saw that the story that must be told is the story of people interacting with their environment. Yes, I at least touch on most technical aspects of the work, but I have tried to hold such to a minimum and merely reference the more complete works.

My hope is to portray the *totality* of landscape, the myriad of both physical and human factors that have wrought changes on the Hill Country and that, in turn, have been affected by the complexity of both positive and negative feedbacks. With this, I'm put in mind of Barry Commoner's first "law" of ecology, that everything is connected to everything else (Commoner, 1971). Too often, we see things in a restricted, isolated, even myopic way, and my intention here is to pull as many of these diverse strands together as possible. In turn, I have borrowed from many fields of academic endeavor: hydrology, geomorphology, soil and water conservation, agricultural science, and rural sociology, to name a few. I hope that I have done them justice.

Rome was not built in a day, nor can a deep and careful landscape history be dashed off in a short time. So here is a book with not just the almost 40 years I've given it, but with a total of almost 80 years of data gathering since USDA started the process in 1933. I hope the book will serve to (1) teach about hydrologic-fluvial processes and landscape change, especially in the Hill Country, (2) give pause to those who see only the destructive side of humans, and (3) inspire coming generations to study other landscapes in equal detail. I also wish to show that the developments in the Hill Country have both national and international significance.

Just a word about the data collected by Dr. Happ and his US Department of Agriculture co-workers and eventually bequeathed to me. First and already mentioned were the extremely systematic, well-organized, and precise survey and borehole data that I was able to use for Coon Creek, Lake Marinuka at Galesburg, Wisconsin, and to some degree in the Whitewater River in Minnesota. Second, there were considerable textual materials such as crop and land use reports and academic theses. Third, and to the point here, was a wonderful collection of photos, probably 300 in all, most of which I recovered from the attic of the South Agricultural Building in Washington, DC, 1974–75 but that are now in the National Archives. Most of these were technical photos of erosion and fluvial phenomena, varying from stratigraphic cross-sections of accumulated sediment to broad landscapes. But of greatest interest to me, and to the reader I believe, are old landscape photographs, mostly of streams and stream-oriented phenomena, which allowed time-lapse photography of landscape changes. While this USDA collection was itself remarkable, I was later able to about triple the number of such photos from archives, local residents, and many other sources. In my view, these photos are one of the most valuable types of evidence and documentation offered so I'm using a lot of them. There are several reasons for this. First, many of the landscape changes are so profound that it might be difficult for some people to believe them without seeing them. Second, most anything can occur as an anomaly but the *many* photos show a replicated pattern much like a statistical trend. Third, since I give the exact location of each photo using the US Land Survey system, it should be useful to future researchers in documenting the effects of climate and/or land use changes on both streams and vegetation. Some photos go back to the 1880s and many date from c. 1900, so we are now looking at the second century. I know of no other region in the world with such a strong historical documentation of physical (and cultural) landscape changes.

While I have included many historical photos, they are only a small fraction of the ones I hold. Choosing among the photos was one of the most difficult things I had to do, and I regret that the rest are not here. Perhaps it may be possible in the future to place them in a repository or put them on a website.

# Biography

Stanley W. Trimble is Emeritus Professor of Geography at the University of California, Los Angeles, where he became a member of the faculty in 1975. Among his professional interests are historical geography of the environment and human impacts on hydrology, including soil erosion, stream and valley sedimentation, and stream flow and channel changes. His regional interests are the humid U.S. states and western and central Europe. Trimble studied physical science and mathematics at the University of North Alabama and earned a B.S. in 1963. Taking an Army ROTC commission, he spent 2 years as an intelligence research officer and served with the 101st Airborne Division from 1964 to 1965. After a year teaching in Europe, he earned M.A. and Ph.D. degrees in geography from the University of Georgia in 1970 and 1973, respectively.

Dr. Trimble was a research hydrologist (adjunct appointment) with the U.S. Geological Survey from 1973 to 1984 and a visiting professor at the Universities of Chicago (1978, 1981,1990), London (University College, 1985), Vienna (1994, 1999), Oxford (1995), and Durham (1998). Among courses taught by Trimble are hydrology, geomorphology, soil and water conservation, and historical geography of the American environment. He has also taught courses in environmental geology/hydrology for the U.S. Army Corps of Engineers, and he is a hydrologic/geomorphologic consultant for several agencies. Additionally, he has served on research committees concerning watershed management for the National Research Council. His awards include a Fulbright to the U.K. and the Mel Marcus Distinguished Career Award from the Association of American Geographers. Trimble was co-Editor-in Chief of *Catena*, a journal of hydrology, soils, and geomorphology, 1996-2006. He has many published articles, ranging from several in *Science* to several in historical journals, including the *Journal of Historical Geography* and *Agricultural History*. His books include *Man-Induced Soil Erosion on the Southern Piedmont* (Soil and Water Conservation Society, 1974, 2008), *Environmental Hydrology* (with Andrew Ward, CRC, 2004), and (Editor) *Encyclopedia of Water Science*, 2 Vols. (CRC, 2008). Since 1978, Trimble has owned and managed a 200-acre farm in middle Tennessee. His personal interests are classical music and jazz, classical and early American architecture, and English landscape gardens.

# *Acknowledgments*

It would be impossible to thank all those who have helped me in this effort over the past 39 years. As I indicated earlier, I owe Dr. Stafford Happ a tremendous debt, but I also owe a huge debt to many other employees of the USGS, USDA, and many other governmental agencies at all levels who have always been interested and often generous with financial and logistical support. Likewise, personnel at libraries and archives ranging from local to national were more than helpful. In particular, I thank the National Archives in Washington, DC, and the state archives and libraries of Wisconsin, Minnesota, and Iowa. The late Herman Friis of the National Archives was always an enthusiastic supporter of this work and was of indefatigable assistance there.

I'm grateful to the many agencies that provided funds and logistical support. Most important are the US Geological Survey (USGS), USDA–Natural Resources Conservation Service (NRCS), National Science Foundation (NSF), Wisconsin Department of Natural Resources (DNR), Wisconsin State Highway Department, UCLA Faculty Senate, and National Geographic Society.

I owe a great debt to my family members who were always supportive. This support ran from helping me in the field, to plotting data, to simply being willing to live in somewhat primitive conditions all summer, to enduring long drives to and from the field, sometimes several times during the summer. My older daughter, Alicia, still remembers driving back and forth from Los Angeles in a huge 1971 Ford LTD with piles of field equipment tied to a roof rack, looking all the while like the Beverly Hillbillies. One great pleasure was having my younger daughter, Jennifer, help me in the field all summer of 1993 and 1995. My wife Alice did all of the above plus tolerating my students and me in cramped quarters and trying to keep order. To begin to appreciate this, one must know that at day's end, we often would be so filthy with swamp muck that we would have to hose ourselves down and remove most of our clothes out in the yard. Indeed, my boots were almost never dry, even with two pairs. But happily, what my family seems to remember best are the waterskiing and cookouts on the Mississippi River late afternoons and Sundays. Nevertheless, neither daughter elected to follow in my footsteps and become a fluvial geomorphologist and perhaps understandably so. Alice was also of indispensable help in preparing and proofreading the book manuscript.

Many others are owed thanks. In all, probably 75–100 students helped me in the field, most without pay. Some have moved on to professorships or other professional positions and now seem to remember fondly their time in the Hill Country. Local landowners gave access to their land, and many older citizens gave useful information and sometimes wonderful historical photographs. From 1974 to 1979 my field car was a yellow 1964 Volkswagen bug with a homemade red wooden roof rack to carry all our field equipment. I mention this because it was immediately recognizable by the local citizenry, and it got to the point that we no longer had to seek permission to go onto land—people knew who we were and what we were doing so we pretty much had the run of the place. Some local citizens took an interest in the project and in us. There were many, but I do want to mention Burt and Janice Lee of Coon Valley, Wisconsin, whom we met the first year and who have been thereafter helpful as well as dear friends. This also goes for Don and Joan Doll of Stoddard.

I'm also grateful to UCLA and the Department of Geography there which gave me important support for Hill Country work from 1975 onward when I moved from the University of Wisconsin to UCLA. I particularly want to thank the two outstanding illustrators, Chase Langford and Matt Zebrowski, who did the excellent work evidenced herein and our management services officer, Kasi McMurray, who has a genius for cutting through red tape and getting things done. Their assistance in the later stages of the book preparation was indispensable.

And there's one more outstanding debt of gratitude. In early June 1977, I visited several agencies in Madison, including the State Library, the Public Service Commission (for railroad records), State Highway Department (for old bridge plans), and the Department of Natural Resources (DNR) for the records of dams and old mills. The DNR official holding them, Edmund M. Brick, was a rather no-nonsense hydraulic engineer who dutifully asked me why I wanted these records. When I told him that it was to date *buried* milldams and other buildings in western Wisconsin, he simply did not believe me. The very next week he showed up in the Hill Country to see for himself. The experience so moved him that he devoted much of the rest of his career and retired life working to make sure people appreciated the improvements in the Hill Country and he continues to write on the subject (Brick, 2010). But more than just enthusiasm, he has gathered information, helped me with surveys, done many surveys himself, elicited research grants, and even personally provided financial as well as logistical support. I owe him much.

Finally, I thank the many scientists and other scholars who over the past 39 years have encouraged me, critically read papers in preparation, acted as a referee, provided information, invited me to lecture or publish in some outlet, and encouraged or even provoked me. It's always dangerous to name people, but I do note (in no particular order) Sir Ron Cooke, Andrew Goudie, Bob Meade, Ron Shreve, Tim Burt, Des Walling, Claudio Vita-Finzi, John Costa, Tom Dunne, Karl Butzer, Michael Conzen, Dick Sartz, Gert Verstraeten, Angela Gurnell, Peter Houben, Greg Nagle, Olav Slaymaker, Reds Wolman, Pierre Crosson, Bill Renwick, Ken Gregory, Kevin Scott, Frank Trainer, Richard Hadley, Ian Douglas, George Foster, Terry Toy, Sam Skemp, Tim Beach, Janet Hooke, Peter Wilcock, John Boardman, Andy Ward, Peter Haff, Ken Gregory, Stan Schumm, Jim Knox, Will Graf and Ken Potter. Of these, I must single out Michael Conzen of the University of Chicago who, as a human historical geographer, has followed this work for 34 years and has always been supportive of it and its eventual publication. Four days with him in the region in September 2011 were especially insightful and encouraging. And when the book was in pageproofs stage Robert H. Meade went through the book with a fine-toothed comb. The improvements were significant.

# Introduction

## Natural and human effects on the land

How do landforms change with time? One may think of the natural *forces* acting on the landscape to make it change: gravity and the water from rainfall are two major ones that we commonly see. In turn, the earth normally offers *resistance* to the forces in the form of the strength of soil or rock and/or the resistance given by vegetation (Figure 0.3). For example, vegetation, with its extensive root system, may counter the force of gravity that would otherwise move soil or rock down slopes. Or vegetation may lessen soil erosion in several ways. One is that the leaves of the plants protect the soil from raindrops, thus reducing the force. Additionally, the biomass of the plant and other organic material slows the flow of water over the soil, giving it more time to infiltrate, again reducing the force. Yet another is that over a long period, plants can improve the soil by adding organic material and thus encouraging pedofauna, or animal life such as termites, earthworms, and moles. In both cases, the soil is made stronger, increasing the resistance, and also is made more permeable, allowing rainfall to infiltrate, thus reducing the force on the soil from overland flow. It will be seen that if resistance is as great as the force, no change will take place, a concept always best articulated by W. L. Graf in his many works. But when the force is greater than the resistance, earth material will be moved and geomorphic work will be done. Moreover, the *form* taken by the portion of earth in question may be traceable to the *process* that formed it. This is basic geomorphology, the study of the earth's landforms and the processes that cause change in these landforms.

Of course, humans can alter these relationships. In the Hill Country, Europeans cleared much of the natural forest and prairie and then set much of that to the plow. On the prairie, they broke the sod and exposed deep rich soil that had been forming for centuries. The initial effect was to subject the soil to the erosive effects of rainfall. But since the original qualities of the soil were still intact, there was little erosion in the early years and crop yields were good. This helped lead to the mistaken idea that the soils, especially the prairie soils,

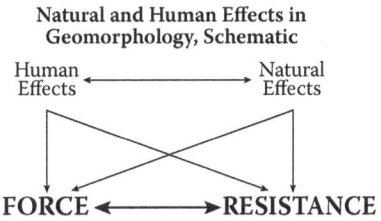

*Figure 0.3* Natural and human effects in geomorphology. (From A. Ward and S. Trimble, *Environmental Hydrology*, 2nd ed., CRC, Boca Raton, 2004. With permission.)

were "inexhaustible." But with continued depletion of the nutrients by crops, oxidation of organic material by exposure to the air, and reduction of the pedofaunal population, soil structure with its high permeability was destroyed. Thus, more overland flow (force) was directed across slopes with lowered resistance. The result was significantly increased runoff and soil erosion from each rainfall event. Later, domestic grazing animals, especially cows, became important across the region. Grazing led to changes of both force and resistance on the landscape. Trampling of the soil by many heavily loaded small hooves compacted the soil greatly, forcing more water to flow off, and over, slopes and thence into the stream channels. At the same time, the hoof disturbance, as well as the removal of vegetation by grazing and browsing, reduced the erosional resistance of the slopes and of steam channels. The result was greatly enhanced erosion of both slopes and channels.

Poor land use also affected stream flow in the Hill Country. Water that leaves the landscape as overland flow is water that will not be available either in the soil or in groundwater. Expressed another way, water that infiltrates the soil is no longer an erosion hazard. And, as a bonus, it is then available to the soil and often even to groundwater by deep percolation. Thus, decrease infiltration capacity as described above not only can cause soil erosion but it can also decrease soil moisture and lower groundwater reserves. This results in decreased flow from springs and seeps, which maintain baseflow, the flow of streams between rains. Conversely, enhancing infiltration in turn enhances baseflow of streams.

## *Two conceptual models of runoff*

The principles described above allow us to understand just how important is the infiltration capacity of the soil in the runoff process. This allows us to understand the two basic conceptual models of runoff: the variable source-area concept (also known as the Hewlett model, Ward, 1975) and the limited infiltration concept (also known as the Horton model).

**Hewlett Model.** This model assumes that the soils of a stream basin have infiltration capacities greater than the intensities of most rainfall events. Thus, the rain is infiltrated and then moves laterally through the soil to a lower part of the basin, usually near a stream or valley. There, the water accumulates until that low area is saturated. At that time, water begins to flow across the surface, but only near the stream. As the rain continues, the saturated area increases in area upslope, thus giving a longer run of overland flow. But in permeable basins, rarely would the entire basin ever have overland flow. The generic name of this concept, "variable source-area," refers to the area producing overland flow. The virgin forest or prairie as found in the Hill country was the ideal scenario for the Hewlett model. Because overland flow is so limited in area, the opportunity for soil erosion is limited.

**Horton Model.** This model assumes that soils have limited infiltration capacity, usually as a function of land management. When rainfall intensities are greater than the infiltration capacity, overland flow begins. Unlike in the Hewlett model, overland flow can occur at any point in the stream basin, even in the most upslope areas. Thus, runoff water accumulates as it flows downslope in a cascade, greatly increasing the erosive force on the landscape and increasing the area affected. The Horton model would therefore apply to the landscape as modified by historical agriculture with plowed cropland and heavily grazed areas.

Very simply put, the period from settlement to the 1930s is the transformation of the hydrologic landscape from the Hewlett model to the Horton model, which we will consider the *destructive phase*. Then from the late 1930s to the present, we see the partially successful attempt to transform that hydrologic landscape back to the Hewlett model. This we call the *conservation phase*.

## Predicting runoff based on land use

Just how much can runoff change with different land uses? Expressed another way, how much hydrologic impact can humans have? Scientists, especially from the US Department of Agriculture (USDA), have been working on this question for almost a century. While their models now are sophisticated and complex, we can get a very good idea from their simple models. For example, what happens with, say, 4 in. (10 cm) of rainfall on a modestly wet soil? With forest or prairie in good condition, runoff might range from zero to less than one inch (2.5 cm) depending on the soil (Ward and Trimble, 2004, Chapter 5). But from poorly cultivated or heavily grazed land, runoff from the same storm might range from about 1.5 (4 cm) to over 3 in. (8 cm)! Moreover, since more water is moving across the surface as overland flow, rather than moving through the soil, it moves much more rapidly so that the peak flows become much greater and more destructive to both slopes and to streams. All this will be demonstrated for the Hill Country.

## Predicting soil erosion

Like all natural and physical processes, the causes of soil erosion are complex, but they can be generally understood by use of a simple model, the **Universal Soil Erosion Equation** as formulated by Wischmeier and Smith (1978). It has since been greatly modified, becoming much more sophisticated, and now used all over the world, not only for agriculture but also for urban and recreational areas. The model was devised to predict the actual mass of soil eroded from a field in tons per acre per year or tonnes per hectare per year but is presented here in a more conceptual manner:

$$\text{Annual soil erosion} = \mathbf{R} \times \mathbf{K} \times \mathbf{S} \times \mathbf{L} \times \mathbf{C} \times \mathbf{P}$$

where:

**R** is the *erosive power of rainfall*. It is the product of the kinetic energy and intensity of each rainstorm, summed for an average year. The value of R varies with climate and varies by more than 25 times across the United States, being highest in the southeastern United States, high in the Midwest, and lowest in the arid West and the marine Pacific Northwest. Within the Hill Country, there is only moderate variance with the higher values to the south.

**K** is the *inherent susceptibility of soil to erosion* and depends on several factors, including texture (grain size), percent organic material, and structure. There is roughly a 4 times difference between the most and least erodible soils, but most agricultural soils are relatively similar in their erosional characteristics, especially in the Hill Country.

**S** and **L** refer to *declivity and length of slopes*, suggesting the potential energy supplied by the physical terrain itself. Anyone who has ever gone down a hill on skis, bike, skates, or even a coaster wagon intuitively knows the potential energy of both slope declivity and slope length. As we will see, this turns out to be one of the most important factors of erosion in the Hill Country: soil erosion tended to be most severe in the area of greatest relief and steepest slopes.

At this point, it will be observed that the foregoing factors are all natural. These compose the natural stage on which humans act. The following factors show the power of the *human hand* in causing or preventing erosion within the context of the natural factors above.

C reflects the protective effect of *vegetative cover*. Of course, the natural dense vegetation of forest and prairie gave the most protection, but these mostly were replaced by agriculture. Row crops like corn and soybeans, which normally leave much bare soil exposed, can be about 1000 times as erosive as virgin forest or prairie. Most crops and land uses fall within a much smaller range. Among common crops in the Hill County, for example, row crops such as corn and soybeans are about three times as erosive as small grains such as wheat, oats, barley, and rye, and about 10 times as erosive as hayland. However, these relative erosive values can be radically changed by conservation practices, **P**, discussed below.

**P** refers to *conservation practices*. For example, simply plowing on the contour rather than plowing up and down slope can reduce erosion by about half in regions with modest rainfall. Strip cropping, where erosive crops are alternated with grass strips all on the contour, a practice widely seen in the Hill County since the late 1930s, can reduce erosion even more. Other useful practices are crop rotations that leave grass on the soil for several years, leaving crop residues such as cornstalks on the field and fertilizing. These conservation practices are important to the eventual recovery of the Hill Country.

## Background to this study

Humans have ruptured the hydrologic system many times during the course of history, and the effects have usually been severe on human habitation. The ruin and burial of civilizations of the ancient world has been described by several investigators (Bennett, 1939; Laudermilk, 1953; Judson, 1963; Vita Finzi, 1967; Butzer, 1974; Goudie, 1997, 2006), but the new world has also had some dramatic cases of erosion and sedimentation (Gilbert, 1917; Happ, et al., 1940; Gottschalk, 1945; Trimble, 1974). There appears to be a perception, even among some geographers, that such phenomena are endemic, if not restricted, to arid-to-humid tropical and semi-tropical climates with intense rainfall patterns and/or very fragile ecosystems. However, even deep continental locations are not immune if slopes are steep, precipitation is great and intense, and land use is abusive and prolonged enough, all characteristic of the Upper Mississippi River Hill Country (Figure 0.1). Although the region may have been glaciated earlier, only part was during the last (Wisconsin) period, so the landscape is mostly stream-carved, and the regionalization is based on local relief (>200 feet or 60 m). The region is generally described as a submaturely to maturely dissected plateau with local relief of about 500 feet (150 m), steep valley sides, and narrow valleys. It is underlain by consolidated sedimentary rocks—older, coherent to very friable sandstone and shale, and younger, fairly resistant dolomite. The rolling uplands are capped by Pleistocene loess, and the valleys contain a series of sandy terraces that are probably of Pleistocene age (Martin, 1965). Most upland topsoils have a silt loam texture.

Accelerated erosion and sedimentation, primarily from agriculture, were noted by early soil surveys (e.g., Gray et al., 1929) and when later investigated in detail by the US Forest Service, the University of Wisconsin, and US Department of Agriculture (e.g., Bates and Zeasman, 1930; Bates, 1931; McKelvey, 1939; Happ et al., 1940; Adams, 1942; Kunsman, 1944), the problem received national attention (Bennett 1939). One of the most dramatic findings was the partially buried village of Beaver in the Whitewater River Basin, which understandably drew much attention (SCS, 1942b; Siebenaler, 1955; H.B. Johnson, 1976). Also reported was sedimentation in the Galena River at Galena, Illinois, which transformed a spacious navigational waterway and steamboat landing into a narrow, sediment-choked stream (Adams, 1942; Brown, 1948). Flooding in both locations had become more frequent and severe, both because of sediment-filled channels and from increased storm runoff from eroding agricultural land.

However, Beaver and Galena were far from being isolated instances, and the problems suffered at those two locations were, in fact, widespread. The hydrologic problems affected not only villages but also isolated farmsteads, roads, and bridges, water-power sites, and agricultural production. Moreover, the degree of damage differed with location, depending on what stream processes were and are in progress, and there is a general system to this spatial variation. But the good news is that since the 1940s, agricultural land use practices have improved dramatically with commensurate improvements in the natural environment. This study is an attempt to tell the whole story.

## Overview of the problem: physical and cultural causes

In reading a book, it is often useful to know where the story is headed. Thus, the mind does not detour so much but rather is able to keep matters in perspective. This short summary is offered in that spirit.

Settlement of the region by Europeans began in southwestern Wisconsin during the 1820s. The big draw at that time was minerals, especially lead, and little agriculture developed there until the 1840s (Brown, 1948). Meanwhile, agricultural settlement was taking place in northeastern Iowa. By 1850 the southern part of the region was thickly settled, and by 1880, the population distribution was widespread. The initial agricultural occupance in much of the region was the growing of wheat, which peaked about 1880 (Brown and Nygard, 1941; H. B. Johnson, 1976). This was replaced by dairying and a more diversified agriculture. Soils which had already been strained by continuous wheat cultivation were now subjected to corn without adequate crop rotations. This further depleted organic content, destroyed soil structure, lowered infiltration capacities, and generally deteriorated the good original hydrologic characteristics, all being further reduced by the resulting erosion of topsoil. Meanwhile, the number of cattle, sheep, and other grazing animals increased rapidly. Pastures and woodland were increasingly overgrazed, thus reducing vegetative cover and compacting the soil.

The increased runoff, in conjunction with decreased vegetative resistance, caused severe erosion on upland fields and perhaps more important, caused extremely severe channel erosion on hillsides and in headwater tributary channels. (McKelvey, 1939; Happ et al., 1940; Knox, 1972, 1977, 1987, 2001, 2006; Trimble 1975a,b, 1976a, 1981, 1983, 1999, 2009a; Trimble and Lund, 1982). The tributary channels were themselves made more vulnerable to erosion by continuous grazing, especially by cattle (Happ et al., 1940). Increased flooding covered tributary floodplains with erosional debris while the channels themselves were trenched and widened, increasing downstream sediment loads (Happ et al., 1940; Knox, 1972, 1977, 1987, 2001, 2006); Trimble, 1975a,b, 1976a, 1981, 1983, 1999, 2009a; Trimble and Lund 1982). Downstream in the main valleys, floodplains and stream channels were being aggraded at a rapid rate. This accretion was augmented significantly by huge gullies, many in the slopes of the high Pleistocene Terraces, which were generally located along the broader main valleys in much of the region. These deleterious processes generally appeared late and developed rapidly. Accelerated erosion and sedimentation must have been in progress from the time of early settlement, but it was about the turn of the 20th century or shortly thereafter before they got out of hand and became disastrous (Sartz, 1975; Trimble and Lund, 1982; Trimble 1983). Detailed investigations and measurements in the Coon Creek Basin, Wisconsin, for example, show that floodplain accretion during the two decades between 1920 and 1940 was very rapid. During the 1930s, it exceeded 12,000 tons per year for every square mile of drainage area (4,200 tonnes/km/yr), or a vertical accretion rate of about 6 in. or 15 cm per year (Trimble and Lund, 1982, Figure 10). Further investigations indicate that these rates were probably typical for most streams directly tributary to the Mississippi River, but other valleys may have filled at somewhat lower rates (Happ, 1944; Knox, 1977, 1987, 2006; Johnson, 1976; Magilligan, 1985). It was during this period, the first half of the 20th century, that the greatest damage occurred to human settlement in the Hill Country.

In contrast, accretion rates for several centuries before agricultural occupance in the Hill Country were so low that a deep Mollisol had developed on floodplains, a process that would have required centuries of stable environment (Owen, 1847, 1852; McKelvey, 1939; Happ et al., 1940; Knox, 1972, 1977, 1987; Trimble, 1975; W. C. Johnson, 1976; Trimble and Lund, 1982). W. Johnson (1976) estimated this primeval accretion rate to have been only about 0.03 cm/yr while Knox (1987) gave a similar value of 0.04 cm/yr. Settlers were able to build their homes, barns, and villages on low terraces, and in some cases on higher portions of the floodplain, apparently with little fear of flood damage. This condition lasted until as late as World War I (1914–1918) or so in some areas before such locations became untenable.

Developments since c. 1940 have been far happier. An organized soil conservation program was started in the Hill Country during the mid-1930s and by the 1970s, soil conservation measures were almost ubiquitous in the region (Trimble and Lund, 1982). The transformation of the landscape was striking: rectangular fields were changed to contour patterns (Johansen, 1969; Trimble, 1975a; H. Johnson, 1976; Trimble and Lund, 1982). Raw, gullied slopes and hillsides were mostly healed and are now covered in proper crop rotations, good pasture, or generally ungrazed forest. The soil conservation program has done its job; soil erosion rates of 1975 were estimated to be less than one-fourth those existing in 1934, and valley sedimentation rates were only 2% to 4% of the previous ones (Trimble 1999, 2009). Resurveys of floodplain profiles in 1989–1993 showed that the most recent annual vertical accretion rate was only 0.2 in. (0.5 cm) per year (as compared to about 6 in. [15 cm] per year in the 1920s and 1930s), and moreover, the area affected was much less than had been previously covered (Trimble, 2009a). Additional surveys through 2005 have all suggested continuing low accretion rates. Although he did not establish more recent rates, Knox found that 1940–1985 sedimentation rates in the main valley of the Galena River were less than 20% of those existing in the decade 1930–1940. Decreases in the size of stream channels and their greatly improved condition are strong evidence that high stream discharges are less common than the destructive phase that occurred during the first part of this century (Trimble, 1975b, 1999; Knox, 1977, 1987; Magilligan, 1985). However, sediment yield from larger streams tributary to the Mississippi River does not appear to have decreased greatly since the 1930s. This is partially because upstream channel banks are eroding where possible, thus removing some of the modern sediment deposits and transporting the material some distance downstream (Trimble, 1975a,b,d, 1981, 1983; Knox, 1987). Thus, despite the great decrease in upland soil erosion rates, sediment accumulation rates in the navigation pools of the upper Mississippi River have not decreased (Jackson et al., 1981).

*chapter one*

# The physical region and primeval landscape

It is the physical landscape on which humans impose their cultural landscape and activities, thereby changing parts of that physical landscape. Much of that original or primeval physical landscape such as climate and physiography at the regional scale has changed only slightly since settlement by Europeans. As we will see, these factors are highly influential because they supply potential or even kinetic energy. But part or even much of that original settlement landscape, such as vegetation and stream morphology, has been greatly changed, and only by knowing the original state can we deduce what has happened. We can find no better description of the primeval landscape of the Hill Country than that given by the noted early explorer and natural scientist, David Dale Owen, c. 1845, in northeast Iowa along the Upper Iowa River:

> We find the luxuriant sward clothing the hill-slope even down to the water's edge. We have the steep cliff, shooting up through it in mural escarpments. We have the stream, clear as crystal, now quiet and smooth and glassy, then ruffled by a temporary rapid, or, when a terrace of rock abruptly crosses it, broken up in to a small, romantic cascade. We have clumps of trees, disposed with an effect that might baffle the landscape gardener [an allusion to the 18th- and 19th-century creators of English landscape gardens such as Lancelot "Capability" Brown and Humphry Repton], now crowning the grassy height, now dotting the green slope with partial and isolated shade. From the hilltops, the intervening valleys wear the aspect of cultivated meadows and rich pasture-grounds, irrigated by frequent rivulets that wend their way through fields of wild hay, fringed with flourishing willows. Here and there, occupying its nook, on the bank of the stream, at some favorable spot, occurs the solitary wigwam. On the summit-levels, spreads the wide prairie, decked with flowers of the gayest hue,—its long, undulating waves stretching away till sky and meadow mingle in the distant horizon (Owen, 1852, pp. 65–66).

Not content with his eloquent descriptions, Owen (1844, 1847, 1852) made near-photographic quality sketches of the region's fluvial landscapes (Figure 1.1). For a similar but somewhat less colorful and articulate description of the region in 1817 by Stephen Long, a U.S. Army officer (see Kane et al. 1978, p. 57). However, when Long returned to the region in 1823, he had apparently found his muse when he described the streams and valleys of northeastern Iowa:

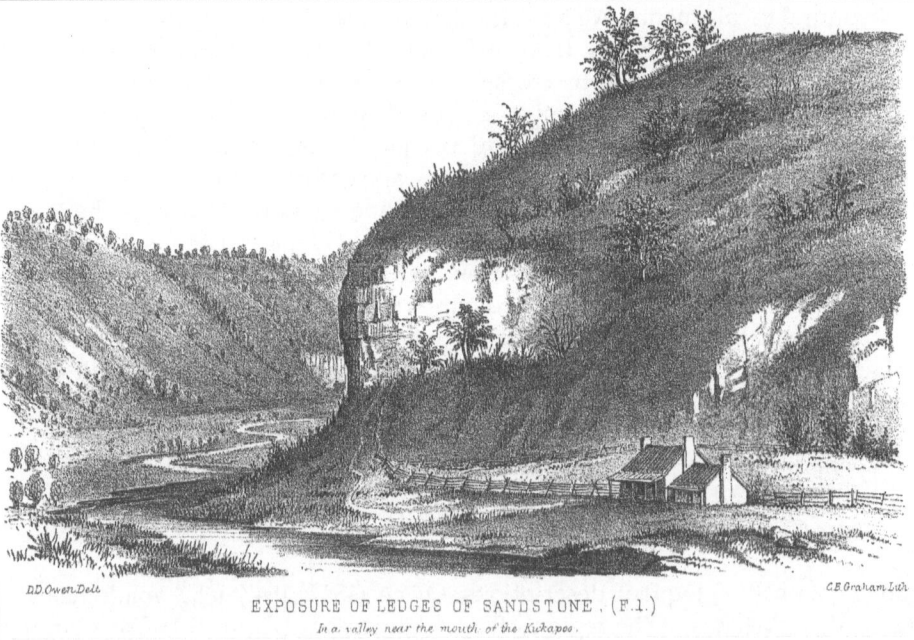

*Figure 1.1* Top: A sketch of the northeastern Iowa landscape by the famed naturalist David Dale Owen, 1852. A similar sketch appeared in Owen (1847) that had slightly more detail but the Native Americans were not included. Bottom: A sketch of the Kickapoo River Valley close to the junction with the Wisconsin River. Note the settler's house and farmstead with rail fencing (Owen, 1847). In all his sketches, Owens portrays the stream bottoms as mostly treeless, but this might have been for the sake of clarity.

### Chapter one:   The physical region and primeval landscape

> ... valleys of water-courses sunk far below the general surface covered with a rich carpet of grass and flowers, and exhibiting their crystal streams meandering in graceful folds conspire to give both beauty and majesty to the landscape. (Kane, et al., 1978, p. 146)

Another early view of the region was given in 1854 by Edward Daniels, the first state geologist of Wisconsin:

> About one-third of the surface is prairie, dotted and belted with beautiful groves and oak openings. The scenery combines every element of beauty and grandeur— giving us the sunlit prairie, with its soft swell, waving grass and thousand flowers, the somber depths of primeval forests; and castellated cliffs, rising hundreds of feet, with beetling crags which a Titan might have piled for his fortress. (cited in Martin, 1965, p. 84)

So what did the settlers themselves think of the primeval landscape? Apparently relying on accounts of his parents, Alex Siebenaler as an elderly man wrote in 1955:

> When the first white men came, they found it [the Whitewater River Valley, Minnesota] just a little short of paradise. They found about everything they needed for a living. A beautiful small river, fed by clear sparkling creeks and springs, meandered its way down the valley. . . . Catching fish was no problem, if anyone was hungry for fish, they just fished. If they wanted game, they went hunting and usually came home with all they needed. The woods gave them all good timber for their buildings and plenty of good, clean water was close at hand for human and livestock use. Most important of all, the soil was rich and fertile, some of the best in the world, a deep, dark silt loam on which almost anything would grow and produce an abundant harvest.

What is the Upper Mississippi River Hill Country region, and how is it delineated (Figure 0.1)? It might be defined on the basis of bedrock or glacial geology. The region has a fairly uniform and simple, relatively flat-lying geology and is sometimes called the "Paleozoic Plateau" for its common geologic age. A fairly complete geologic cross section for the areas of greatest relief is that found in Houston County, Minnesota (Figure 1.2), but some other areas may have a somewhat different cross section, particularly the more maturely dissected areas in the northeastern part of the region for which parts of the upper strata shown in Figure 1.2 are missing.

From a high point in the vicinity of Houston County, Minnesota, the strata dip to the south, north, and west. There is no significant faulting or folding. This geology is important in controlling the largely erosional topography featuring dendritic drainage and ridge and valley topography with rolling upland and maturely to sub-maturely dissected topography. This description characterizes much of the landscape north of the Wisconsin River and most of that immediately along the Mississippi River, a topography that in relative terms might be termed *hilly* or even *rough* because it has steep slopes and presents limited level to gently rolling areas suitable for agriculture. Somewhat more *mature*, rolling, and even muted with less relief in some places are the Cambrian

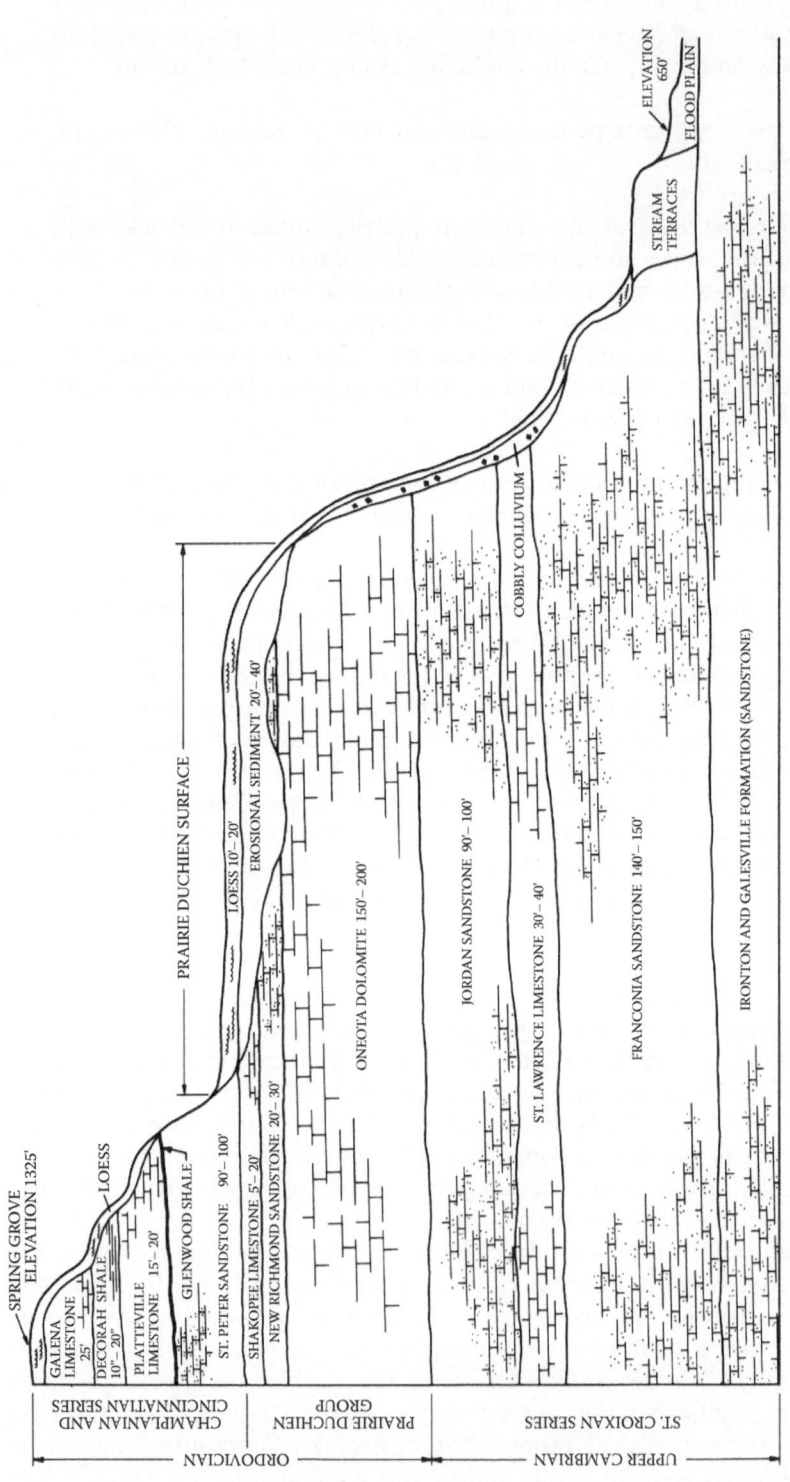

*Figure 1.2* Geologic cross section of Houston County, MN, drawn in a west to east direction. (USDA, NRCS, *Soil Survey of Houston County, Minnesota*, 1984.)

sandstone hills in the northeastern part of the region, where the overlying and more resistant limestones and dolomites are missing. On the western margin and the plateau lying south of Military Ridge (just south of the Wisconsin River), the landscape could be better characterized as "youthful," being a plateau incised by streams but presenting far more level to gently rolling land. Most of the region was covered by a few feet of Pleistocene loess, much of which remains, and this silt is the basis for the most productive soils of the region.

The two most commonly found rock strata in the central and southern areas, the Prairie du Chien (Oneota) dolomite and the St. Peter sandstone, are Ordovician in age. The sandstone is porous and friable and can be a source of sediment where exposed in a road cut. It is also an aquifer and feeds upland springs. The more resistant dolomite often crops out along steep valley walls and forms the escarpments and picturesque rocks on the high bluffs along the Mississippi River that sometimes fancifully appeared as castles and fortifications to early explorers. Below the Ordovician are the mixed Cambrian strata. The Trempeleau formation consists of sandstone, siltstone, and shale and appears at the surface to the northeast of the region, giving a more muted relief (Zakrzewska, 1971). The resistant Jordan sandstone is the most conspicuous member of this formation and sometimes forms cliffs along valley walls. The Franconia shale/sandstone and the Dresbach sandstone form the lower valley slopes with the sandstone being the important aquifer. Its springs influenced the location of many farms and homes in the region, especially east of the Mississippi River. The lower slopes are a complex of loess and older colluvium, now often covered with historical colluvium. The old, dark alluvial soils are now covered with up to more than 15 feet of historical alluvium. More detailed geologic descriptions are given in Trowbridge, 1921; Chamberlain and Salisbury, 1885; Shaw and Trowbridge, 1916; Martin, 1932, 1965; Bates, 1939; and Trewartha and Smith, 1941. Most USDA soil surveys also have excellent geologic information for the particular county or area surveyed.

## Physiography

The region is often called the *Driftless Area*, and it is true that the inner part of it, for some unknown reason, escaped at least the last (Wisconsin) stage of glaciation. But whether or not it was ever glaciated has little bearing on this study. What really makes the region distinctive from the glaciated and gently rolling surrounding area is its hilly topography and steep slopes. And because this study deals with soil erosion and consequent sedimentation, it is the land surface configuration that becomes paramount.

Perhaps the best available quantitative description of this hilly topography is the map of local relief created by Trewartha and Smith in 1941 (Figure 1.3). It is based on topographic relief within a rectangle of about 17 mi$^2$ (44 km$^2$). Note that the threshold relief to be included in the region is 200 ft (67 m) but that the relief increases toward the Mississippi River to more than 600 ft (183 m) in some areas. Trewartha and Smith (1941) also attempted to come up with a map of slopes. Using a threshold average slope of 5%, they outline the region in a pattern similar to the average relief map above (Figure 1.3). Moreover, the degree of slope shown on their map also accords well with the amount of relief. Unfortunately, data were inadequate to complete the slope map for the entire region. Other studies (Read, 1941; H. Johnson, 1976; Argabright, et al., 1996) outlined the region in a similar way (Preface, Figure 0.1).

By far the best qualitative depiction of the physiography of the region is the diagram drawn by Trewartha and Smith (Figure 1.4), which shows the landscape as it might be seen from near space on a perfectly clear day. As already noted, the mature dissection of

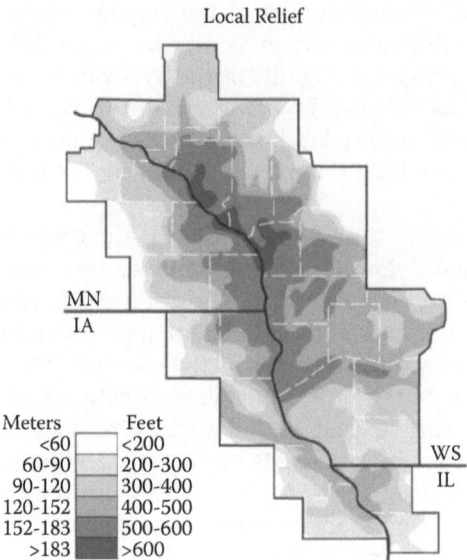

*Figure 1.3* Local relief in the Hill Country. (Modified from Trewartha, G. and Smith, G.-H, *Annals, Assoc. Am. Geogs.*, 31, 26, 1941.)

landscape or "hilly" quality with only remnants of the original rolling upland is clearly greatest between the Wisconsin and Chippewa Rivers in Wisconsin. Scattered but large areas of such hilly land can also be seen in Minnesota, Iowa, and Illinois. These are areas of high soil erosion potential as a function of slope. The major exception is the northeast edge of the region, part of what is often termed the "Central [Wisconsin] Plain," which is largely glacial outwash or old glacial lake beds. Much of it is wetland or sandy or both. Most of this is poorer agricultural land and, as will be seen in the following analysis (Chapter 2), was relatively unused for agriculture.

Hamlin Garland (1860–1940) was a writer and novelist born just north of La Crosse, Wisconsin, who had a profound appreciation of the natural landscape and gave many vivid descriptions of the early Hill Country. The Central Plain was described well by Garland (1926, p. 211) in the trek of his character Richard Graham across the area in the late 1850s:

> ... the land became trackless. Singular castle-shaped ledges of rock rose out of the level valley floor which was covered for many miles with groves of dwarf-oaks alternating with marshes and thickets of willow and birch. All settlement was left behind, for the soil of this region was sandy and poor. ...

By contrast, the area of Wisconsin south of Military Ridge as well as large areas in the western parts of Minnesota and Iowa show a "youthful" landscape with a level to rolling upland surface that is dissected by only the larger streams. Again, Garland (1926, p. 222) vividly describes the vision of this region in western Houston County, Minnesota, in the late 1850s. His characters are traveling westward and have just traversed and left the hilly country immediately west of the Mississippi River:

Chapter one: The physical region and primeval landscape

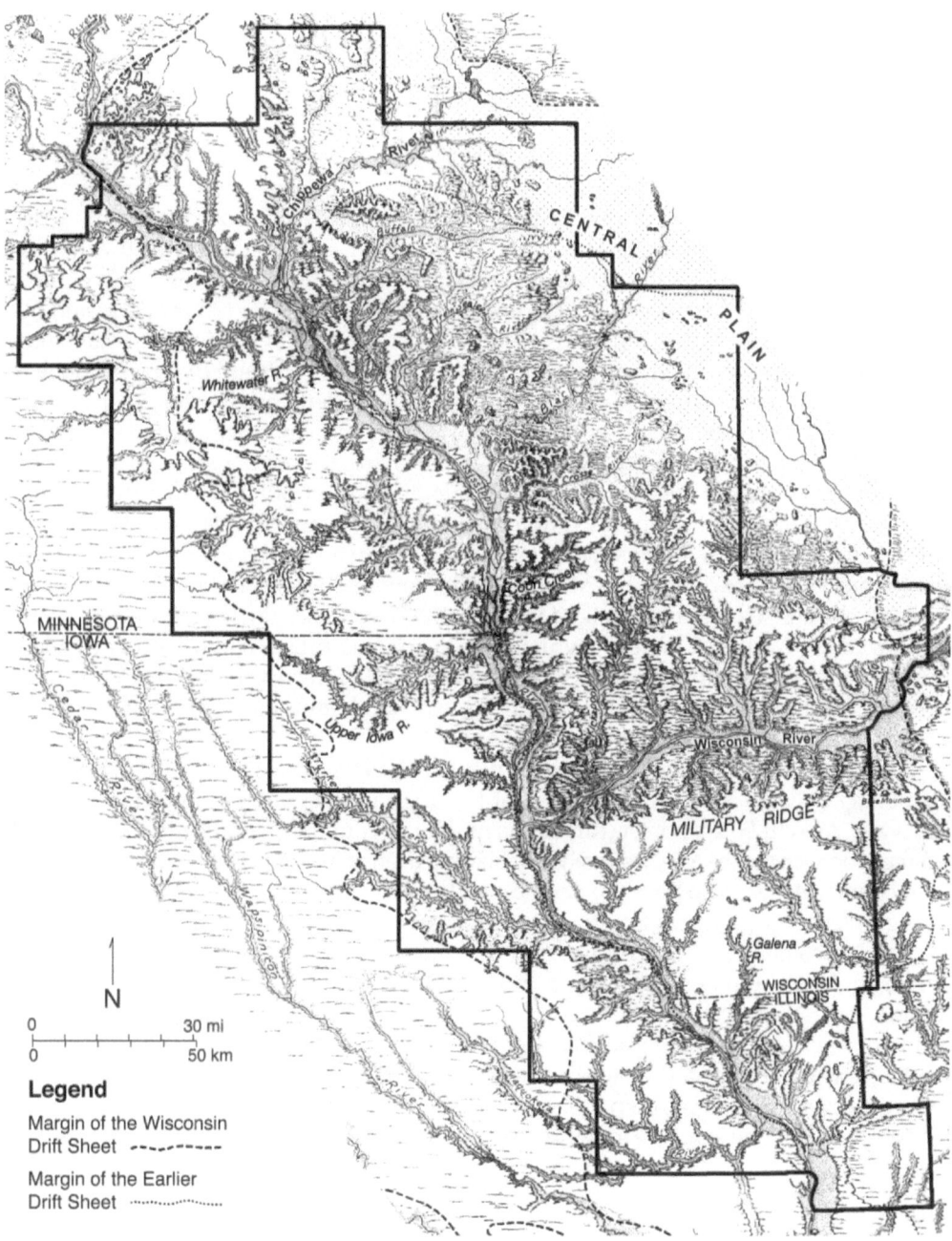

*Figure 1.4* Physiographic diagram of the Hill Country. (Modified from Trewartha, G. and Smith, G.-H., *Annals, Assoc. Am. Geogs*, 31. foldout map 1941.) (Refer to Figure 0.1 for more detailed place names.)

> Illimitably rolling to the west and northwest, the grassy ocean outspread, radiantly green, brilliant with flowers, laced with streams, and islands with groves of trees, with only an occasional settler's cabin lying like a lonely sail against a green wave. Here were the ideal homesteads of their dreams. Here were the farms without a hill or stone...

Because of the greater area of low slopes, these are generally areas of lower soil erosion potential. As we will see in Chapter 2 and later chapters, these differences of physiography had profound influences on the agricultural and erosional history of the region.

Perhaps the most striking feature of the Hill Country is the wide and deep valley of the Mississippi River (Figure 1.4). Carved out by glacial meltwater, along with much of the Hill Country, during the Pleistocene epoch, it is now as much as 4 mi (7 km) wide and 500 ft (150 m) deep, usually with steep bluffs to either side. The glacial stream was much deeper, but sediment has since filled much of the valley. Often the bluffs are lined with huge rock outcrops, cliffs, and fortress-like formations that reminded early travelers of the Rhine River with its castle-studded gorge in Germany. Before the construction of many low-head dams in the 1930s to create a 9 ft (2.7 m) navigation channel, the mainstream river was usually only a few hundred feet wide. Geomorphologists would call this a manifestly "underfit" stream because there is no way that the present stream could have carved out the huge valley in which it flows. Readers wishing to know more about the formation of the Upper Mississippi Valley are directed to Martin (1965), Knox (1985), and Fremling (2005).

With the construction of navigation dams, the river was transformed into a series of shallow lakes (<25 ft or 8 m deep), which usually are about 15–25 mi (25–40 km) in length (Bogue, 1990). The lakes may extend from bluff to bluff directly behind the dams but farther upstream, much of the area, mostly the old floodplain, has been transformed into areas of swamp and marsh termed "backwaters."

As was the case with the Mississippi River, many of its tributaries are likewise underfit, often being only 20–50 ft (6–19 m) wide but flowing through valleys as much as a mile (1.6 km) wide. As will be seen later, this underfit quality can have significant implications for sediment delivery to the Mississippi River. Having been adjusted vertically to the Mississippi River over time, they also have the deep fill and, mostly upstream of Prairie du Chien, have the high Pleistocene terraces found along the master stream, but of smaller extent.

The Wisconsin River is also underfit, having also been a glacial meltwater outlet. It still struggles to transport sand from melted glaciers, and it is a textbook example of a braided stream, the configuration a stream takes when it is overcharged with bedload sediment like sand or small gravel, a scenario that we will revisit later in this study with the debris from accelerated historic erosion. The Wisconsin also has a very broad valley with wide, sandy floodplains and terraces. These are used for agriculture but require irrigation because of the sandy, "draughty" soils, which do not hold moisture between rains.

## Climate

The region lies in an area of humid continental climate (Strahler, 2010). Lying in a continental area at higher midlatitudes, the region experiences extremes of temperature with warm to hot summers and cold, long winters. Summer temperatures sometimes exceed 100°F (38°C), but winter temperatures can plunge as low as −40°F (−40°C). Annual precipitation averages about 30–34 in. (75–85 cm), but there can be great variation between consecutive years. There is a decided summer maximum of precipitation when temperatures are warm to hot and the air is moist and unstable. The summer combination of warm to hot temperatures, plentiful moisture, and long days is perfect for growing crops. In winter, the low temperatures permit little moisture in the air, and precipitation then falls largely as snow, which may stay on the ground for long periods.

In summer, warm, moist air from the Gulf of Mexico is pulled up to the region by advection. Because the air is so moist and there is such a high temperature gradient between the air at the surface and the air aloft, the air is often atmospherically unstable so that violent thunderstorms are common. While not as severe as those in the southeastern United States, the size and intensity of these storms are much greater than those found in the old countries of Western Europe. The storms so exceeded what most Western Europeans had experienced that their violence actually frightened the early settlers (H. Johnson, 1976). This heavy rainfall is a basic principle in understanding soil erosion in this region and, indeed, for the eastern United States (Trimble, 2011). The problem was that rainfall intensities were greater than the infiltration capacities of soils after being used for agriculture. While virgin soils under forest or grass could absorb most rainfall events, soils used for agriculture eventually lost much of that ability, a concept that will be expanded on later. Rainfall that does not infiltrate normally becomes overland flow and erodes the soil. Most settlers, even "old Americans" from the northeastern United States, had never experienced such violent storms, and, in any case, they did not have the technology to deal with them so that they might mitigate overland flow and erosion (Trimble, 2011).

The contrast between rainfall intensities in the Hill Country and Western Europe is much greater than most people, even geographers, may realize (Figure 1.5). Because data are incomplete for Western Europe, the northwestern United States, which has a similar west coast marine climate, is used as a surrogate with supplementary data from the UK lowlands, the agricultural heart of Britain. It will be noted that the 100-year storm in the Hill Country is about twice as great for longer storm durations of up to 24 hours but for

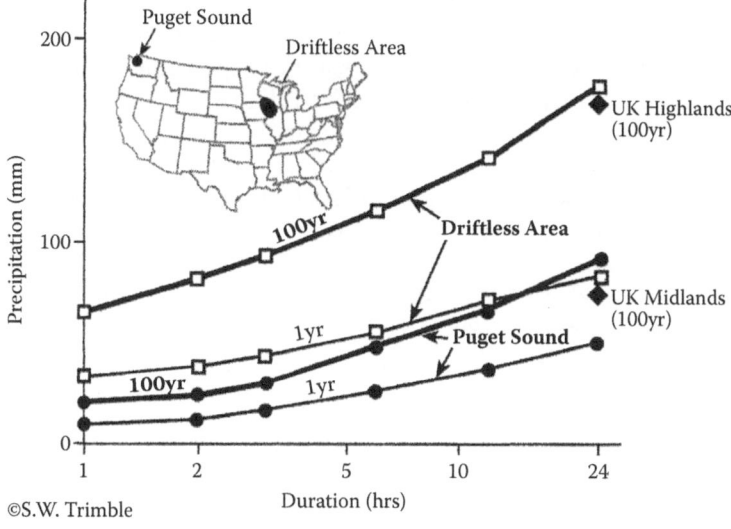

*Figure 1.5* Frequency-magnitude relationships for precipitation events, Hill Country or Driftless Area (humid continental) and Puget Sound and United Kingdom (marine west coast). (From Trimble, S., in Bennett, S., and Simon, A. (eds.), *Riparian Vegetation and Fluvial Geomorphology*. American Geophysical Union, Washington, 2004.) Note that the 24-hour, 100-year event is about 2 times greater in the Hill Country while the 1-hour, 100-year event is 3 times greater.

shorter durations is about 3 times greater. For smaller storms like the 1-year storm presented, the contrast is similar; storms are 2 to 3 times greater in the Hill Country than in western-most Europe. Note, however, that 100-year storms in Britain for all durations are about equal in magnitude to *1-year* storms of the Hill Country! Indeed, they are actually greater in the Hill Country for the shorter-duration events. This means that storm rainfall totals equaled or exceeded on the average every *year* in the Hill Country are equal to storms equaled or exceeded only once per *century* in Britain. While continental locations in Europe would experience somewhat higher intensities than Britain, they would still be moderated by the marine influence.

There are two extremely significant features of these larger rainfall events. The first is that much more erosional force is directed to the soil. Recalling from the Introduction that the erosive power of rainfall, **R,** is essentially the annual sum of the product of total kinetic energy and intensity of rainfall for each event at any location, it will be seen that R must be much higher for the Hill Country than for Western Europe. We do not have an R value for Europe, but we can use the Puget Sound lowlands as a surrogate value as in Figure 1.5, and by this approach, we find that the R value in the Hill Country is about 6 to 7 times that of the Puget Sound, meaning that 6 to 7 times more soil erosion would be experienced, everything else being equal (Troeh et al., 2004).

The second significant feature of this excessive rainfall in the Hill Country is the potential to cause overland flow and greater runoff. We may see this demonstrated by comparing rainfall intensities to the infiltration capacities of soil. Soils are categorized into *hydrologic groups* depending on their sustained infiltration capacities with the A group (8–12 mm/h) being the highest and D (<1 mm/h) being the lowest. When we then compare rainfall intensities to soil infiltration rates (Figure 1.6), the *excess* becomes readily apparent, and it is this excess that can, especially under conditions of human disturbance of the soil, rapidly flow off the surface, creating rills and gullies on slope and flooding downstream. Hence, the contrast of the *excess* between the United States and United Kingdom can be much greater than just the contrast in event magnitudes (Figure 1.5).

Note that the excess is even greater in the southeastern United States. This excessive rainfall was a damaging force that Europeans had not experienced, and they therefore had no agricultural practices to effectively manage it. Not surprisingly, the Southeast also experienced severe erosion, perhaps even greater than the Hill Country (Trimble, 2011).

There is a small difference of R within the Hill Country. For the northernmost part of the region, the value is less than about $3 \times 10^4$ ft-tons per acre-year ($6 \times 10^6$ j/ha-yr.) but in the extreme south, the value exceeds $3.5 \times 10^4$ ft-tons per acre-year ($7 \times 10^6$ j/ha-yr.). Thus, from the effect of climate alone, one could expect about 10–15% more erosion to occur in the average year in the southern area of the Hill Country given the same physical and land use conditions.

A recurring question is climate change: has there been any during the period of historical agriculture? More specific to this study is rainfall: has it become more or less erosive over time? While the historical data to answer this question are not as detailed and complete as one would like, it appears that the climate as measured by annual rainfall was wet in the mid-19th century, becoming somewhat dryer as the 20th century began and became driest in the 1920s to 1930s. Since then, the trend has been decidedly wetter (Figure 1.7). While the annual rainfall is not in itself a good predictor of soil erosion, the big, erosive rainstorms appear to be more common in wet years. Thus, we may assume that rainfall was most erosive at about the time of settlement, became less erosive as the 1930s approached, and has become more erosive since then.

# Chapter one: The physical region and primeval landscape

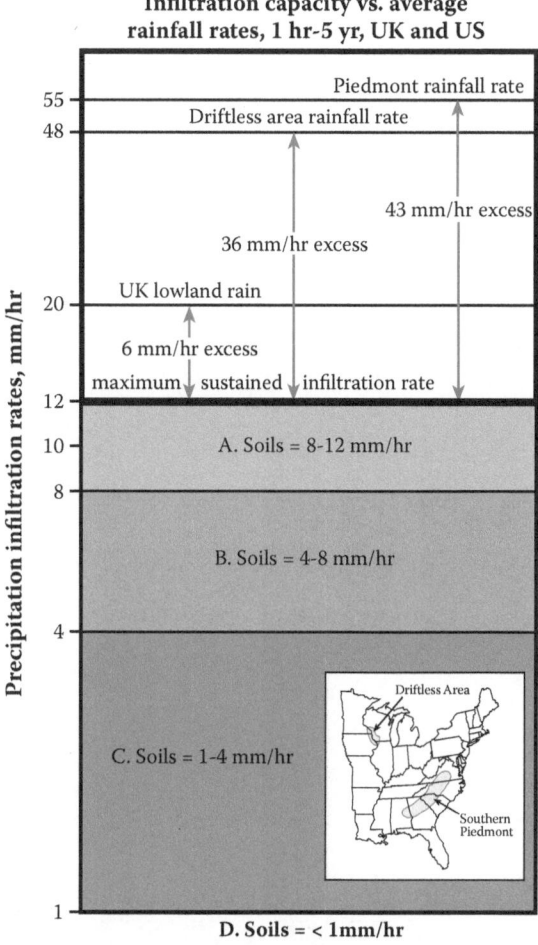

*Figure 1.6* Infiltration capacities of soils versus average rainfall rates, United States and United Kingdom. The difference is excess rainfall that flows off the surface as overland flow. Note that it is much greater for the United States than for the United Kingdom. (Modified from Trimble, S., in Brunn, S. (ed.), *Engineering Earth*. Springer, Berlin.)

Several studies support the climatic trends described above. Knox et al. (1975) show the annual flood series for the Upper Mississippi River at Keokuk, Iowa, c. 1850–1973, which features greater but declining floods in the 19th century with the declines continuing to about 1920. Then after about 1940, the floods gradually increase, reaching the mid-19th-century levels by the 1970s. Unfortunately, this is an imperfect indicator for the Hill Country alone because it includes such a large area, mostly outside the Hill Country, and because it includes regional snowmelt events. But the real question is the occurrence and magnitude of large and extreme rainfall events: are those increasing as indicated above? The first—and benchmark—analysis to quantify the occurrence of such events, a true magnum opus, was by David Hershfield (1961). Working with about 40–60 years of rainfall data, he established rainfall frequency-magnitude relationships for most of the United States. For the Hill Country, he found the 100-year, 24-hour event to be about 4–5 in. (10–12.5 cm). Over three decades later, Huff and Angel (1992) combined the more recent data

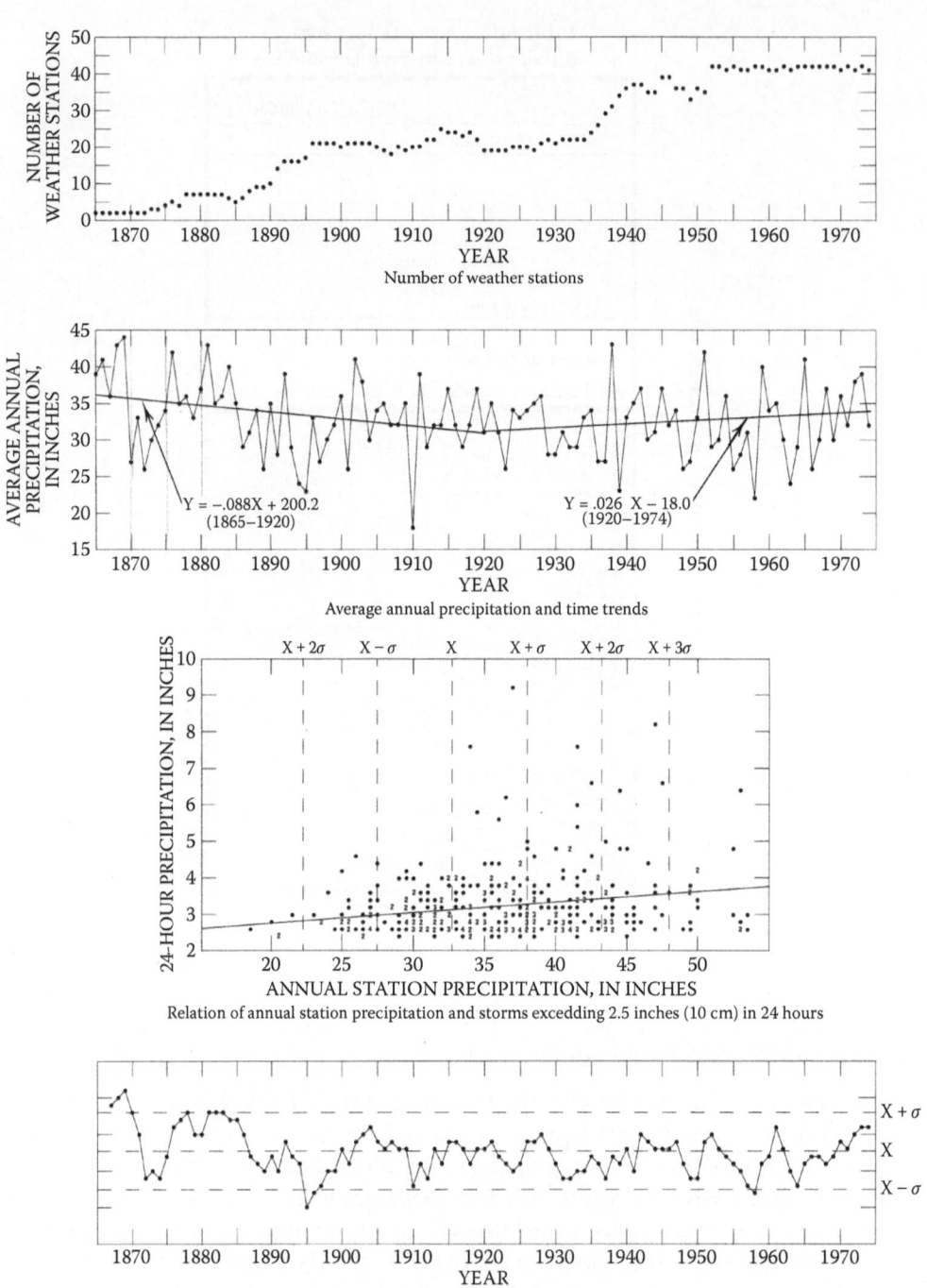

**Figure 1.7** Historical precipitation trends in the Hill Country (from Trimble, S., and Lund, S., *US Geol. Surv. Prof. Paper.* 1243, 1982). The data show a nadir of precipitation in the early 20th century and increasing wetness and storms since 1930. See also the annual flood series for the Upper Mississippi River for the period c. 1850–1975, which shows a similar pattern (Knox et al., 1975). (Continued)

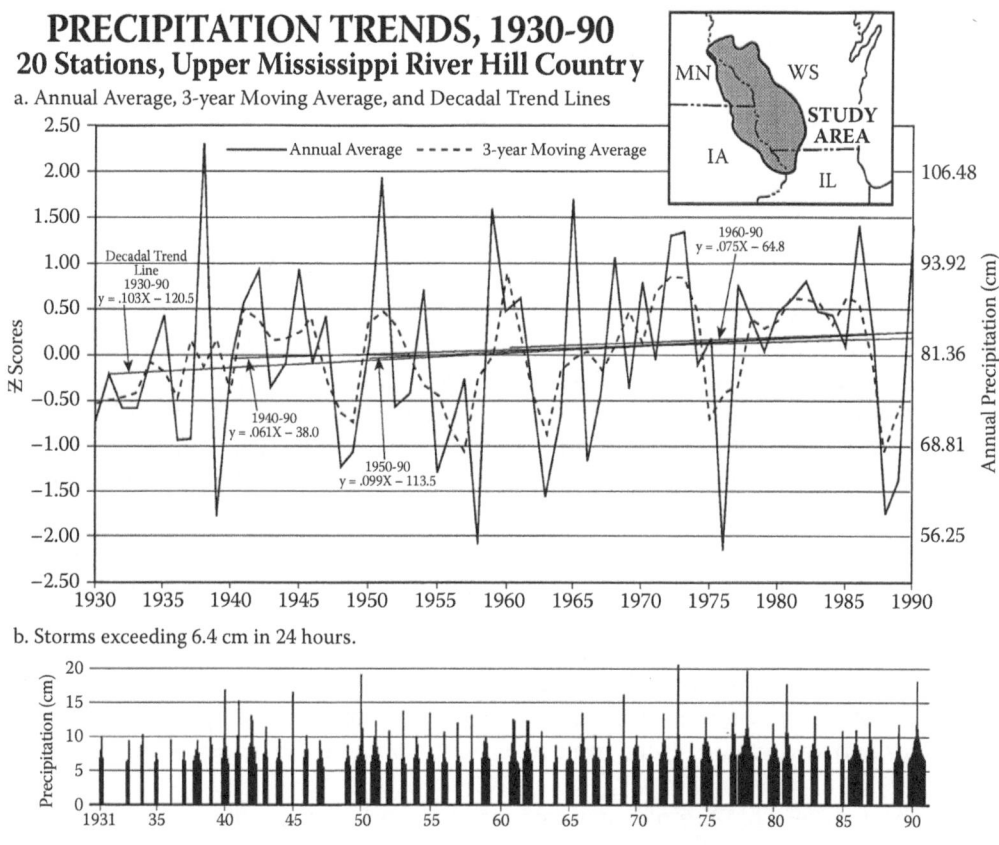

*Figure 1.7 (Continued)* Historical precipitation trends in the Hill Country (Trimble, unpublished). The data show a nadir of precipitation in the early 20th century and increasing wetness and storms since 1930.

in their analysis, and they found the 100-year, 24-hour event in the Hill Country to range from about 6 in. (15 cm) in the north to over 7 in. (17.5 cm) in the south! In other words, they found the 100-year, 24-hour event to be almost 40% larger than previously thought. Or perhaps it had increased during this period. By comparing the older 1900–1940 data to that of the newer 1940–1980 period, they found as much as a 20% increase in size of storms (for a given duration and return period) in Illinois where data were abundant. However, data limitations for parts of the Hill Country made the comparison there uncertain. More recent analyses available from USDA-NRCS indicate the 100-year, 24-hour event to be about 5 in. (12.5 cm) in the north and east of the Hill Country increasing to slightly over 6 in. (15 cm) in the south and west. So while the data may be somewhat confusing, they certainly indicate that storms for a given return frequency are getting larger. This conclusion appears to be underlined by the occurrence in 2007 and thereafter of several extreme storms of up to 18 in. (45 cm) of rain in 24 hours. The return frequency of such storms is still uncertain, but they are thought to be about 500–1000 years. The final chapter of this book considers the erosional effects of one such storm in August 2007.

These apparent climate changes loom very large in this study because, as we shall see, they are *negatively correlated with the changes on the landscape*. In particular, the strong wet trend since the late 1930s (Figure 1.7) is negatively correlated with the great decrease of local flooding and erosion since then. Thus, we can discount climatic amelioration as the cause of the improvement. Stated in geomorphological terms, while the *forces* on the landscape have become progressively greater since the 1930s, the *resistance* to those forces has been even greater so that less geomorphic work (erosion) is being done.

## Vegetation and soils

As a gross regional generalization of primeval vegetation, the more level to rolling uplands tended to be in prairie, hillsides tended to be forest, and valleys were varied. Prairie grassland in the Hill Country is defined as having less than one tree per acre (Stroessner and Habeck, 1966). It was part of the huge prairie archipelago of the Midwest (Lapham, 1844; Butterfield, 1881; Blanchard, 1924; Trewartha, 1940a; Smith, 1975. See Kuechler, 1975, for a map of the archipelago). Prairie grassland has often been characterized as the most rich and diverse biome. Plants include little and big bluestem (*Schizachyrium scoparium* and *Andropogan gerardi*), side-oats grama (*Bouteloua curtipendula*), and several other grasses (Mirk, 1997). Flowering plants such as western sunflower (*Helianthus occidentalis*) were abundant and put on magnificent shows, especially in the spring and autumn. Plants grew rank and were sometimes taller than a person on horseback. Insect, bird, and small animal life was diverse and rich, perhaps the most diverse on earth.

Some early maps of the region clearly delineate the prairies (Friis, 1969). Later, most such maps were synopses or compendia of the original land survey notes and maps (e.g., Wisconsin Board of Commissioners, 1848). Land surveyors normally recognized the quality of soils and always noted the presence of prairie both on the maps and in their written notes.

McMaster (1893, p. 10), a native of upstate New York and one of the earliest settlers in Galena, Illinois, described in 1834 his delight over a spring ride across the mostly unsettled uplands north of Galena:

> Some fifteen miles was over a broad rolling beautiful prairie, without any settlement, and we had only a dim track most of the way. I saw on my way a number of deer bounding away over the prairie. This was the first large prairie that I had seen. Being in June it was covered with masses of bright flowers. I enjoyed the ride intensely.

Hamlin Garland had a strong affection for the prairie. His family migrated westward in the late 1860s, and he described the prairie of northern Iowa at the western edge of the Hill County:

> Each mile took us farther and farther into the unsettled prairie until in the afternoon of the second day, we came to a meadow so wide that its western rim touched the sky without revealing a sign of man's habitation. . . .
>
> The plain was covered with grass tall as ripe wheat . . . we looked about us in awe, so endless seemed this spread of wild oats and waving blue-joint. . . .
>
> Far away, dim clumps of trees showed . . . and no living thing moved except our cattle and the hawks lazily wheeling in the air. My heart filled with awe as well as wonder. (Garland, 1917, pp. 81–82).

Why would prairies have existed in an area in which forest is the optimal or climax vegetation and would be expected to cover the region? The general answer is fire, which would have killed tree seedlings (and some older trees too), but would have created the ideal propagation conditions for grass. Some fires were naturally caused by lightning during dry periods, but it appears that Native Americans intentionally burned the region (Curtiss-Wedge, 1917; Marks, 1942; Curtis, 1959; Stroessner and Habeck, 1966; Pyne, 1997). Beltrami (1828), an early traveler in the region, states that the Indians "set fire to the brushwood, so that the surface of all the vast regions they traverse is successively consumed by the flames" (p. 176). He further gives this violent account of such a fire in northwest Iowa:

> The flames towering above the tops of the hills where the wind raged with much violence, gave them the appearance of volcanoes, at the moment of their most terrific eruption; and the fire winding in its descent through places covered with grass presented an exact resemblance of the undulating lava of Vesuvius or Aetna (Beltrami, 1828, p. 177)

The immediate intention of setting fire to the landscape was sometimes to herd large game for hunting (Marks, 1942). But it appears that there was a longer-term goal of perpetuating the grass cover, which large game, buffalo in particular, preferred. In turn, large animals browsed the area, often killing off tree seedlings, helping to perpetuate the prairie. There is some conflict in accounts on the presence of browsing animals at the time of European settlement. Bunnel (1897) reported that buffalo were common east of the Mississippi River on the prairies of Illinois and Wisconsin. Stephen Long, a US Army officer traveling up the Mississippi River in 1817, stated that buffalo were found along the Buffalo River in Buffalo County, thus the name (Kane et al., 1978). Coues (1895, p. 58), who edited the 1805–1807 travel journals of Zebulon Pike, states that there were "plenty of buffalo" in the region until the building of Fort Snelling in 1819 at what is now Minneapolis–St. Paul. Calvin Fremling (2005), a regional naturalist of note, indicates that there was a significant population of elk, buffalo, white-tailed deer, and antelope but that by the 1850s, elk, bison, and antelope had been eliminated from the eastern side of the Mississippi River. Curtiss-Wedge (1917) reported that buffalo had disappeared from the region before European settlement, but elk were present in Trempealeau County as late as 1865 and deer until the early 1890s. Further south, Smith (1975) reported that elk had disappeared by the 1830s. While these accounts may have some disagreement, what they all agree on is that there were large numbers of grazing and browsing animals in the region before European settlement. For a detailed look at early wildlife in Iowa, see Dinsmore (1994).

The result of fire and animal browsing was open grassland in some areas and scattered trees in others. The remaining trees were usually the more fire resistant, especially oak and hickory. In Iowa County, Wisconsin, the original land surveys found 75% of the trees to be oak (Stroessner and Habeck, 1966). For Trempealeau County, Wisconsin, Curtiss-Wedge (1917, p. 70) comments on the scarcity of trees at the time of settlement and the presence of brush:

> Here and there in secluded places along the hills were forests, but generally the country was untimbered and covered with brush and wild grass, which was burned over each year by the Indians.

The presence of this brush growing on prairie soils was also noted by Edwards et al. (1928). The "brush" is probably regrowth of woody vegetation on previously burned areas (Curtis, 1959). One may still see the modern analogue: the same type of growth on fields previously cultivated or grazed but now unused. Much of this regrowth is hardy pioneer species such as sumac (*Rhus* spp.) and poplar (*Populus* spp.) and is generally a great nuisance to the farmer. This regrowth is often rapid and vigorous, so much so that in the absence of fire, present-day farmers intending to reuse a field after the lapse of a few years must use rotary cutters to suppress the brush. An early brand of mechanical rotary cutter pulled by a tractor was Bush Hog® (Bush Hog LLC, Selma, AL), so most farmers now generically refer to any similar cutter as a "bushhog" and the process of using it as "bushhogging." One might consider fire to have been the "bushhog" of Native Americans. Indeed, fire for this purpose is reentering the management scenario in modern American farming and is now often recommended as a management tool by the USDA, especially for crop set-aside programs such as the Conservation Reserve Program (CRP).

Forests were mostly typical northern deciduous hardwoods. On north-facing slopes, which were cooler and more moist, were found trees such as sugar maple (*Acer saccharum*), beech (*Fagus grandifolia*), and basswood (*Tilia americana*). On slopes that received more sun, trees tended to be more xeric in nature, with bur oak (*Quercus macrocarpa*) often dominating in the dryer forest areas. To the north of the region, northern coniferous and mixed forest was found, especially on northern slopes.

Many of the steep southern and western slopes were often treeless and termed "goat prairies," an allusion to the steepness as well as the absence of trees. These were noted by Owen (Figure 1.8). The mostly treeless slopes were the subject of ecological studies in the early 20th century (e.g., Marks, 1942), which generally concluded that the lack of trees, other than small juniper (*Juniperus virginiana*), was due to xeric edaphic conditions: that is, the sandy soils derived from the residual sandstones were too shallow and droughty to hold sufficient water to sustain deciduous trees over drought periods (Leopold, 1935).

*Figure 1.8* Sketch by David Dale Owen (1852) showing what were later called "goat prairies," where the sunnier southern and western hill slopes were often bare of trees. While he terms these "northern and southern," they are more likely northern (forested) and western (treeless) in this sketch.

But this assumption was often wrong (Sartz, 1959). After the suppression of fire, cedar (juniper) slowly increased. And with the reduction or cessation of grazing in the later 20th century, deciduous trees began to appear. Many or even most of these former goat prairies are now thickly forested, as we will later see.

For detailed vegetation descriptions and analysis of the entire region, see Trewartha (1940a), wherein he used the original plats of survey to reconstruct vegetation patterns. Marks (1942) did an extremely detailed study of Coon Creek, Wisconsin. A more general map of primeval vegetation for the entire United States is by Kuechler (1975).

Under long-term grass cover, prairie soils, known now as Mollisols, were formed on the loess cap. A common soil series is the Tama. Such soils were deep, fertile, high in organic carbon and basic nutrients, and had rich faunal activity, including ants, earthworms, and small mammals. All of these factors provided a strong structure that allowed high infiltration rates, even after being used for crops. In his 1844 report, Owen (p. 61) comments on the high organic content of soils in the region. One soil analyzed by him (1852, p. 67) had an organic content of 8.2%. The root mass was so thick and strong that large plows pulled by several oxen had to be developed to "bust the sod."

Soils formed under forest were known as brown forest soils under early soil classifications but later and currently as Alfisols. Common soil series in the Hill Country are Fayette and Dubuque. These soils were not as deep, rich, organic, structured, basic, and fertile as the prairie soils. While perhaps not as pervious as prairie soils, the presence of tree roots and old root channels along with pedofaunal activity nonetheless gave these forest soils high infiltration capacities when first cleared and cultivated.

While soils develop very slowly in response largely to climate and vegetation, vegetation sometimes can change over much shorter periods. Thus, there was in some places a mismatch between the soils and the vegetation existing at the time of settlement. That is, a soil long established in grass and producing a prairie soil might have recently grown up in brush or even forest, given the absence of fire or other disturbances. The literature on this is somewhat conflicting and confusing. Marks (1942) indicates that most Indian fires occurred in the century after 1750. Nesbit (1973) states that John Muir suggested that hardwoods spread in the region after European settlement with the cessation of Indian fires. Blanchard (1924) further indicates that some of those trees were planted by the settlers, and the fact that they prospered gave credence to the idea that fire had formerly suppressed trees. Smith (1975) reported that while prairie was widespread and had been reportedly created by fire, he had never seen any charred wood (c. 1838), suggesting that it had been a long time since fires occurred. Conversely, Marks (1942) reported finding fire scars on large trees, suggesting that fires had continued.

Vegetation on bottomlands (floodplains and terraces) was highly varied. Early reconnaissance and exploration accounts mention forests in some place and grasslands elsewhere. Lapham (1846) reports trees growing in bottoms as being soft (silver) maple, birch, and elm. Along the Zumbro River, bottomlands were described as being about one-fourth "prairie and meadow, the balance heavily timbered with elm, ash, oak, basswood, butternut, and black walnut" (Warren, 1867, p. 58).

The streambanks themselves were often lined with trees, sometimes even when the bottoms were in grassland (Butterfield, 1881). This may have had implications for streambank stability. Later studies (e.g., Trimble, 1997, 2004; Lyons et al., 2000) suggest that, at least at present, forested banks of lower stream reaches in the region are less stable than grassed ones. However, when one closely examines the several sketches of bottomland done by Owen (1844, 1847, 1852), no more than 20% of streambanks appear to have trees and most sketches show much less (Figures 1.1 and 1.8). While it is possible that Owen deleted riparian trees for the sake of clarity in his sketches, we should also consider the account of Long who in 1823 characterized the valleys as a "rich carpet of grass and flowers" (Kane et al., 1978, p. 146) but also later pointed out that many trees grew there also.

Nearly all bottomland soils were apparently fertile and, even late in the 19th century, farmers often grew crops there and left them stored on the fields well into the autumn (Leopold, 1935; McKelvey, 1939), suggesting that they did not flood often at that time. Bottomland soils were usually dark with organic material and in some cases were Mollic in nature, indicating long-term grass cover. Owen (1847, p. 37) described the region as having a "rich, black soil, covered as far as the eye can reach, even down to the very edge of the small streams, with a thick and high growth of prairie grass." Later, in 1852 (p. 502), he described bottomland soils found near Lansing, Iowa, as "rich, dark, sandy loam." In his earliest report on the region, he reported a bottomland soil on the Platte River as having 11%–15% organic matter (Owen, 1844, p. 58). These soils, when examined in more recent times, were generally unstratified, indicating very slow, non-episodic accretion and/or long-term bioturbation had been in place for a long time. Gross (1973) describes these so:

> ... the buried [alluvial] soils have surface horizons with accumulations of illuviated clay suggesting older well developed soils (i.e., Mollisols). The buried soils formed on floodplains during apparently extended periods of vegetational, geological and climatic stability. (p. 1)

Indeed, the mature nature of bottomland soils allowed the interpretation that these were the presettlement soils. Based on pedological principles and cultural artifacts, this is the justified assumption made by Happ and his colleagues when they began sedimentation studies in the region in the 1930s (McKelvey, 1939; Adams, 1940, 1942; Kunsman, 1944; Happ, 1944; Happ et al., 1940; see Trimble, 2008). Thus, by boring through historical sediment to the old soils, the depth of sediment accretion since European settlement, often several feet, could be rapidly ascertained. However, there were exceptions where an old soil could not be found by borings (Kunsman, 1944). Knox (1972, 1977, 1987, 2006), W. Johnson (1976), and others added much to this work by establishing radiocarbon dates on these soils showing that they were indeed old and that they had accreted very slowly.

Wetter bottomlands had a dark, organic nature (Histosols). The original plats of surveys mention forests, grasslands, and in some places, especially near the Mississippi River, marshlands. Swamp and marsh were apparently uncommon south of the Wisconsin River. For Grant County, Lapham (1846, p. 240), an early Wisconsin naturalist, reported that "there is neither swamp, lake, or stagnant water of any kind in the county."

Stream terraces, or relict floodplains lying at a somewhat higher elevation, were common in many valleys, usually perched 3–20 ft (1–6) above the floodplain although some were higher. These were fluvial (stream dissection of an earlier, higher floodplain created by streams) in upstream valleys. In many cases, two to four sets of terraces were found at different levels along streams. In the northern part of the region along the Mississippi River, generally north of the Wisconsin River, were found high terraces, usually lacustrine (stream dissection of lake sediment deposits) in nature, and lying 10–30 ft above the floodplain. These were formed from material deposited during periods of natural river damming in the Pleistocene. As the river later eroded part of this material away, it left terraces which were sometimes large and high enough for cities to occupy. Indeed, they were often the settlement sites for native Americans and later for European settlements. Examples are found at Prairie du Chien and La Crosse, both in Wisconsin, and Wabasha and Winona,

Minnesota. Typically, these high terraces were formed of sandy material but were covered with several feet of silt on which a fertile soil, often a Mollisol similar to the floodplain soils, was usually found. As will be seen later, these terraces, especially where found along tributaries to the Mississippi River, are important to this study because many later eroded and gullied so easily and often furnished vast quantities of sandy sediment into streams and onto floodplains (see book cover).

Stream terraces were important to human settlement because they provided a generally flood-safe zone for agriculture, farmsteads, villages, and roads. As will be seen later in this study, many of these terraces, especially the lower alluvial ones, are now buried by historical sediment.

The role of Native Americans has come up several times. The region was somewhat thinly occupied by several tribes, including Sioux, Chippewa, Sacs (also called Sauk), Fox, Kickapoo, Pottawatomie, Menominee, and Winnebago (Paullin and Wright, 1932). Fremling (2005) points to the great decline of Indian populations by disease in North America, suggesting that the pristine appearance of the middle United States was due in part to this decline. My reading of early travel accounts suggests little extensive agriculture by Native Americans other than on the level, high river terrace locations like Prairie du Chien, Lacrosse, Winona, and Wabasha. Blanchard (1924) also mentions large Indian fields at Cassville and Muscoda, Wisconsin, both valley sites. And Owen (1847) mentions cultivation of the Upper Iowa River floodplain by a small band of Winnebagoes. More detailed information on the location of Indian tribes as well as Indian land cessions for the region is found in Paullin and Wright (1932) and Burke (2000).

As we will see, much evidence of Indian agriculture along streams has been covered with modern sediment. This makes recovery of such evidence quite difficult for archaeologists (e.g., Stanley et al., 1985).

As already indicated, the main role of Indians in changing the landscape was by fire. And as we have seen, not only did they change the vegetation, they indirectly changed the soil. By the use of fire to increase prairie, they inadvertently increased the area of the more productive prairie soils. Farmers in the Hill Country and, indeed, much of the north-central United States, can thank the Indians for some of their present agricultural productivity.

## *Streams*

Within the Hill Country, streams were mostly clear at the time of settlement, indicating little erosion of any kind. The only exception seems to be the Pecatonica River, which was noted as being "muddy" by both Long and Lapham, the noted Wisconsin naturalists, although the Lapham description might be derivative to Long (Lapham, 1846, p. 187; Kane, et al., 1978, p. 140). Apparently the widespread fires, the very limited soil cultivation by Indians, and the movement of large animals had minimal effect in most areas. Generally, the only coloration noted was from organic acids where streams had their origins in swamps, the latter being found mostly in the northeastern part of the region. Five lines of evidence are offered here to demonstrate that streams were mostly clear and transporting very little fine sediment at the time of European settlement, this condition, indicating very little erosion of either uplands or streambanks.

1. Well-developed floodplain soils. As already discussed, floodplains almost everywhere had a dark, well-developed, generally nonstratified soil. This evidence tends to preclude either significant vertical accretion on the one hand, or significant lateral migration of streams with the resulting, generally coarser, lateral accretion. Thus,

little sediment transport could have been occurring in streams. The major exception to this generalization was along the Chippewa River, which was transporting sand, and Owen (1852, p. 56) reported the source of this sand was the stream erosion of old high terraces. At some much earlier time, sand transported from the Chippewa River was massive enough to dam the Mississippi River and create Lake Pepin at the mouth of the Chippewa. The Wisconsin River also carried a heavy sand load, but this came from the formerly glaciated areas farther upstream.

2. Radiocarbon dates in older alluvium. Professor James Knox, of the University of Wisconsin, and his students have extensively dated the older alluvium. For example, W. Johnson (1976) established a date of about 2500 years before present at a location about 30 in. (75 cm) below the old floodplain surface, and other dates he found were in accordance. More recently, Knox (2001) has given a general rate for primeval tributary floodplains of about 0.02 cm (0.008 in.) per year.

3. Well-developed and well-vegetated upland soils that suggest little surface erosion. Under natural conditions, these deep, well-structured forest and prairie soils in their natural state apparently had enough infiltration capacity for all but the greatest precipitation amounts and intensities (Sartz, 1961, 1976). Thus, most stormflow would have been subsurface or throughflow (Ward, 1975; Ward and Trimble, 2004), with surface flow generally restricted to channel areas. Experimental watershed research in the region shows that relatively undisturbed forest and grassland generally have little surface flow (overland flow) and practically no erosion (Sartz, 1976). Many tributary streams draining about one-half square mile (1.3 km$^2$) had no channels at the time of European settlement, indicating minimal surface flow, which, over time, would have eroded channels.

Available evidence suggests low response of streams to rainfall events in presettlement times. Even as late as 1900, people were still building houses on low terraces, presumably with little fear of flooding based on 50 or so years of observation. The fact that tributary channels were so small and narrow, typically narrow enough that a person could leap across (McKelvey, 1939), is strong evidence that floods were minimal. In 1934, an early settler near Coon Valley, Wisconsin, told Aldo Leopold that until the 1870s, streams generally remained within their channels and farmers stacked wheat near creeks (Johnson, 1991, p. 60). McKelvey (1939) reported much the same thing. All this is not to say that floods did not occur in the early period. Johnson (1991) reports a flood near Coon Valley in 1881 that drowned seven people. It is important to note that even the best land use such as found on the primeval landscape cannot totally control floods.

While it is true that the magnitude of many potential flood events may be mitigated, extreme storms will produce large floods, no matter what the land use and management might be. And as seen in the discussion of climate, the region was in a period of wetter climate during the early settlement period.

4. The presence of brook trout (*Salvelinus fontinalis*), which tolerate little suspended sediment and flashy streamflow (Trowbridge, undated; Winchell, 1884; Bunnell, 1897). Owen (1852, p. 502) described "streams in whose sparkling waters are to be seen an abundance of fine trout" (see also Owen, 1847, p. 35). In Winona County, Minnesota, in the 1850s (*City of Winona and Southern Minnesota*, 1858), the following description was given:

> The streams flowing through these valleys, particularly the Rollingstone [creek], swarm with large speckled trout, a source

of pleasure here as well as a most certain indication of a healthy country.

Stream channel bottoms were usually sand or gravel (Smith, 1975), which would suggest little fine sediment and also good spawning areas for the brook trout. Even in the downstream reaches of the Zumbro River, the bottom, as seen through 1–10 ft of water, was fine sand and gravel. The water was so clear that the interbedding of sand and gravel could be described (Warren, 1867, p. 59).

5. Contemporary accounts. Early explorers of the region, including scientists, uniformly praised the clarity of the streams. Long described the streams as "crystal" (Kane et al., 1978, p. 146). "Crystal" is also the term used by Daniels (1854, p. 12). David Dale Owen, the early explorer and scientist, described stream bottoms as seen several feet beneath the surface of the water. At one place on the Upper Iowa River, Owen was even able to describe a bottom of "stiff blue clay" in "very deep water" (Owen, 1847, pp. 35, 37, 58). Many of these stream bottoms were usually sand or gravel (Smith, 1975). In Vernon County, Wisconsin, the Bad Axe River was described thus: "The waters of that river have ever been cool, clear, and sparkling and bright, and the trout that darted through its crystal waters, very large, lively fellows, and of a superior flavor" (Brunson, 1884, p. 228). Since that 1884 description was written in past tense, it may be just the memory from an earlier time. An early Army Engineer report indicated that the Upper Mississippi River was still clear enough in the 1860s that stream bottom dune formations and vegetation could be easily observed, even in Lake Pepin, which was 30 ft deep (Warren, 1867, pp. 11, 17, 59, 61). While sand moved in the Upper Mississippi River, there was apparently little fine sediment. The same report stated that "its waters are clear and there is no mud deposited by it. The bed and bottom are nearly all sand" (Warren, 1867, p. 313). The report emphasized that sediment would not be a problem in the operation of navigation locks, which were being proposed at that time for the Mississippi River and several of its tributaries. The same report stated that before settlement, the river was in "permanent condition in which little sand was brought in and moved along, and the water flowed uniformly clear and with a darkened color, indicating the slow drainage and abundance of vegetative growth," and even in the 1860s, "there is very little material in suspension in the waters of the upper Mississippi. What material there is is dragged by the current along the bottom [sand transported as bedload]" (Warren, pp. 19–20). Finally, the original land survey notes have many references to clear streams (e.g., Wisconsin Board of Commissioners of Public Land for Wisconsin, 1848).

I note that some streams in the glaciated area to the northwest of the Hill Country were indeed transporting fine sediment on occasion as evidenced by their turbidity. The French explorer Joseph Nicolet noted in September 1838 that the Minnesota River downstream of St. Peter, Minnesota, was "without transparency" and suggests this was a permanent condition (Bray and Bray, 1976, p. 111) although this would seem to conflict with the above observations of Warren (1867). Farther to the southeast in Dakota County on September 15, 1838, Nicolet mentions crossing "several small and muddy rivers" (p. 122). The latter were in the headwaters of the Cannon River, which flows through the northern tip of the Hill County and into the Mississippi River at Red Wing, Minnesota. Thus, it seems probable that at least some fine sediment was brought to the northwestern part of the region by tributaries originating in the glaciated area to the west and north.

## Conclusion

The Hill Country was a rich and beautiful land that the European settlers found to be a virtual paradise in many ways. But as we shall see in the next chapter, they wasted little time placing their cultural and physical imprint on it.

*chapter two*

# European settlement and changes of land use

Before land could be properly taken up by settlers, it had to be obtained from the Indians by treaty (Read, 1941), surveyed into parcels suitably sized for farming, and then suitably organized to permit the formation of townships and counties. With this done, public land could pass into private hands, a process termed a "land patent." Nearly all of the Hill Country was surveyed under the US rectilinear land survey whereby the land was surveyed into rectangles, at least as closely as could be done on the surface of a sphere (Thrower, 1960; H. Johnson, 1976, 2010). Stemming from the ordinances of 1784, 1785, and 1787 and largely from Thomas Jefferson's Enlightenment ideas of creating small agricultural landholders, the rectilinear survey became a model of Enlightenment thought in terms of its logic, fairness, and reproducibility (Figure 2.1). Note that the basic unit of survey is the *section*, which was 1 square mile or 640 acres (2.6 km$^2$), but which could be subdivided into many rectilinear permutations. An adequate farm for the Midwest was thought to be about 160 acres (64 ha) or a quarter section, but, depending on the wealth and intentions of the patentee, various-sized options were available, sometimes with all the land lots not contiguous.

One of the debatable results of the rectilinear survey from a sociological viewpoint was to perpetuate the model of scattered farms and farm residents. And from an environmental viewpoint, its effects were also often problematic. For example, roads were often run along survey lines, which meant that a road might cross a stream several times, whereas a road allowed to follow the landscape might avoid the stream altogether (Thrower, 1960).

The imposition of such a geometric scheme on the steep terrain of the Hill Country created other problems. Roads running up and down steep slopes invited severe erosion of both the unpaved road surface and the ditches to either side, the sediment often being routed directly into a stream. However, roads in the steeper parts of the Hill Country usually followed much more pragmatic routes of milder slopes, avoiding crossing of streams where possible. This was not done on environmental grounds but rather to provide milder slopes for the animal-drawn conveyances of the day and to avoid the expense of building and rebuilding bridges. A comparison of 19th- and late-20th-century county road maps will show that some roads were changed from survey conformity to topographic adjustment (Thrower, 1960).

But perhaps the worst aspect of the survey system as imposed on the Hill Country was that it encouraged farmers to lay out their fields in rectangles and to plow straight rows, even when directed up and down slopes. This culture of "plowing a straight row" was established before the Hill Country was even settled and was still dominant in much of the United States as late as the 1950s. As we will soon see, this culture later slowed the adoption of contour plowing and contour strip layouts.

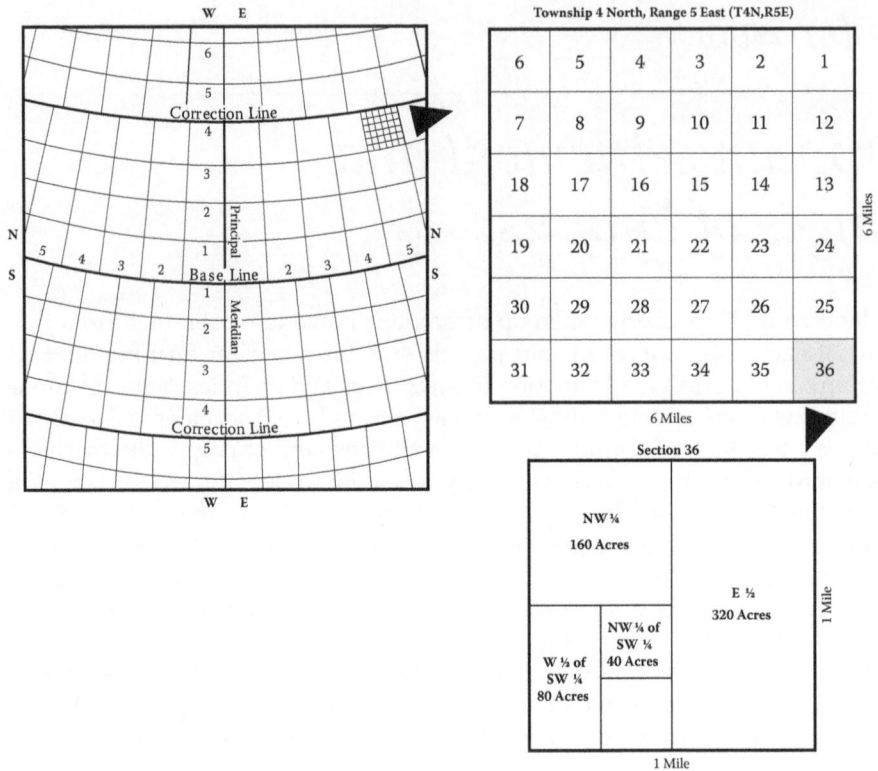

*Figure 2.1* The US Rectangular Survey System (redrawn from Johnson, H., *Order upon the Land*, Oxford, 1976). Thus, the location of the NW quarter of the one square mile unit at the lower right of the diagram (termed a "section") would be northwest quarter of Section 36, Township 4 North, Range 5 East, abbreviated NW ¼, Sec. 36, T4N, R5E. In the Hill Country, tracts as small as 40 acres (16 ha) could be purchased, and tracts were not always contiguous.

## Early settlement

Drawn first by the lure of lead deposits, intensive and permanent European settlement began in the 1820s south of the Wisconsin River in the area centering around Galena, Illinois, but soon spread northward into what is now Grant and LaFayette Counties, Wisconsin (Trewartha, 1940b; Conzen, 1997). Even before the land was ceded by the Indians there in 1829 (Read, 1941), hundreds of miners moved into the area, staked claims, and started mining operations. By the early 1830s, the mining activity had spilled over to the Iowa side, which was ceded by the Indians northward through most of Clayton County in 1832 (Reid, 1941; for a fictionalized but seemingly accurate portrayal of this period, including Indian wars and cessions in the region, see Derleth, 1940). Dubuque had become a town, and, according to the US Census, settlers had spread westward into the Iowa hinterlands by 1840. There was now a considerable population to feed, and agriculture had begun on both sides of the river, but even so, the most densely populated area of the mining area around Galena only had about 5 people per square mile (2 people/km²). North of the Wisconsin River, only a few people were found, mostly along the Mississippi River terraces in the vicinity of Prairie du Chien and Winona (Trewartha, 1940b). By 1850, the population density around Galena had increased to over 10 persons/mi² (4/km²) but

Chapter two:   European settlement and changes of land use    25

*Figure 2.2* Sketch of the Lead Region (from *Harper's New Monthly Magazine*, 1853). Note the barren, almost lunar-like quality of the landscape and the scattered mine pits, or "diggings."

was still generally less than one person per square mile north of the Wisconsin River even though that area was ceded by the Indians in 1837 (Read, 1941). The poorer, sandy areas of the northeast part of the Hill Country, the "Central Plain," were not even settled. Most of the Minnesota area was not ceded by the Indians until 1851 and consequently had few settlers except along the Mississippi River and the southern border with Iowa, part of which was ceded by the Indians in 1846 (Paullin and Wright, 1932).

Early mining around Galena was done at the surface, termed *diggings*, and the spoil was left on slopes, the irregular piles being particularly vulnerable to erosion. Later, subsurface mining and the processing of the ore left even greater spoil piles, which also eroded. Contemporary descriptions of the landscape suggest little beauty, and a sketch done for *Harper's Magazine* in 1853 shows an almost moon-like terrain with obvious gashes and no trees (Figure 2.2).

While many activities—mining, road building, construction, forestry—contributed to erosion of the landscape, the overwhelming cause of erosion and sedimentation over the historic period, however, was agriculture. Even in the lead mining area, later studies (Adams, 1944) indicated that most sediment came from agriculture. Indeed, it was the agriculture that eventually attracted most of the settlers into the region. By 1840, settlers had moved well beyond the mining activity, and by 1850, they were rapidly moving out over much of the Hill Country, most with the intention of farming. Steamboat navigation began on the Upper Mississippi River in 1823, and even before that there had been downstream raft traffic to transport regional products, especially timber. With this excellent transportation artery, small towns such as Guttenberg, Prairie du Chien, Lansing, La Crosse, Winona, Wabasha, and Red Wing flourished and became market towns where farmers could sell their products and buy supplies, further promoting agriculture.

Before proceeding further with a detailed analysis of land use for the region, it would be appropriate to give a historical and spatial overview of agriculture to help the reader keep matters in perspective and context.

## Overview of historical agriculture in the region

Based on the historical trends of improved land for agriculture, there are two major agricultural regions in the Hill Country (Figure 2.3), and they are strongly influenced by the physiography and soils. "Improved Land" is the US Census classification for land that has been converted from forest and prairie to agriculture: cropland, pastures, orchards, and farmsteads. Given a choice, farmers will improve the more level and fertile parts of their farms, that is, the land that will give the best economic return.

The first of these (Region I) is located on the most suitable area for agriculture, that is the area with the most level or gently rolling land. As explained in Chapter 1, moreover, these level-to-rolling upland areas are more likely to have the deep and fertile Mollic soils formed under prairie. Referring to the maps of slope and physiography from the last chapter (Figures 1.3 and 1.4), these areas are generally south of the Wisconsin River (actually south of Military Ridge) and also several miles west of the Mississippi River. Here, agriculture expanded rapidly, and much of the land was "improved" or cleared of the native vegetation and much of that was first planted to wheat, the early commercial crop. The rapid expansion of settlement and agriculture in Region I is demonstrated in the six counties of Minnesota. Opened for settlement after the Indian cessions in 1853, most of the land was occupied within four years (Colby, 1924). This topographic predilection should not be surprising: John Hudson, a noted agricultural geographer of the Midwest (2010, p. 195), states that "The history of American agriculture has been witness to an almost constant shift of farmers toward the flattest, richest, grassiest land available." While it may be difficult for many to understand why people would want to leave the beauty of the

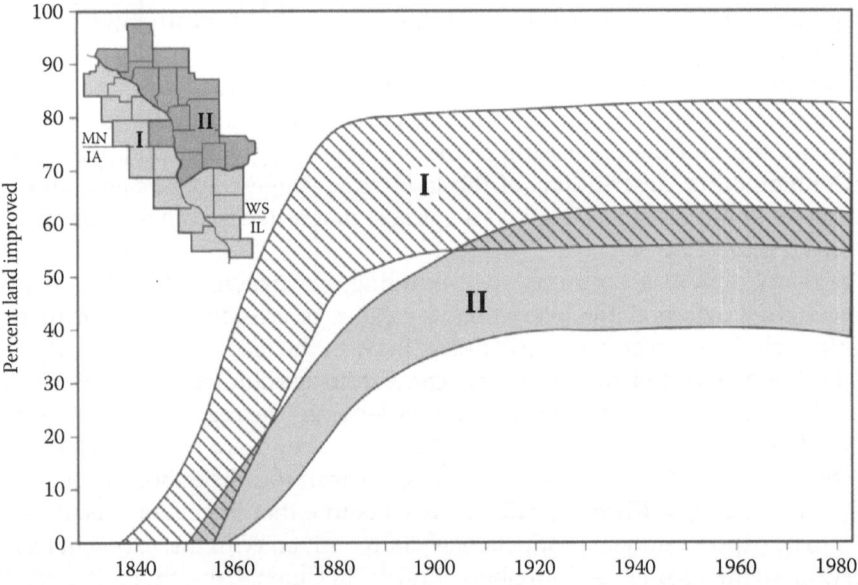

*Figure 2.3* Regionalization of the Hill Country based on historical trends of agricultural land use.

rolling and well-watered forested land farther east and settle on the almost treeless prairie, Garland (1917, p. 42) emphasized the difficulty and frustration of tilling steep, rocky slopes and, conversely, the perceived ease of tilling level, stoneless plains. While wheat eventually disappeared from Region I, specialized commercial crops, mostly corn and soybeans, have continued. Although crops and other agriculture evolved over time, crop farming has remained the overwhelming use of the land.

An interesting footnote to rapid agricultural uptake of Region I is that at least some early settlers of the Midwest tended to avoid the prairies. Many thought that trees, and especially nut-bearing trees, were the indicators of fertile land, and these were largely absent on the prairie. Additionally, prairies could not provide wood for houses, fences, fuel, and other uses, and the sod was difficult to break (Leverett and Sardeson, 1919). Indeed, Loehr (1939) points out that as many as 20 oxen were required to pull the breaking plows. However, once the fertility of the prairies was widely recognized, the prairies filled up rapidly and most were taken by 1880 (Curtis, 1959).

Garland (1917, p. 104) eloquently describes his mixed feeling about the breaking of the prairie :

> . . . as I saw the tender plants and shining flowers bow beneath the remorseless beam, civilization seemed a bad business, and yet there was something epic, something large-gestured and splendid in the "breaking" season. Smooth, glossy, almost unwrinkled the thick ribbon of jet-black soil rose upon the share and rolled away from the mold-board's glistening curve to tuck itself away upside down into the furrow behind the horses' heels, and the picture . . . gave me pleasure in spite of the sad changes . . .

Further, Garland (1917, pp. 99–105) is clearly disturbed by the process of breaking, plowing, and dragging (breaking up clods), and, being an observant naturalist, he goes into some detail about how each plant and animal is affected.

The second region (Region II) is the more dissected or hilly part of the region lying mostly north of the Wisconsin River and east of the Mississippi River. This area, less suitable for agriculture because of the steeper surface configuration and generally less fertile soils, was populated later, and agriculture evolved more slowly and to a lesser degree than in Region I as shown by the expansion of improved land. Moreover, some who settled this hilly area were dissatisfied and moved to Region I. Again, Garland states (1917, p. 42) that his father disliked farming in the hilly, stony land north of Lacrosse, Wisconsin (Region II), and could hardly wait to move farther west to the level prairie land. He states (1917, p. 42) that "it irked [his father] beyond measure to force his reaper along a steep slope, and he loathed the irregular little patches running up the ravines beyond the timbered knolls, and so at last like many another of his neighbors, he began to look to the west as a fairer field for conquest." Indeed, he moved his family to Winneshiek County, Iowa (Region I), in the late 1860s. While specialized commercial crops evolved in Region II also, the nature of farming was much more diversified and has remained so.

Whatever the merits Region II had for farming, Garland (1917, p. 78) wistfully looked back on the area as a frontier Arcadia:

> As I look back upon my life on that woodland farm, it all seems very colorful and sweet. I am re-living days when the warm sun, falling in radiant slopes of grass, lit the meadow phlox and tall tiger

> lilies into flaming torches of color. I think of blackberry thickets and odorous grapevines and cherry trees and delicious nuts which grew in profusion throughout the forest to the north. This forest which seemed endless and was of enchanted solemnity served as our wilderness. We explored it at every opportunity. We loved every day for the color it brought, each season for the wealth of its experience, and we welcomed the thought of spending all our years in this beautiful home where the wood and the prairie of our song did actually meet and mingle.

Arguably, a third region, or perhaps subregion, could be identified. As mentioned earlier, this would be the sandy, often wet, and less fertile Central Plain area in the extreme northeast part of the Hill Country where relatively little agriculture was pursued. Much of this area was the old lake bed of glacial Lake Wisconsin (Martin, 1965). Agricultural progress in this area would have depended on massive drainage, and even with that, the acidic and sandy soils were not naturally productive (Whitson et al., 1923). According to a map in Ostergren (1997, p. 138), parts of eastern Eau Claire, Jackson, and Monroe Counties by 1920 had not even been settled by Europeans. However, some of this area is a Winnebago Indian Reservation and is used for growing cranberries. Because this sub-region is composed primarily of parts of only three counties, it appears better to fold it into Region II.

Note that much of Region II was naturally forested, especially in the more northern areas. In many places, these trees had to be cut and removed to clear fields for agriculture (Rohe, 1997). Michael Williams (2010), the eminent historical geographer of American forests, states that clearing an acre of forest for agriculture required about 32 man-days, whereas breaking the sod and plowing an acre of prairie, as in Region I, required only 1.5 man-days! The difference was perhaps not as great in the upper Midwest, where it cost only about twice as much to clear timbered land (Jarchow, 1949). While there may be several reasons for Region II developing more slowly, the removal of trees looms large. Even after being cleared, the steeper slopes of Region II would have been more difficult to plow than the more level fields of Region I, which allowed larger equipment pulled by several animals. The stoniness of some slopes just added to the problems.

In addition to clearing land for agriculture, there was also commercial timbering in some areas, especially the headwaters of the Wisconsin, Kickapoo, Black, Trempealeau, and Chippewa Rivers. Again, Garland (1917) gives a good description of this early commercial timbering and the downstream conveyance of the logs. The process of cutting trees, as it was practiced then with logs being pulled from the forest by animals, probably created little erosion and sediment. But the plowing and ensuing agriculture that usually followed often had dire results.

Towns such as Black River Falls, Eau Claire, and La Crosse early on had commercial sawmills. Since most of these were water-powered and thus located on streams, sawdust and other wood waste was simply released into the stream to be carried away. A US government survey in 1867 described the early effects of timbering and agriculture in the northern and eastern parts of the Hill Country:

> The ploughing of the prairie, felling of forests, erection of mills and other causes have already begun to change the former state of things. The water is no longer as clear and dark as it used to be, and more sand has accumulated in the stream, and a noticeable quantity of saw-dust and chips from the lumber mills of the Mississippi,

St. Croix, Chippeway [Chippewa], and Wisconsin [Rivers] is also deposited along the banks.... We must, therefore, expect a gradual rise ... in the bed of the upper Mississippi, though it may be of so little effect to be disregarded. (Warren, 1867, p. 17)

This latter observation on the anticipated effects of humans again underscores just how clean the water was before settlement. These changes came slowly at first. Even after over 40 years of agriculture, a flood of 3-year frequency farther downstream on the Mississippi River at Keokuk, Iowa, had a sediment concentration of only 100 parts per million (Warren, 1867, p. 321).

Geographic regions are simplified intellectual constructs, models used to portray concepts of spatial functions through time. Figure 2.3 is a simplified model. Note that the time functions of "percent improved land" have some overlap, in that they are not mutually exclusive. This is largely because the data are compiled from the US Census of Agriculture, which is based on a county, or political, basis. If they were based on physiographic and soil units, they would have been seen as more mutually exclusive. Note that the only county west of the Mississippi River to fall mostly into Region II is Houston County, Minnesota. Looking back at the map of relief (Figure 1.3) and the physiographic diagram (Figure 1.4), it will be seen that Houston County is visibly more hilly or "rougher" than the rest of that western region. And further, it will be seen that all up and down the river, the areas near the river are physiographically "rough." Thus, they fit more comfortably into Region II than Region I to which they are politically attached. Hence, one should associate Regions I and II with physiography and soils rather than political boundaries. Indeed, Blanchard (1924) convincingly contrasts agricultural conditions in the "rough" areas as opposed to the more level sections of southwestern Wisconsin south of the Wisconsin River.

With these simple regional models in mind, a better idea of the historical spread of agricultural land use can be seen from the construction of isopleth maps over the historical period (Figure 2.4). The 1860 map shows a general northward march or "wave" of land improvement coming from the nodal area in the lead country to the south, but the "wave" becomes distorted, being deflected to favor the more level and rolling western side of the Hill Country, the areas more favorable for agriculture. Meanwhile, the eastern side north of the Wisconsin River (Region II) languished, whereas the area to the south was rapidly becoming a mature agricultural landscape. To place all these changes into a national context, one is directed to an excellent series of historical crop maps by Paullin and Wright (1932).

By 1880, the western margin in Iowa and Minnesota had caught up with the south, but Region II generally continued to trail far behind. This is especially pronounced to the far northeast, especially in Eau Claire, Jackson, and Monroe Counties, Wisconsin, where the marshland and less productive sandy soils are found. Also to be noted is that higher values extend eastward across the Mississippi River into the rolling lowlands of Trempealeau County, Wisconsin, while low values extend westward into the steep hills of Houston County, Minnesota. An "island" of particularly low values for the entire historical period is seen in Crawford County, Wisconsin, an especially dissected area with great relief. By 1900, Region I has changed little while Region II continues to see increases of improved land. These general patterns exhibited in 1900 persist and were even accentuated over the following century and into modern times, with 1980 being similar to 1900, the difference being that improved land had generally increased in Region II up to about 1960 but decreased slightly since then as shown in Figure 2.3.

Nearly all counties in the Hill Country had more than 90% of the land area in farms at maximum extent. That is, the land had been purchased with the presumed intention of conducting agriculture of some kind. The exceptions were, as might be expected,

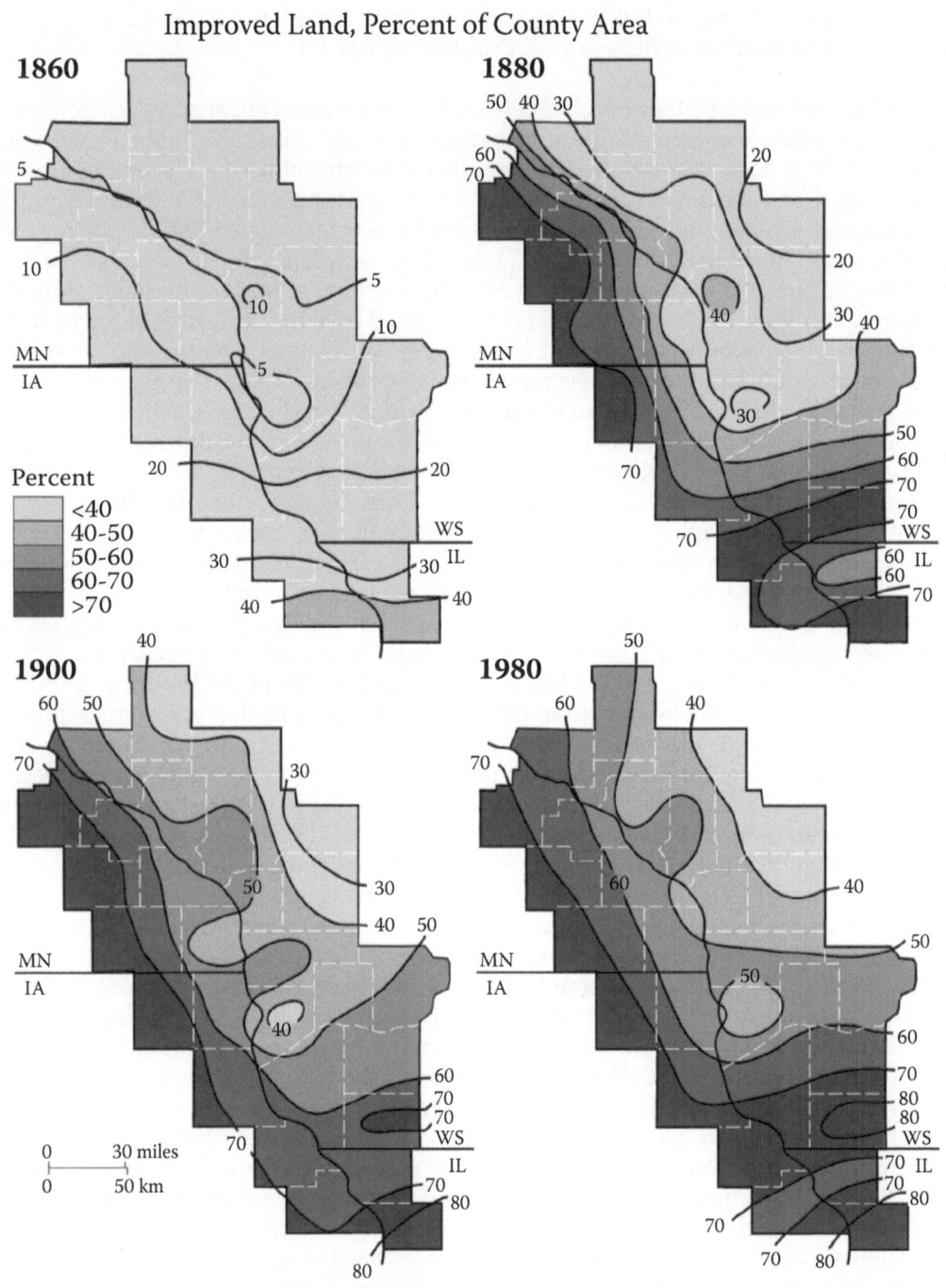

*Figure 2.4* Improved land, 1860, 1880, 1900, and 1980 (US Census of Agriculture).

Monroe, Eau Claire, and Jackson Counties (Wisconsin) in the extreme east of the region. For reasons of land and soil suitability already mentioned, only 80%, 75%, and 60%, respectively, of the land areas were taken up as farmland in those counties. Land in farms was slightly greater in Region I than in Region II. In almost every county, land in farms has decreased over the past century or so, sometimes by as much as 10%. The reasons are usually urban expansion, rural housing, highway rights of way, and public land acquisition such as for parks, reservoirs, recreational areas, and wildlife refuges, mostly at the expense of "land in farms" as opposed to active farmland. Improved land has also generally decreased but not as much as land in farms. This suggests that the land reverting from agriculture is generally less suitable for agriculture: farmers are selling off land less desirable for agriculture.

One might well also consider "unimproved land" as well as improved land in this regional study. Generally, it remained unimproved because it was too steep, rocky, rough, infertile, or wet to pay the farmer to clear and use it. Most would be in forest, brush, swamp, or perhaps "goat prairie." While it was not mapped here, the proportion of unimproved land in each county can be easily estimated for each county by subtracting the percentage improved from the average proportion of land in farms (about 90%). From this it will be seen that Region II had 40%–50% land unimproved, while Region I normally had only about 20%, a very significant difference in land use that underlined the physical difference in the two regions. By the early 20th century, the "unimproved" category would also have included land once improved but so depleted or eroded that the farmer had simply let it revert to brush or forest and no longer considered it "improved." In some cases, land became so eroded that the farmer had no intention of ever clearing and using it again. In the South, such land was often termed *abandoned* (Trimble, 1974), but this term was rarely used in the Hill Country.

Another way to demonstrate these land use trends is to show two sample counties, each being more or less representative of its region. For Region I, Winneshiek County, Iowa, well demonstrates what happened in the west (Figure 2.5). The land was rapidly patented, and over 90% was put into farms by 1880. Almost as rapidly, the farmland was improved, reaching 76% of the county by 1880. This means that only about 14% of the county was left unimproved. Small grains, mostly wheat, covered almost 35% of the county in 1880 but were rapidly reduced thereafter, covering only about half as much in 1890. At the same time, corn was rapidly expanding, presumably displacing small grains on the better fields. Corn and other row crops (mostly potatoes), covered about 19% of the county in 1930 but over 34% in 1980, with soybeans comprising about one-seventh of that. Soybeans were more important toward the northwest in the Hill Country. In Goodhue County, Minnesota, for example, soybeans were about one-fourth of the row crops in 1980. To the south in Region I, wheat and, later, soybeans were not as important as seen in the western part of the region.

That Region I continued to burgeon and maintain greater agricultural output is again not surprising. Already noted is the level, fertile land more suitable for crops. To that, one might consider how much easier it would be to get crops to market over level or rolling land instead of having to traverse the often steep slopes and cross the many streams of Region II. But there's even more. Hudson (2010, p. 188) states that " the better the land, the greater the crop acreage and the finer the 'mesh' of the town-and-railroad network." That is, the advantages of Region I would have been synergistic: more agricultural production begat more market towns and a denser transportation system. By 1890, most farms in the six-county area of Minnesota were within ten miles of a railroad (Colby, 1924), and by 1916, only small areas in the Wisconsin portion of Region I were more than five miles from

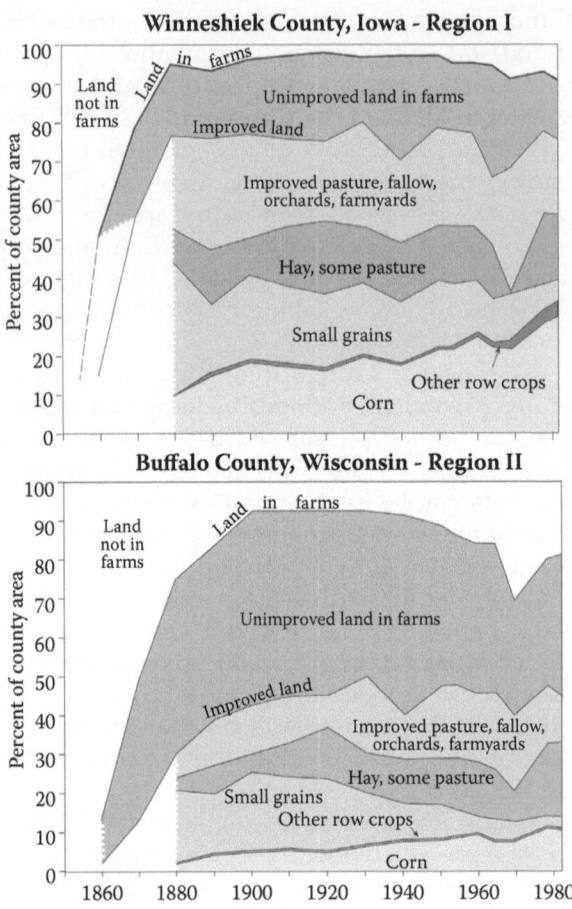

*Figure 2.5* Agricultural land use, 1860–1982, Winneshiek, Co., IO (Region I), and Buffalo Co., WI (Region II). US Census of Agriculture.

a railroad (Blanchard, 1924). To all these advantages, one might add that the more level land was much more favorable for tractors and large agricultural equipment when they appeared (Gray et al., 1929).

For Region II, Buffalo County, Wisconsin, is typical (Figure 2.5). There are both similarities and contrasts with Region I. For example, note that the land was patented almost as rapidly as in Winneshiek County and to almost the same proportion (over 90%), but the improvement of farmland followed much more slowly and to a lower level, exceeding only 40% by 1890. Thus, unimproved land was about three times as great in Region II as in Region I. Wheat and other small grains were never commercially important as they were in Goodhue County, and about half of reported small grains was oats grown for farm animals. Corn and other row crops (mostly potatoes) were never dominant, remaining below 10% until the grain boom of the 1970s. As in the southern part of Region I, soybeans were not important. Hay and pasture were important in both Regions I and II. But by far the major difference between Regions I and II was the amount of unimproved land. At least part of this unimproved land might have been former farmland, temporarily abandoned because of depletion or erosion, and being allowed to grow up in brush. Winneshiek

County averaged only about 20% in unimproved land while Buffalo County had almost 50% until after c. 1940. As previously noted, this area was often forest, brush, and goat prairie and was usually grazed with deleterious results that will be discussed later.

While it is now unfashionable in the field of geography to suggest physical determinants of human economic activity, the relationship between extensiveness of agriculture and landforms/soils appears manifest over historical time in the Hill Country and, indeed, in much of the United States. This study demonstrates the influence of landforms at the regional scale, but earlier studies did that at a more local scale. For the area in Wisconsin south of the Wisconsin River, Blanchard (1924) showed how much less intensive was agriculture in the highly dissected area north of Military Ridge than on the relatively smooth backslope to the south (Figure 1.4). At an even finer scale, Schafer (1932) used different townships in Grant County, Wisconsin, to dramatize the difference. In 1880, the values of both land and farm commodities were four times greater in a rolling prairie township than in a "rough" one, but that great differential had decreased by 1905 because of more dairying in the rough township, agricultural activity which could utilize steep slopes. Of course, there are other influences operating, including markets, ease of transportation, local climate, patterns of landholdings, education and ethnicity, so we must simply point to the "influence" of the landforms in this case. But the relationship between level land and high value is unmistakable.

## Historical crops and other agricultural land use

### Crops

In pioneer settlement, the first crops tend to be subsistence in nature. But even at that stage, when markets were available, cash or market crops were grown. Steamboat traffic began on the Mississippi River in 1823, and several market towns such as Galena, Dubuque, Guttenberg, Prairie du Chien, La Crosse, Winona, Wabasha, and Red Wing, among others, sprang up so that markets were often quite close, thus promoting market crops. The later building of railroads gave even better market accessibility for both selling and buying.

In the Hill Country, wheat was the first big cash crop, rolling across the region with settlement, and known as "King Wheat" in some areas (H. Johnson, 1957, 1976). Because the census of agriculture reported crop acreages beginning only in 1880, it captures only the end of the wave (Figure 2.6). While wheat had been an important market crop all across the region, nowhere was it more important than on the prairie of Minnesota, especially in the western parts of Goodhue, Olmstead, Fillmore, and Winona Counties, where the wheat crop peaked in 1877 (H. Johnson, 1976). While perhaps no one became rich from growing wheat, a very hefty profit was made from it, and this wealth is suggested by some of the fine churches built in the area, especially in Minnesota, during this early period. Farther south in Winneshiek County, Iowa, even with the limitations of the census, it is clear that much of the land being cleared or "improved" went directly into wheat (Figure 2.5). By 1865 in Olmsted County, Minnesota, only 12 years after the beginning of settlement, 65% of all cultivated land was in wheat (Colby, 1924). In Goodhue County, wheat occupied 81% of all cultivated land by 1875 (Colby, 1924). But competition from better wheat lands to the west, declining markets, declining yields, cinch bugs, and wheat "rust," a fungal disease, caused its virtual disappearance from most of the region by 1900 (Robinson, 1915, Figure 2.6).

The effects of the cinch bugs on the crop of 1879 and 1880 were dramatized by Hamlin Garland (1917, p. 229) when the wheat crop was almost destroyed:

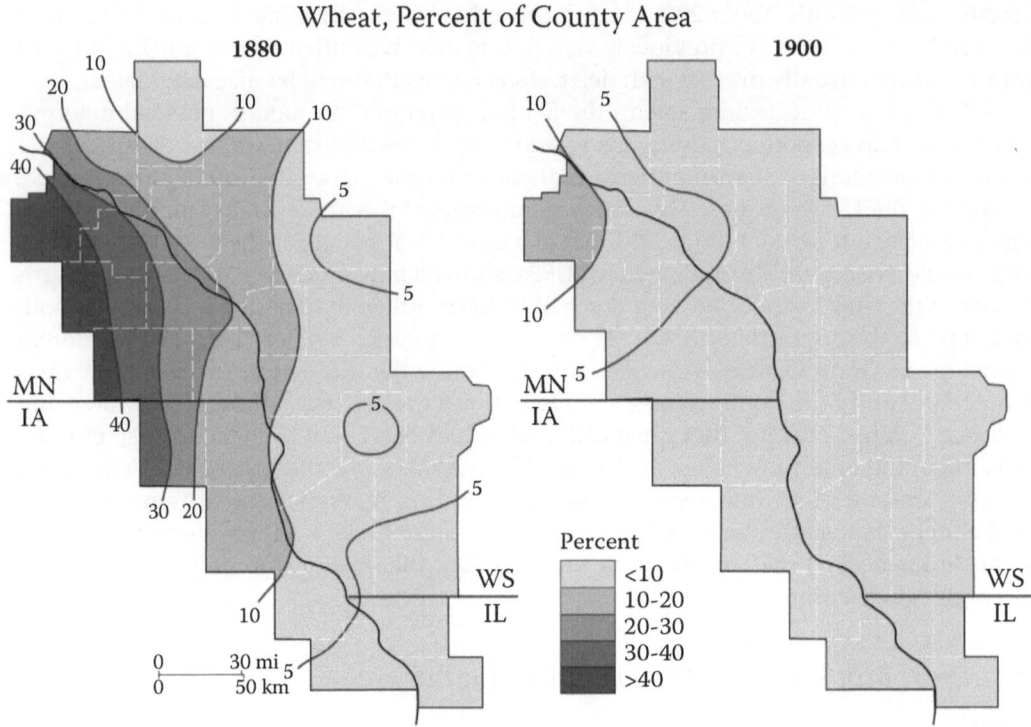

*Figure 2.6* The decline of wheat cultivation. Proportion of land in wheat, 1880 and 1900 (US Census of Agriculture).

> The harvest of '80 had been a season of disgust and disappointment to us, for not only had the pestiferous mites devoured the grain, they had filled our stable, granaries, and even our kitchens with their ill-smelling crawling bodies—and now they were coming again in added billions. By the middle of June they swarmed at the roots of the wheat—innumerable as the sands of the sea. They sapped the growing stalks till the leaves turned yellow. It was as if the field had been scorched, even the edges of the corn showed signs of blight.

The precipitous decline of wheat is portrayed in a graph by Colby (1924, p. 216): Goodhue County produced about 30,000 bushels of wheat in 1875 but only about 6000 in 1880! The other counties of Minnesota experienced similar declines By 1880, only the western tip of the region was still active, and even that soon diminished (Figure 2.6).

This wave of wheat growing probably caused little soil erosion. Because it is so much more thickly planted, it gives the soil much more protection, and the **C** (vegetative cover) value is normally well less than half of clean-tilled row crops such as corn. However, plowing and the cultivation of wheat did partially deplete the soil of nutrients, expose the soil organic material to oxidation, decrease faunal life in the soil and allow compaction and breakdown of soil structure, thereby making the soil more vulnerable to erosion. Writers on the period were especially critical of the emphasis on wheat, with one calling it an "assault on fertile new lands" (Johnson, 1991, p. 2).

Wheat and other small grains have continued in the Hill Country, but they have declined greatly with time (Figure 2.5). Oats, in particular, were important for feeding horses used as farm draft animals, but the advent of the tractor in the 1920s virtually meant the eventual demise of oat growing. As with wheat, these small grains usually caused minor soil erosion, and they made the soil more vulnerable to erosion with the later growing of corn.

In a sense, corn (*zea mays*) replaced wheat in the Hill Country. The area of corn continued to increase during most of the historical period. Some was sold as a cash crop but much was also fed to animals, especially pigs as we will see later. Looking at the spatial diffusion of corn (Figure 2.7), the nodal area is the south with more than 25% of some counties being planted to corn in 1880 while most of the north had less than 5% at that time. However, the expansion is to the western part of the region, with large acreages of wheat being replaced with large acreages of corn by 1900 (Figure 2.7). Corn continued to expand in the northwest and, by 1940, was as dominant as in the south so that Region I became almost uniform in this respect. Starting in the 1940s, an additional row crop, soybeans, was introduced and by 1980, was grown in all counties. Whereas soybeans generally comprised less than 10% of the row crops in the more hilly Region I, they comprised up to 20% in the more level areas of Region II, especially Goodhue and Olmstead Counties in Minnesota. And as will seen in Figure 2.7, corn and soybeans covered more than 30% of some areas to the west and south by 1960. These areas are now part of the "corn belt" of the United States (Hudson, 1994). Yet, just a short distance to the east in the hillier area of Region II, a few counties had less than 10% of their total area in corn and none had as much as 20%, again demonstrating the influence of terrain and soils. In the 1970s occurred the "grain boom," which affected Region I much more than Region II (Figure 2.7). While an additional 1%–3% of land in Region II was put into corn between 1960 and 1980, Region I counties saw as much as 9% of their land area being converted to corn during that period (Figure 2.7). And the expansion of corn continued well into the 1990s (Argabright et al., 1996).

Corn and, to a much lesser extent, soybeans have been and are the staple crop of the Hill Country. While both are sold as market crops, corn has also been important for feeding animals, especially swine. But as described in the Introduction, corn and soybeans are row crops (planted in rows with more soil exposed to direct impact by rainfall) and permit more erosion than most other crops. Without proper management, moreover, row crops of the early 20th century continued and even accelerated the deterioration of the physical qualities of soils already seen under wheat, especially the decrease of soil organic material, the breakdown of soil structure, compaction, decreased infiltration, and decreased resistance to erosion. The widespread nature of these crops as shown in Figure 2.7 means that much soil of the Hill Country has been exposed to the debilitating nature of row crops along with the intense rainfall of the region. This explains some of the disastrous erosion that will be examined later.

Tobacco is another row crop that was traditionally grown in parts of the Hill Country, mostly in Wisconsin north of the Wisconsin River. Much of the tidewater area of Virginia and Maryland was severely eroded by the extensive growing of tobacco during the colonial and early national periods (Gottschalk, 1945). In the Hill Country, however, it was normally grown on level land, was intensively cared for, and was of small acreage, so that despite being a row crop, tobacco probably caused little soil erosion, although it might have had the effect of shifting other erosive crops to steeper slopes. Very little tobacco is presently grown.

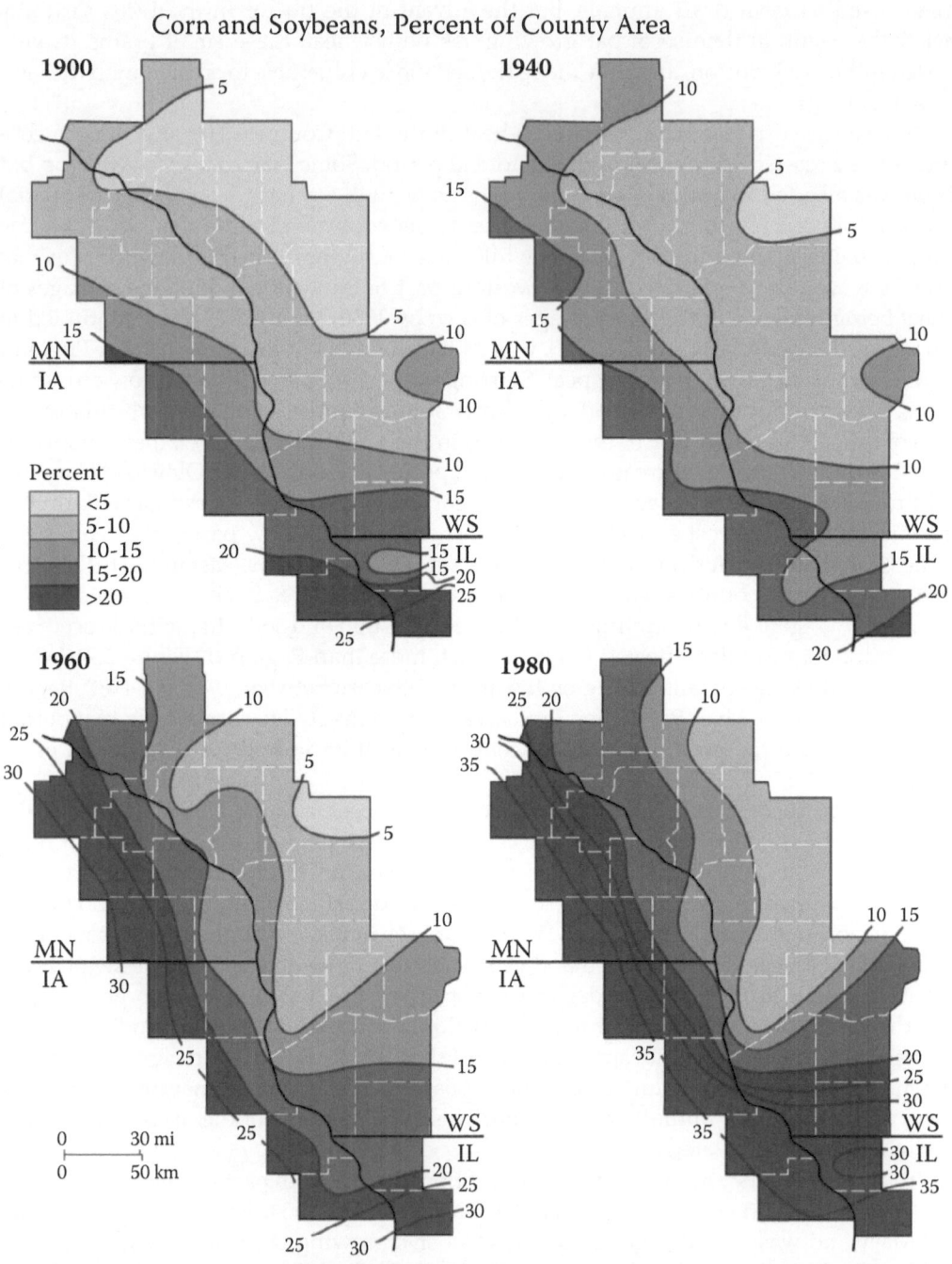

*Figure 2.7* Corn and soybean cultivation. Proportion of land in corn and soybeans, 1900, 1940, 1960, and 1980 (US Census of Agriculture).

## Grazing and animal husbandry

Animal husbandry has always been an important part of agriculture in the Hill Country. As already suggested, draft animals, especially horses, were important from the beginning for cultivation and for transportation. Dairying later became important and, indeed, the Hill Country, especially Region II, became part of the US "Dairy Belt." Fresh milk was important around larger towns and cities, but the making of butter and cheese allowed dairying in even more remote areas. Sheep were also grown in many areas. Considerable area was devoted to the growing of hay so that these animals would have winter subsistence. Improved grass pastures were also developed for them.

These three domestic animals, horses, cows, and sheep, were important to the later destruction of the landscape because they were allowed to extensively graze. To be sure, some were kept on improved pastures, but many were allowed to simply forage on less used land, especially forest and brush. To a very large degree, this meant steep, erodible "unimproved land" where animals were often forced to browse the leaves rather than graze grass. While the comparative biomass consumed, that is, grass, on pastures versus woody vegetation on unimproved land, is uncertain, the nutritional value of grass is much greater. And with continued browsing, small trees were killed, leaving only scattered larger trees in a park-like environment but with only a poor and discontinuous grass cover. Animals often desperately moved through the woods trying to get enough to eat. As one old diary farmer told me in 1974 with understandable exaggeration, "For a cow to get enough to eat in the woods, it would need a mouth three feet wide and move 30 miles per hour." Adding to the difficulty was that the grazed unimproved land was on steep slopes so that animals were also expending more calories offsetting their meager woodland diets even more. One wag professor at the University of Wisconsin termed the wooded hillsides of the Hill Country during this period as "bovine gymnasia" (Johnson, 1991, p. 132).

It is difficult to overestimate the environmental damage done by heavy grazing of steep forestland and overgrazing of pastures (Trimble and Mendel, 1995). The primary problem is that the weight of heavy animals is carried on four small hooves. This means a pressure of many pounds per square inch even when the animal is merely standing. Watching an animal walking, however, will show that all the weight is sometimes transferred to one or two hooves, especially when the animal is climbing a slope. And the steeper the slope, the harder the animal must push and thus the greater the pressure. The result is highly compressed soil, especially on steeper slopes. Such compressed soil has very low water infiltration capacity, resulting in the overland flow of water creating soil erosion and flooding (Trimble and Mendel, 1995). Experiments in the region showed that ungrazed oak forest soils had an infiltration capacity of 7.46 in. (18.94 cm) per hour. Grazing those same slopes reduced infiltration to 0.05 in. or 0.13 cm/h (Stoeckeler, 1959)! Farmers have been known to joke that heavily grazed slopes "shed water like a tin roof" (Cohee, 1934; Leopold, 1935). Without grazing, forested slopes were able to infiltrate much of the runoff from the cultivated fields lying upslope but with grazing, the forested slopes gullied (Curtis, 1966).

Animals on steeper slopes create another problem, the shearing of soil. On level ground, the hooves are normally pushed directly into the ground, resulting mostly in compression only. But on steeper slopes, the compression of the hoof is directed at an angle to the ground, which often allows the hoof to skid, resulting in highly disturbed soil which makes it more vulnerable to erosion. Where animals were penned, as in farmyards, the damage was severe (Figure 2.8).

*Figure 2.8* An eroded hillside from overgrazing, probably 1950s. (Photo credit: R. Sartz.)

Not only did heavy grazing alter the soil characteristics of woodland, it also changed the woodland itself. First, heavy grazing sometimes killed trees because the severe compaction did not allow enough water and air to reach the roots. And because animals were able to reach and browse the understory composed of young trees, they were often wiped out. Thus, older trees dying from disturbance or just old age were not replaced. Additionally, low-hanging limbs were browsed, often eliminating them to a height of 6 ft (2 m) or so. The end result was a scattering of larger trees with no understory or lower limbs, much like the park savanna of the tropics. One old farmer told me in 1974 that during the 1920s and 1930s, he could sit on his porch and see the sun setting between the trunks of trees on a wooded hillside next to his house. The effects on associated resources such as wildlife can only be imagined.

While heavy grazing of steep woodlands can create overland flow and erosion on its own, perhaps its greater role was to impair or even destroy a natural buffer. The woodlands in their natural state had a very high infiltration capacity and, often positioned between the cropland and the streams, were apparently able to reinfiltrate much of the overland flow produced by cropland and pasture on the level to rolling uplands above them (Curtis, 1966). And as sediment-laden water infiltrated, sediment was deposited on the forest floor as colluvium. But with extreme grazing, the flow from the cultivated upland combined with the flow of the grazed hillsides, often producing deep gullies down the wooded slopes. Such gullies were widespread to ubiquitous over the Hill Country by the early 20th century, and their effects are discussed in Chapters 3 and 4.

There is yet another strong hydrologic downside of hillside grazing: Not only does the infiltrated water on ungrazed slopes recharge the soil water, but it often happens that much of the *groundwater* recharge in the region is through geologic formations encountered on hill slopes, especially sandstones. Not only did hillside grazing increase *connectivity* allowing the flow of water and sediment from upland fields to reach valleys, promoting gully erosion and increasing the movement of water and sediment to streams and valleys, but it also decreased groundwater recharge, thereby reducing the flow of springs and the baseflow of streams (Juckem et al., 2006; Juckem et al., 2008).

There were also positive aspects of animal husbandry. First, it required farmers to devote a significant part of their improved land to hay rather than to more erosive cash crops such as corn. Second, manure from confined animals such as horses and dairy cows could be gathered and worked into at least a portion of their fields, thereby increasing the fertility and making the soil more resistant to erosion.

The growth of the grazing animal agriculture started in the late 19th century and has remained important until the present, although there has been some decline over the past 50–75 years in some places (Figure 2.9). The distribution of these three grazing animals has been more widespread and uniform in the Hill Country than the crops seen so far. The denser distributions were in the south at the beginning of the 20th century but then, like the grain crops, gravitated toward the west, where the animal populations in Region I remained about 50% greater than most of Region II. A much more detailed view is given for Vernon County, Wisconsin, in Region II (Figure 2.10). It clearly shows the increase of sheep up to about 1900 and their virtual disappearance since then. Horses (not shown) increased up to about 1920, when tractors became more common (Trimble, 2009a). Both sheep and horses have almost disappeared from the landscape in recent decades. Cattle, however, have increased over most of the period but, increasingly, they are beef cattle rather than dairy cows.

The introduction of dairy farming with its attendant grazing had mixed results in Region II. In this region of limited suitable cropland, the initial result was to give a new source of income to farmers, who could then more efficiently use previously underused, but highly fragile land. This additional income was enough to raise land values to something closer to those of Region I (Schafer, 1932). However, the steeper slopes of Region II made the land much more vulnerable, as the previous discussion above indicates. It seems probable then that this early economic asset later became an environmental liability and arguably was a major factor in the generally greater erosion in Region II.

But mere numbers of animals do not tell the compete story. Greatly improved pastures and hay yields since the 1930s have allowed farmers to remove grazing animals from the steep forested slopes so that grazed forest has decreased radically (Figure 2.11). By 1969, the last year the US census listed grazed forest, most counties had only about 40%–60% as much as in the period 1925–1930, perhaps the peak of forest grazing. Since 1969, moreover, there has been a strong decline of forest grazing as farmers (1) increasingly realize that grazing of forests and woodlands produces little net nutrition for animals, (2) have better pastures and more hay as the result of better agronomic practices, (3) realize that grazing not only can cause more runoff and erosion but can also degrade their soil as well as their woodland resource both for wood production and for wildlife habitat, and (4) recognize the offsite damage from runoff and sediment to their own land and to the land of others downslope and downstream.

Another important animal in the Hill Country has been swine, but their role in causing significant soil erosion is difficult to evaluate. The distribution of pigs was highly correlated with the growing of corn, and some of the counties in Region I had astounding numbers, the number of pigs sometimes being greater than the population of grazing animals (cows, horses, sheep) by a factor of two to three times! In Region II, with much less corn, the number of pigs was more like 20% to 50% of the number of grazing animals.

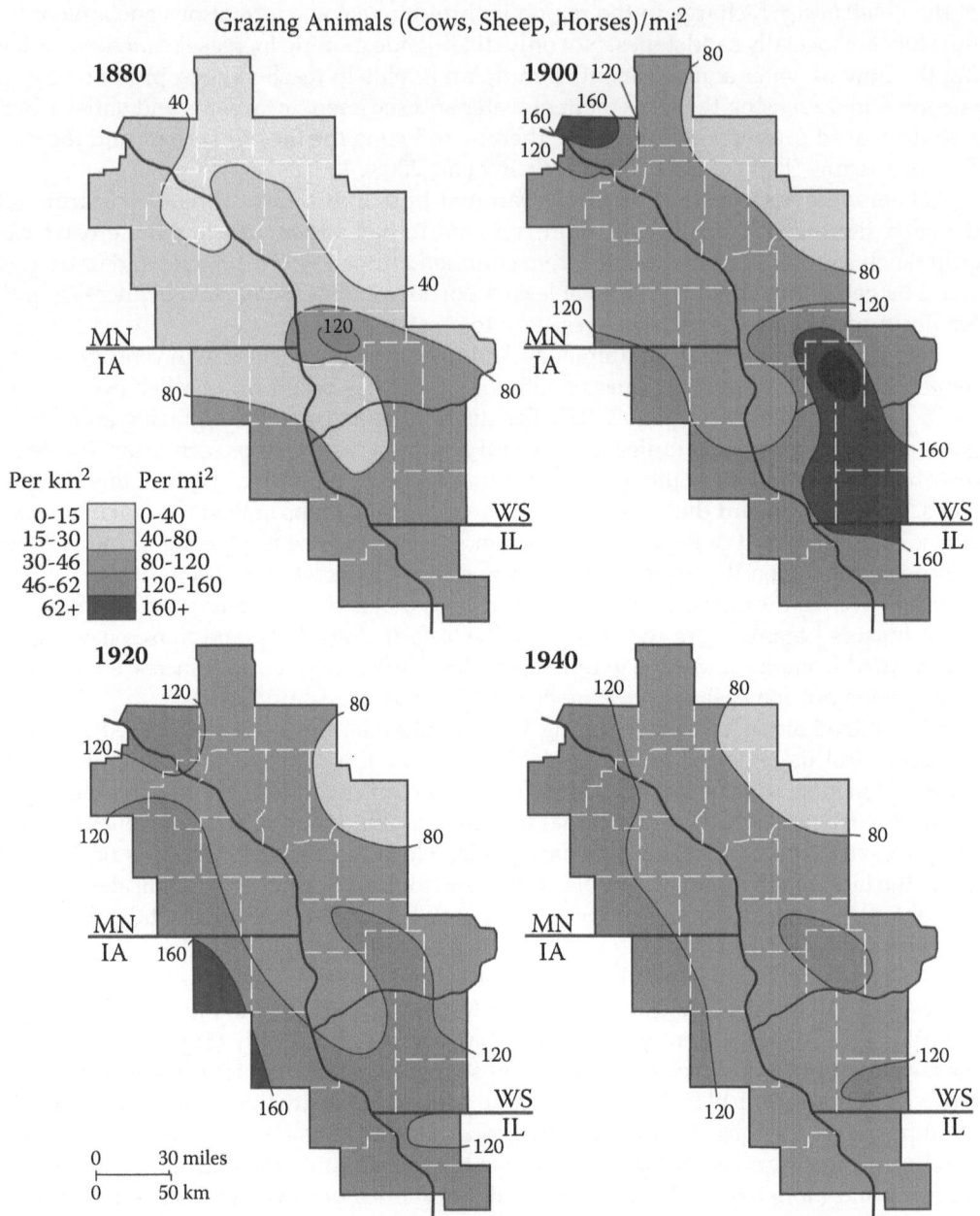

*Figure 2.9* Density of grazing animals (cows, sheep, horses), 1880, 1900, 1920, 1940 (US Census of Agriculture).

But how much erosion did all those pigs cause? Pigs are normally densely confined on a mild slope for good drainage, and the food and water are brought to them there. Given the trampling and disturbance of the soil, the erosion from such pigpens was sometimes locally serious, with both soil and pig offal being swept into the nearest stream with every rainstorm. In Region II, the pigpens sometimes were located on steeper slopes and

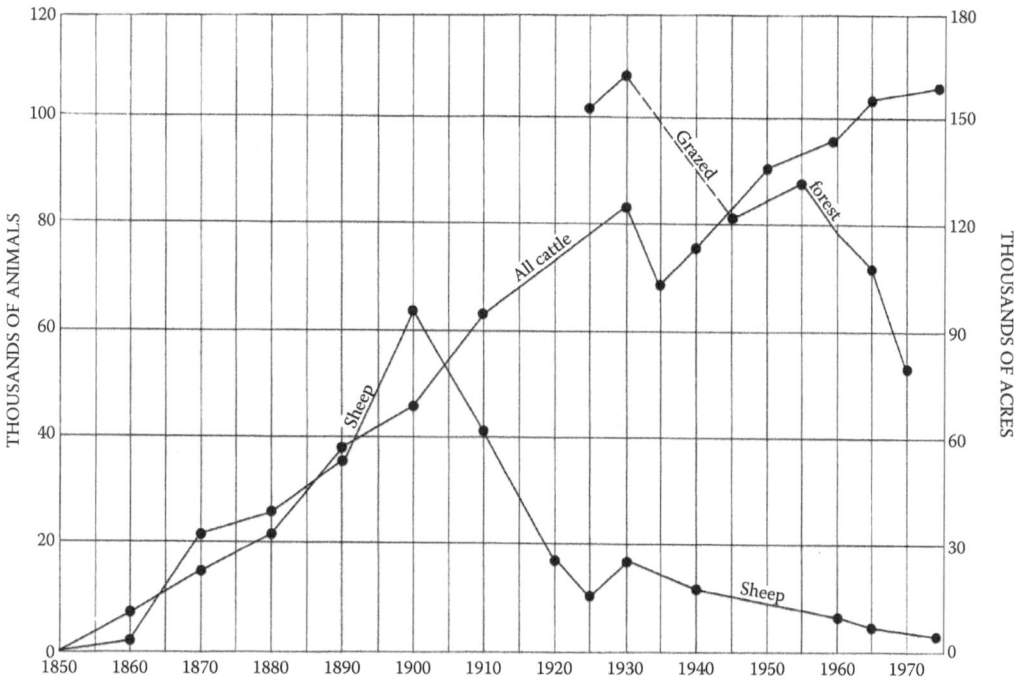

*Figure 2.10* Grazing animals, 1850–1975, and grazed forest, 1925–1970, Vernon Co., WI (Trimble, S. and Lund, S., *US Geol. Surv. Prof. Paper* 1234, 1982).

even into the 1970s, I saw severely denuded hillside plots of 0.5–5 acres (0.2–2 ha) from confined pigs. The only saving grace for the entire Hill Country was that such pens were of relatively small area. Another practice was to sometimes "harvest" corn by turning the pigs into the grain field, a practice termed "hogging off" (Bogue, 1963). Argabright et al. (1996) report that about 10% of corn acreage was harvested this way in the 1930s. They did not differentiate within the Hill Country, but this was probably more common in Region I, where it is still practiced. The erosional effects are probably minor because most corn fields were on the more level land and because the pigs are on the field for only a short period.

## Land use management from the time of settlement to the 1930s

In the upper Midwest, the deep and fertile soils, especially prairie, found by the original settlers were so productive that the concept and terminology of "inexhaustible" soils crept into the perception of farmers and even into the literature. It is thus possible that a somewhat relaxed attitude about careful land management was held by the earliest settlers in the region, many or even most of whom were "Old Americans" or people who had been in the U.S. for more than one generation (Shafer, 1922). While Southern influence was strong in the southern part of the Hill Country (Hancock, 1913), most American settlers were from the northeastern United States, many of whose families had migrated across the United States and had perhaps owned and sold several farms as they "leapfrogged" across the country taking up new land. A good example is the father of Hamlin Garland, who was born in Maine and then migrated from farm to farm westward across the country, eventually winding up in South Dakota (Garland, 1917). Hence, some perhaps perceived cheap

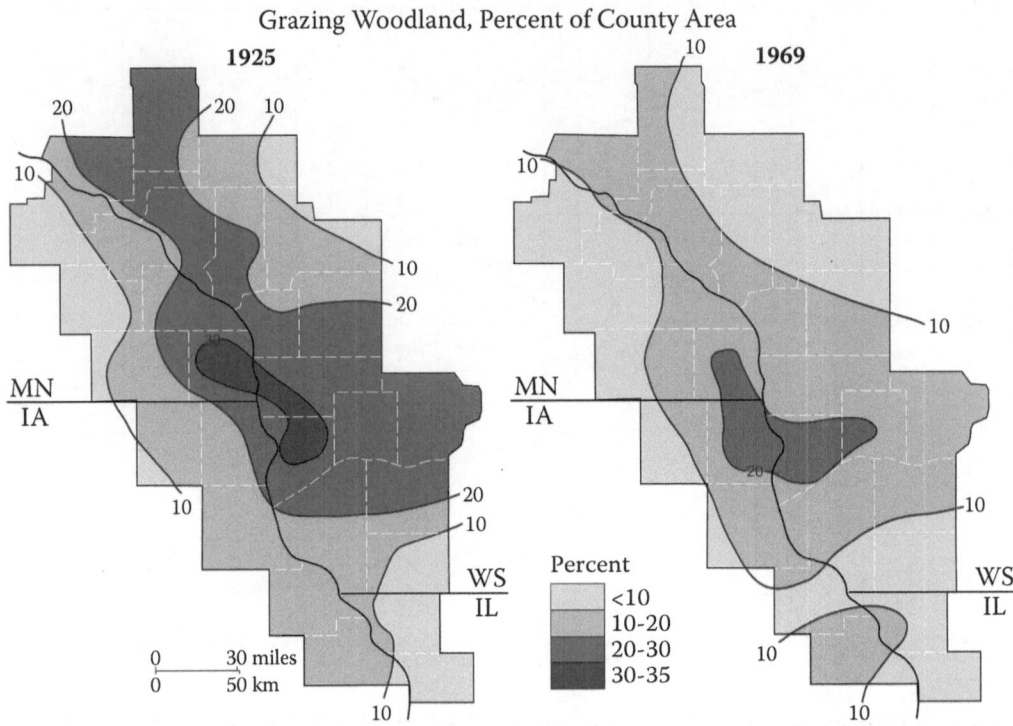

*Figure 2.11* Grazed woodland, percentage of county area, 1925 and 1969 (US Census of Agriculture).

land as a disposable commodity, as was the documented case farther south (Trimble, 1974, 1985). Such attitudes could also be engendered by the cheap price of land (Trimble, 2010a). While early agriculture in the Hill Country was not as erosive as cotton farming down South, it was still perceived in some quarters as "land skimming" (Johnson, 1991, p. 3).

The early "Americans" were soon followed by central and northern Europeans, largely Scandinavians, Germans, Irish, and Luxemburgers. Schafer (1922) contends that Germans were better and more prudent farmers than "Yankees" or old Americans, and, if that is true, it seems reasonable to assume that other Western Europeans were as good as the Germans. Many of these newly arrived Europeans had been small landholders or even tenants in the old country and were "land hungry," their intention being to hold onto this land and pass it on to posterity. Thus, they generally employed the best cultivation methods they knew. The prevailing practice in Europe, which had preserved soils for centuries, was generally the "3-field" method, where fields were on an annual rotation of one year small grain, one year vegetables (peas, beans, lentils, cabbage), and one year fallow. Whether the European 3-field system was ever practiced in the Hill Country is uncertain, but farmers there soon devised a somewhat similar 3-year rotation, which was one year row crop (mostly corn), one year small grains, and then one year of grass for hay which replaced the one year of fallow in the 3-field system (Benton and Russell, 1927; Trimble and Lund, 1982). While this was an improvement, only one year out of three in grass was not adequate to restore the soil qualities that allowed it to resist erosion, especially given the steep slopes and intense rains of the Hill Country. And in some cases, the rotation was even poorer, leaving fields in corn for two years rather than one (Veatch and Orben, 1923). On the other hand, there are recorded instances of a seven-year rotation that included two years of corn, two years of small grains, and up to three years of hay (Whitson et al., 1914), but this was apparently rare.

The 3-year rotation described above was a "best case" in most instances. Argabright et al. (1992) suggest that by 1930 some farmers did little with rotations and, for example, grew continuous corn. This is also suggested by several soil surveys of the period.

As mentioned earlier, a major problem, not just in the Hill Country but in much of the United States, was that crop rows and thus plowing were usually aligned with the cardinal directions as dictated by the rectangular land survey. Indeed, a "plowing ethic" arose in which a farmer was partially judged by the straightness of his plowed rows. On the mild slopes of glaciated country, this usually made little difference but in the Hill Country, it meant that rows were often at steep angles that channeled water downhill, eroding slopes. However, it must be said that some farmers did lay out their fields so that they could plow across the slope in general, but not necessarily on the contour. For example, see the 1934 air photo in Figure 2.12. But detrimentally, corn was often "checked" or plowed both ways (both across and up- and downslope) to control weeds, and this could promote erosion, especially in the very early planting and crop growth stages (Whitson and Dunnewald, 1916). As late as 1929, it was reported that some farmers still plowed up and down slope (Gray et al., 1929).

Thus, the combination of gradual soil deterioration, inadequate rotations, plowing up and down slope, and overgrazing led to increasing runoff and soil erosion, with rills and gullies eventually scarring the landscape (Figure 2.12). As soils became progressively

*Figure 2.12* Air photos of Coon Creek landscape, 1934 and 1967, just north of Coon Valley (SE1/4,T15N, R5W, Vernon Co.). 1934: Note rectangular fields and gully systems extending into upland cultivated fields. (From Trimble, S. and Lund, S., *US Geol. Surv. Prof. Paper* 1234, 1982.) (Continued)

*Figure 2.12 (Continued)* Air photos of Coon Creek landscape, 1934 and 1967, just north of Coon Valley (SE1/4,T15N, R5W, Vernon Co.). 1967: Note contoured and strip cropped fields with no rills or gullies. (From Trimble, S. and Lund, S., *US Geol. Surv. Prof. Paper* 1234, 1982.)

thinner and less well structured, even more water and soil went down the rills and gullies with every rain. Valleys were covered with sediment while small tributaries, flushed with the additional storm flow coming off cultivated and heavily grazed hillsides, began to erode their channels, sending even more sediment downstream.

What drove the soil erosion, of course, was the hydrologic changes. Water that earlier had infiltrated, recharging soil and groundwater, now cascaded off slopes, causing erosion of the slopes and flooding in the streams, a runoff condition termed by hydrologists as *flashy*. But between storms, there was consequently less groundwater available to keep streams flowing (baseflow) so springs and small streams began to dry up. Although this was problematic for people, it also affected wildlife. For example, the original brook trout were largely extirpated by muddy floods followed by low flows.

## A revolution in agricultural land management

The Hill Country was not alone in the United States in suffering severe soil erosion. Indeed, much of the United States had severe soil erosion problems by the late 19th and early 20th centuries (Bennett and Chapline, 1928; Bennett 1939). A major problem was that people, primarily from Western or Northern Europe or at least having acquired that culture, were farming land in environments heretofore not experienced by them. Thus, they had not developed the proper crops, agronomic practices, and engineering techniques to deal with the American environment, especially climate (Chapter 1; see also Trimble, 2011).

Of greatest moment in the humid United States was the amount and intensity of rainstorms as compared to the old country, as discussed in Chapter 1.

But this problem was not only true of recent immigrants but also true of "old Americans." By the early 20th century, it was widely recognized by American agricultural experts that agricultural techniques in the United States, even the techniques employed by those who had been here for decades, were not adequate to cope with both wind and water erosion. Thus, states, followed by the federal government, started erosion control experiment stations with the goal of devising methods to mitigate soil erosion (Meyer and Moldenhauer, 1985; Johnson, 1991; Trimble, 2009b, 2010a, 2011). One of these was the Upper Mississippi River Valley Soil Conservation Experiment Station established in 1931 near La Crosse, Wisconsin (Hays, McCall, and Bell, 1949; Johnson, 1991).

Just three examples of findings and methods from these experiments will indicate their importance:

1. In a crop rotation, one year of grass was found to be inadequate to allow the soil to regain its structure and thus its infiltration capacity and resistance to erosion. Rather, it required 3–4 years of grass. Clearly, longer crop rotations with more years in grass were called for.
2. Contour plowing had been practiced on some steeper slopes in the United States since the time of Thomas Jefferson, who strongly advocated its use. The idea was for the plowed ridges on the contour to act as small dams and hold water until it could infiltrate. In the first place, the problem was that few people in the Hill Country even thought about contour plowing. But even had they done so, the problem would have been that during the very large storms so common in the Hill County, the excess water filled the rows and then overflowed, cascading down the slope and causing severe erosion. And the steep slopes exacerbated the problem. The solution was to install 44–60 ft (14–18 m) wide *contour strips* of alternating land use such as corn, small grains, and grass. In particular, the grass intercepted the water and sediment cascading from upslope, allowing the water to infiltrate and the entrained sediment to be deposited (Figure 2.13).
3. Improved crops were introduced or encouraged. One example is alfalfa (lucerne), which sends down extremely deep roots and encourages infiltration. While it was grown earlier in the Hill County, it was encouraged, became more widespread, and its management became more scientific. Other plants were also improved genetically (e.g., hybrid corn), resulting in much higher plant populations per acre, better growth, higher production, and less erosion.

Figure 2.12 shows how the landscape was transformed by contour strip cropping. Details on the fields, crops, and rotations are given by Trimble and Lund (1982). Figure 2.13 shows some of the changes of land use and management practice in the Coon Creek basin, 1934 and 1975. It will be observed that cropland declined only slightly. Indeed, *land use* changes were relatively small; it was changes of *land management* that were significant. Nevertheless, the small decrease of cropland (about 6%) is significant and deserves comment. Most of this was steeper land that would erode more readily, and thus there was sometimes the conservation motive to remove it from cultivation. But there is another reason: mechanization. A horse and plow could operate quite well on a steep slope, but tractors couldn't, and this was the period when tractors were becoming more common. Tractors and mechanization are sometimes invoked as detrimental to soil conservation (Worster, 2004), and indeed they can be. But they can also make possible better forms of soil conservation such as no-till farming, as we will

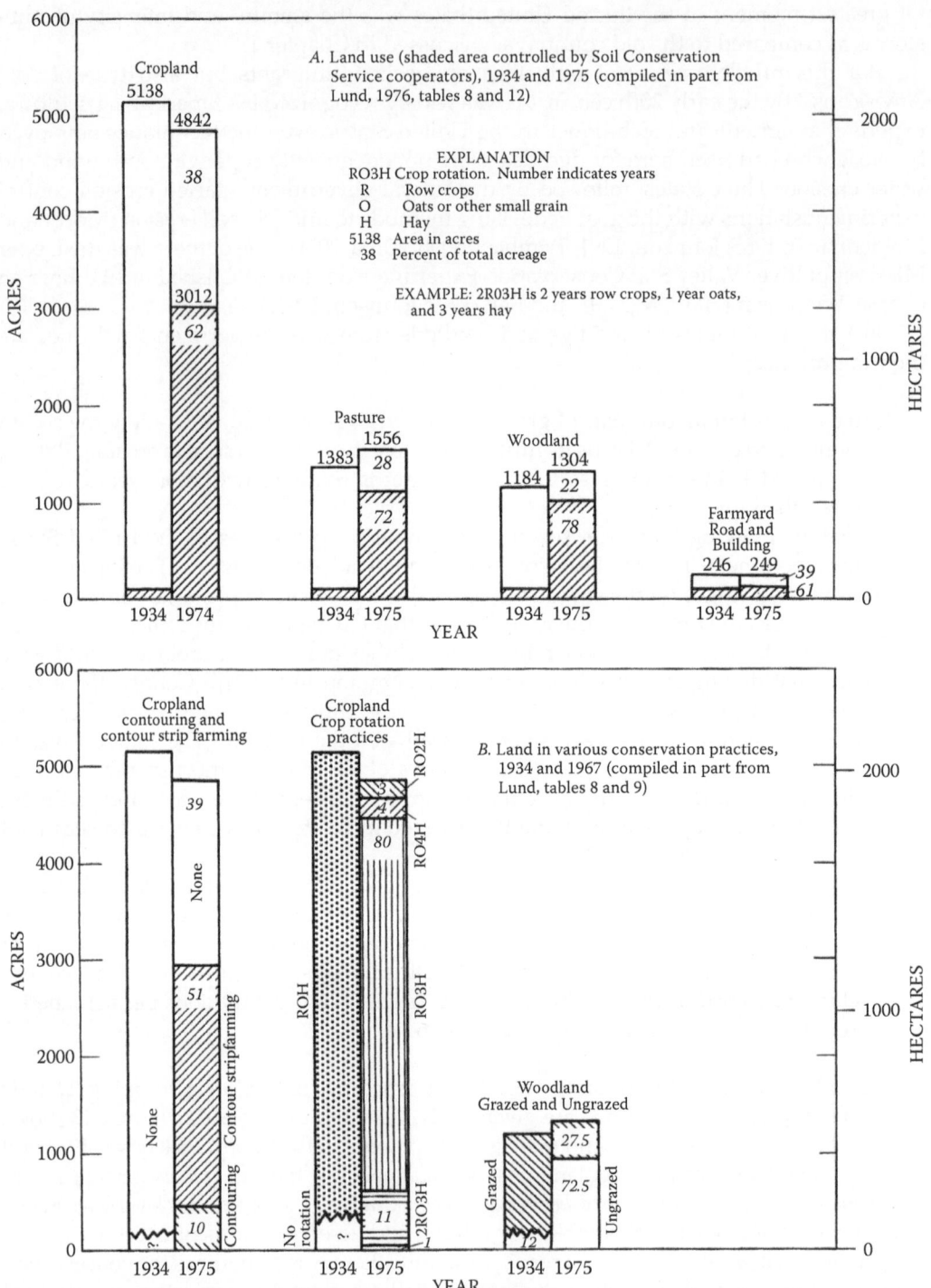

*Figure 2.13* Changes of land use and management in 10 sample basins, Coon Creek, WI, 1934 and 1974. (From Trimble, S. and Lund, S., *US Geol. Surv. Prof Paper* 1234, 1982.)

see later. We should never forget that the worst depredations on American agricultural soils, indeed, world agricultural soils, were mostly done with animal power.

## Conservation agencies

By the mid-1930s, the agronomic methods described above were becoming available, and governmental agencies and programs were instituted to convey them to farmers and put them into use. Specifically, the (Federal) Soil Erosion Service was created in 1933, becoming the Soil Conservation Service (SCS) in 1935, a part of USDA. The SCS (now the Natural Resources Conservation Service, NRCS) brought the newly developed soil conservation methods to individual farmers in many ways. First, there was widespread publicity directed at farmers and other land managers. Second, was the availability of local SCS agents to create individual farm plans and advise on particular practices (Figure 2.14). Third was limited financial assistance for installation of these practices. Fourth was the creation of "Conservation Demonstration Areas" for each agricultural region of the United States. These normally encompassed an entire stream basin of several square miles. Here, modern soil conservation measures were applied to the land with subsidies from the federal government. The concept was that neighboring farmers, seeing success, would emulate these practices. Not so coincidentally, the first one in the U.S. was in the Hill Country, the Coon Creek basin just south of La Crosse, Wisconsin. This project had been envisioned for a long time by Aldo Leopold of the University of Wisconsin and others (Cross and Davis, 1971).

Such demonstration areas were necessary to convince neighboring farmers to invest the time and money to change from the old ways, which were often highly ingrained. A good example of this was changing from straight rows to contour strips, which some farmers termed "crazy–quilt farming" and refused to employ it until convinced of its efficacy by demonstrations (Figure 2.15).

The utility of these demonstration areas is shown by the diffusion or spread of contour-strip farming from the Coon Creek demonstration area to adjacent counties during the period 1939–1967 (Figure 2.15). In 1939, contour stripping was limited to Coon Creek and its immediate environs, but by 1967, it was ubiquitous across the four-county area shown. This portrayal of diffusion shows that the soil conservation revolution was an enduring process and, despite occasional setbacks, was continually improving.

The SCS also assumed the responsibility of soil surveys, which had before 1933 been done by the USDA Bureau of Soils and Chemistry. The newer surveys described the capability of soils to handle intense use and to guide farmers to optimum use and management of soils. Another type of study was the *Erosion and Related Land Use* surveys for some badly eroded areas. One of these in the Hill Country was the Farmersburg-McGregor study in Iowa across the Mississippi River from Prairie du Chien (Perfect and Sheetz, 1942). Another was for Winona County, Minnesota (Brown and Nygard, 1941).

Another USDA agency playing an important role at this time was the Agricultural Adjustment Agency (AAA), later the Agricultural Stabilization and Conservation Service (ASCS, now the Farm Service Agency, FSA), which by financial subsidies controlled the acreage of some crops and in some cases dictated how the crops were to be grown (Helms, 2003). Measuring the acreage of crops required stereographic air photo coverage, and this was done at a scale of 1:8500 for Coon Creek in February, 1934 (Figure 2.12, top). Begun for most of the United States in 1937 and reflown every six years or so, these air photos became important conservation tools because soils, slopes, land use, and management plans could be plotted and annotated directly on the photos. They were originally printed at a scale of 1:15840 (4 in. = 1 mi or 1 cm = 158.4 m), which permitted observation of conservation

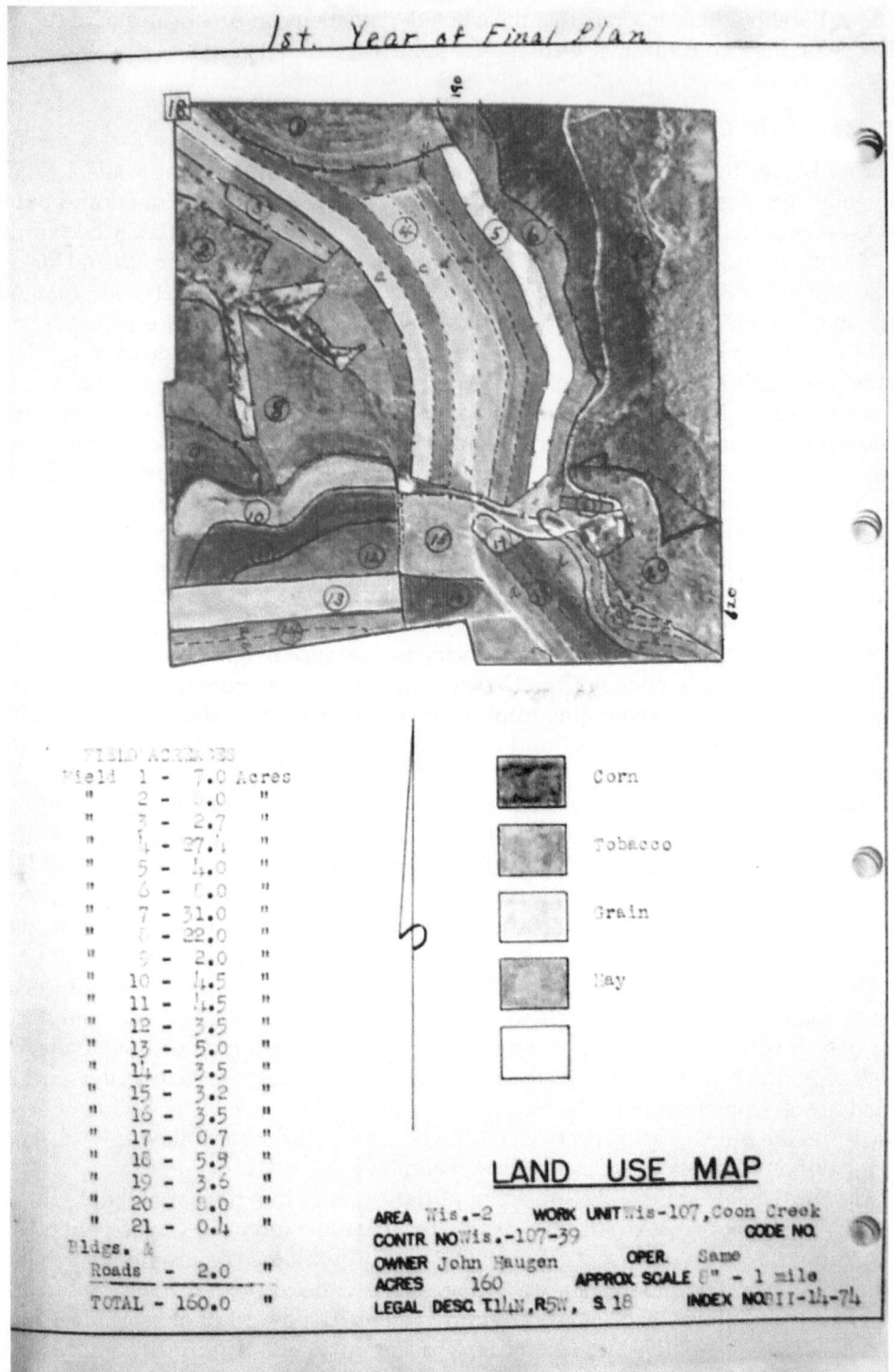

*Figure 2.14* Partial farm plan for John Haugen for his 160-acre farm south of Coon Valley, WI, 1939. (SE1/2 T14 N R5W, Vernon Co.) (Courtesy of Ernest and Joseph Haugen, Coon Valley, WI, Photo credit: Nadine Kleinhenz, July 2006.)

Chapter two: European settlement and changes of land use

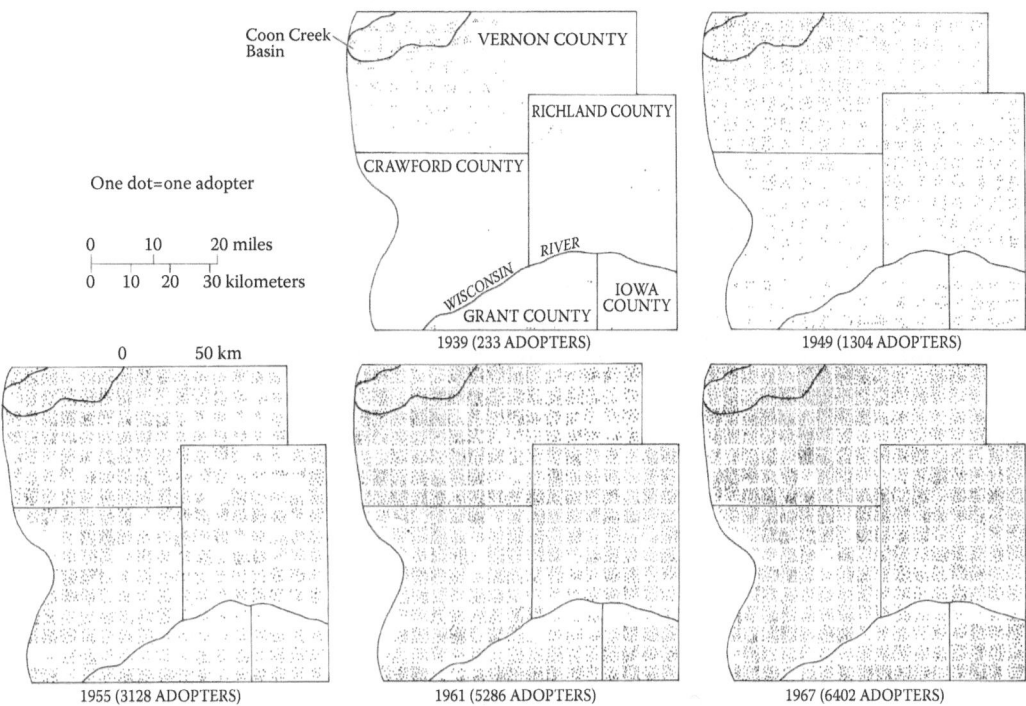

*Figure 2.15* Spread of contour-strip farming from the Coon Creek Conservation Demonstration Area, WI, 1939–1967. (From Trimble, S. and Lund, S., *US Geol. Surv. Prof. Paper* 1234, 1982.)

problems such as gullies and eroding stream banks. Later, soils were delineated directly on air photos at the scale of 1:20,000. By the late 1980s, soil surveys had soils denoted on orthophotomaps, that is, air photos adjusted to coincide perfectly with USGS topographic maps at a scale of 1:24,000 (1 in. = 2000 ft or 1 cm = 240 m).

Yet another federal project of that period was the Civilian Conservation Corps (CCC), a military-type organization that executed construction-type projects in conservation such as building small dams, stabilizing gullies, and protecting eroding streambanks (Figure 2.16). There were several CCC camps in the Hill Country; perhaps the best known is the one at Coon Valley, Wisconsin. Soil erosion problems in the Hill Country as compared to the surrounding glaciated landscape may be shown by the fact that of nine CCC camps established in Wisconsin in 1933–1943, eight were in the Hill Country (Johnson, 1991).

The building of small dams and ponds, started by the CCC in the 1930s, was continued. While many ponds were created by government agencies to help ameliorate local floods, most were built as private farm ponds, usually with technical and financial assistance from government agencies, primarily the USDA. The net result of such ponds is to detain significant sediment that otherwise would have moved downstream (Renwick et al., 2005). They were also used to stabilize gullies.

Other federal agencies were also active in the soil and water conservation enterprise, but somewhat more tangentially. An example is the Works Progress Administration (WPA), which sponsored historical work on various regions of the United States and also compiled a photographic file of soil erosion, now in the National Archives (Record Group 118). The Federal Resettlement Agency and other agencies also produced two classic films that relate to some degree to the Hill Country. These are *The River* and *The Plow That Broke*

*Figure 2.16* CCC project to stabilize high Pleistocene terrace undercut by stream. Exact location unknown but probably Trempealeau Co., WI, late 1930s. SCS Photo.

*the Plains*. With pathbreaking direction and cinematography by the poet Pare Lorentz and a musical score by the prominent music critic and composer Virgil Thomson, these were powerful films for the time. While somewhat overdrawn and perhaps rightly thought of as propaganda, they show remarkable footage of soil erosion and related problems in the United States during the early 20th century, and they did help to get public support for large soil conservation programs.

State and county agencies also played important roles in this revolution of soil conservation. Already mentioned is the fact that state agencies also sponsored and published soil conservation research going back to the late 19th century and some states, notably Missouri, started erosion control experiment stations long before the federal government. Agricultural Extension was part of this, and extension agents were part of the cadre advising farmers. All four states of the Hill Country had strong agricultural programs and were able to give valuable assistance to federal efforts.

Assistance in curtailing erosion and sedimentation sometimes came from unusual quarters, in this case, state, county, and township road departments. Because amplified runoff was often undermining public roads, railroads, and bridges, and sediment from accelerated erosion was sometimes burying the same features, many counties in the Hill Country were spending large proportions of the road budget for maintenance and repairs. Thus, it was in the interest of road departments to mitigate the problem. Since they had no control over land use and management on private land, all they could do was install engineering works and structures that protected the roads—but in doing this, they often brought badly eroding areas such as gullies under partial control.

## The continuing soil conservation revolution

It is important to note that the soil conservation revolution has never ceased in the Hill Country, or indeed in the United States. As shown in Figure 2.14, the new methods eventually spread from nodal points such as Coon Creek to the entire region. Farmers became much more sophisticated and grew into the habit of soil conservation as they realized its value to them. The new techniques were improved over time, and new machinery, along with the increasing wealth of farmers, made it easier to implement the measures. The SCS (now the NRCS) and the AAA (now the FSA) continued to work with farmers. Just between 1982 and 1992 in the Hill Country, grassed waterways increased by 28%, contour strip cropping 11%, and terracing 20% (Argabright et al., 1996). New agencies such as the state departments of natural resources continue to create wildlife habitat and help stabilize streams.

Since the 1980s, there has been another conservation revolution: *no-till* farming. Instead of plowing the soil, seeds and fertilizer are injected ("drilled") into the soil, protected with a deadened grass cover. With no-till farming, there is less soil compaction so that infiltration rates are significantly improved and erosion rates are greatly reduced. Additionally, soils store more carbon as organic material, and pedofaunal life such as earthworms increases. The practice also conserves water. One may perceive this as a soil conservation advancement just as great and significant as that which took place in the 1930s and 1940s, and its effectiveness against erosion will be demonstrated in Chapter 8.

While the technology and equipment have been around for some time, the equipment was expensive and the technology new. But slowly and with little fanfare, farmers have been adopting no-till farming. No exact figures are available, but "conservation tillage," which includes no-till, increased by 37% between 1982 and 1992 (Argabright et al., 1996). While improvements in the landscape, including water quality, over the past 20 years are due in part to the maturing condition of soils from continued conservation, it is very probable that no-till cultivation has played a strong role too. Indeed, it is ironic that as some scientific journals were publishing jeremiads about a soil erosion "crisis" in the United States, including the Hill Country (e.g., Pimentel et al., 1995), matters were actually getting better (Trimble and Crosson, 2000).

Perhaps just as significantly, productivity has increased greatly in the Hill Country. For example, yields of the major crop, corn, in the 1930s ranged from 35 to 50 bushels per acre, or about 2700–4100 liters/ha (Argabright et al., 1996). By the 1970s, USDA soil surveys indicate that corn yields ranged from about 60 to 135 bushels per acre (4,600–10,300 liters/ha). Presently, the average for the steeper and less fertile soils of Region II is about 150 bushels per acre (11,400 liters/ha) while the level to rolling Mollisols of Region I produce about 180 bushels/acre (13,700 liters/ha). However, excellent farmers on the best soils in Region I sometimes get as many as 300 bushels per acre (22,800 liters/ha). At the same time productivity was increasing, the acreage of corn was expanding. One almost has to step back to realize the full significance of all this: in the Hill Country, soil erosion has been radically decreased over the past 75 years while productivity has greatly increased. The impression is often given that "pushing" the land to higher productivity increases soil erosion. But just the opposite is usually true with modern technology: productivity decreases erosion, especially with no-till farming. This happens for several reasons, but the main ones are the denser-growing and more vital plant populations that help protect the soil during the growing season. Then after harvest, the greater residues provide more mulch to protect the soil until the next planting. The residue cover is especially important for no-till farming. Simply put, the higher the productivity, the lower the erosion rates.

Certainly, massive applications of nitrogen and other fertilizers can leach into both ground and surface waters, and are a significant problem. Farmers should avoid overfertilization (National Research Council, 2008), but it should not be forgotten that the high yields enable farmers to greatly reduce erosion. Since many potential pollutants adsorb onto soil particles, a reduction in erosion generally means fewer agricultural pollutants moving into surface waters.

There is perhaps an even greater significance to the increased productivity of cropland in the Hill Country and other favored American agricultural regions. It means that less agriculturally productive regions can greatly reduce agricultural production, and this former cropland can revert to forest or other conservation land uses (Trimble, 2009b). Indeed, from Texas to Maine, millions of acres of forest now cover what was formerly eroding crop land (McKibben, 1995).

## A composite of erosive land use over the historical period

With the general knowledge of land use and management over time now in hand, we may now show erosive land use (ELU) and its changes over time. I do this by weighting the proportion of counties in various land uses and crops (corn, small grains, pasture, grazed forest, etc.) with a factor proportional to the erosive nature of that land use. And as already explained, the erosive role of land use has declined over time. The weighting proportion is of the erosion rate compared to bare soil. These management-weighting factors are from the Universal Soil Loss Equation as explained in the Introduction (Table 2.1). This methodology is similar to that used in Trimble (1974) and Trimble and Lund (1982). Clearly there would be great disparity within each group, and these values are only estimates (Table 2.1).

Thus, a county entirely in corn (1853–1930) would be presumed to be eroding at a rate of about 0.45, or 45%, of bare earth while a county entirely in ungrazed forest at the present time would be eroding at a rate of 0.002, or 0.2%, of bare earth. Of course, each county is a composite of many land uses, and the mapped values reflect that (Figure 2.17). It is important to note here that the values are conservative and reflect *only* land use and management and do not account for slope, soils, or rainfall. In other words, this is a very conservative estimate of the *human input*. Note that it does not extend to the present period of no-till.

The first map is 1880, the first year that the US census of agriculture reported acreages of crops (Figure 2.17). ELU increases to maximum in about the 1920s to early 1930s and then decreases to about half of that maximum value by 1980. Since that time, significant increases in no-till farming would reduce the value of ELU much more, but no attempt is made here to quantify that.

*Table 2.1* Management-Weighting Factors Used to Estimate Historic Erosive Land Use

| Land Use | 1853–1930 | 1940 | 1950 | 1980 |
|---|---|---|---|---|
| Row crops | 0.45 | 0.32 | 0.26 | 0.20 |
| Small grains | 0.20 | 0.15 | 0.13 | 0.10 |
| Hay and pasture | 0.04 | 0.022 | 0.014 | 0.005 |
| Farmstead and roads | 0.20 | 0.15 | 0.13 | 0.10 |
| Abandoned | 0.05 | 0.045 | 0.043 | 0.04 |
| Grazed forest | 0.03 | 0.025 | 0.023 | 0.02 |
| Ungrazed forest | 0.002 | 0.002 | 0.002 | 0.002 |
| Urban | 0.05 | 0.05 | 0.05 | 0.05 |

*Source:* From Trimble, S. and Lund, S., *US Geol. Surv. Prof. Paper* 1234, 1982.

Chapter two: European settlement and changes of land use 53

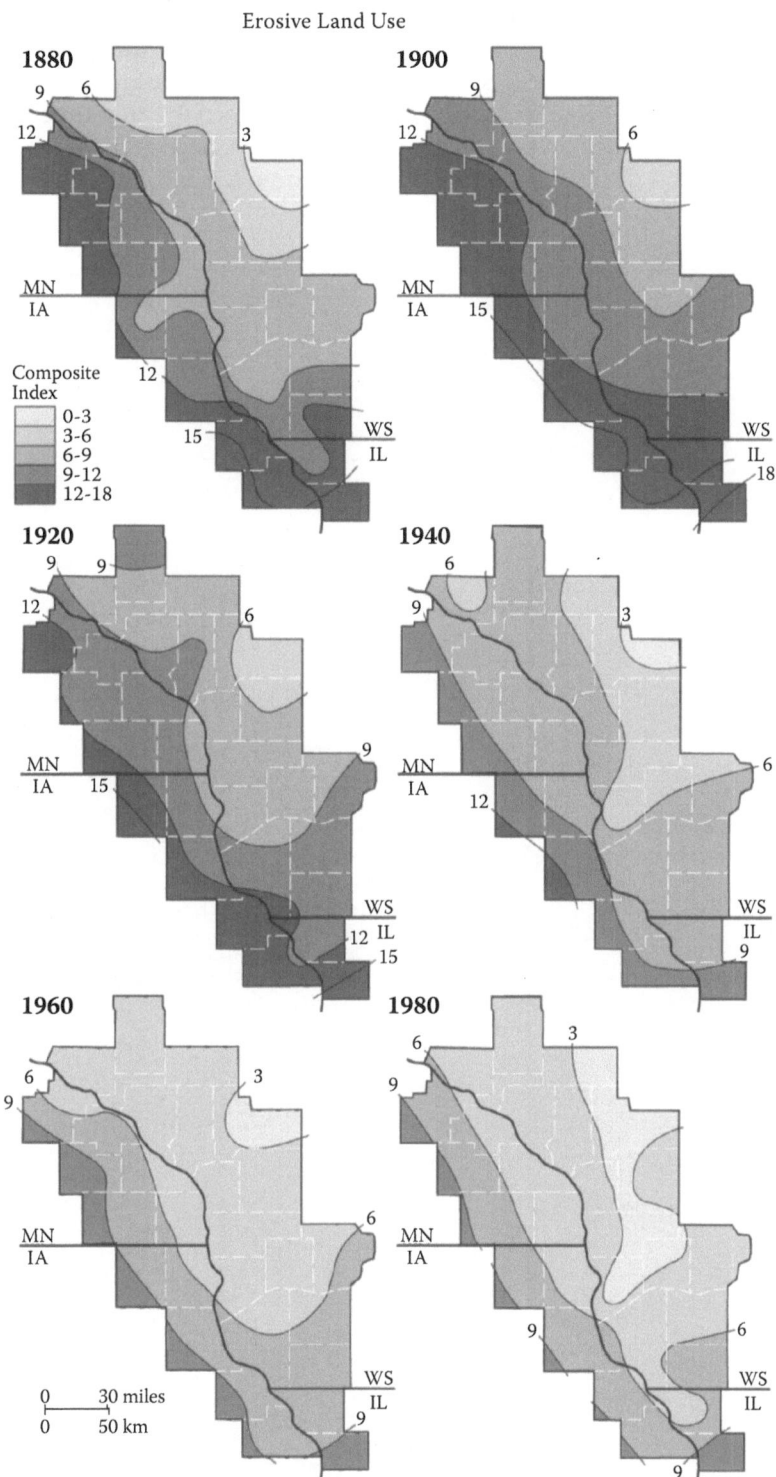

*Figure 2.17* Erosive land use, Hill Country, 1880–1980. See text for definitions and explanation.

For each year, values of ELU in Region I are much greater than in Region II. As might be expected, the minimum values are in the northeast area, a much poorer farming area as already explained. While the maximum spatial disparity of values is over 3 times, the more general difference between Regions I and II is more like two times. While these methods are imperfect, they do suggest that Region I was subjected to much more erosive land use than Region II. In Chapter 3, we will examine the results of this land use and management in the context of landforms and other factors in the Hill Country.

*chapter three*

# The systematic effects of historical agriculture on the physical landscape

We must begin this chapter by reminding ourselves of the buffer or "cushion" that the virgin soils provided. While they were not "inexhaustible" as some thought, they were resilient and their depth, structure, infiltration capacity, and resistance to erosion gave the farmers an initial reprieve from serious erosion. How long was this reprieve? While we have no experimental studies to guide us on this, one to four decades seems reasonable based on the historical record. It would depend on the soil, slope, land use, and land management. Moreover, all of these variables varied widely and differed from place to place even on the same farm. In any case, 50 plus years of cultivation had to have had a deleterious effect on the excellent soil characteristics described in Chapter 1. With time, organic material, pedofaunal activity, and soil structure declined. Knox (2001) showed that organic carbon in a prairie soil had declined from about 5% to 6% in the virgin state to about 1.5% in a long-cultivated soil. The result of the foregoing changes was decreased structural integrity, decreased ability to resist erosion, increased bulk density, decreased infiltration capacity, and more overland flow; all the foregoing increased erosion. And with erosion of the topsoil, the process was accelerating. As discussed in the Introduction, the process involved the hydrologic transformation from the Hewlett Model to the Horton Model.

Cycles of heavier rainfall probably also played some part. Short periods of greater rainfall occurred around the turn of the century (Figure 1.7), and that might have helped bring on an earlier onset of serious erosion. But if so, it merely accelerated a process that was inevitable given the land use and management practices of the previous half century.

## The role of modified hydrology (in particular, the role of rills and gullies)

Even after soil began to erode, there was yet another buffer that delayed the delivery of water and sediment to the small tributary streams: the general scarcity of channels. I refer to both natural channels and human-induced ones such as rills and gullies, which rapidly conveyed water and sediment to streams and valleys. Until the development of these rills and gullies, much of the eroded material could not be efficiently transported by overland flow and would be deposited as colluvium downslope, at the base of slopes, in local concavities, and into the permeable soils of wooded hillsides. But as discussed in Chapter 2, these hillsides were being heavily grazed, thereby gradually losing their ability to absorb excess flows. From these several factors, rills and gullies developed in time, speeding water and sediment off slopes (Figure 2.12). Such gullies were widespread. By the 1930s, the Whitewater River basin in Minnesota, an area of about 450 mi$^2$ (1170 km$^2$), had thousands of gullies. Of these, 2788 were large enough to require an engineering structure such as a dam (SCS, 1942).

Rills and gullies develop as the result of excessive overland flow, that is, the transition from the Hewlett concept to the Horton concept (see Introduction). Water always seeks the

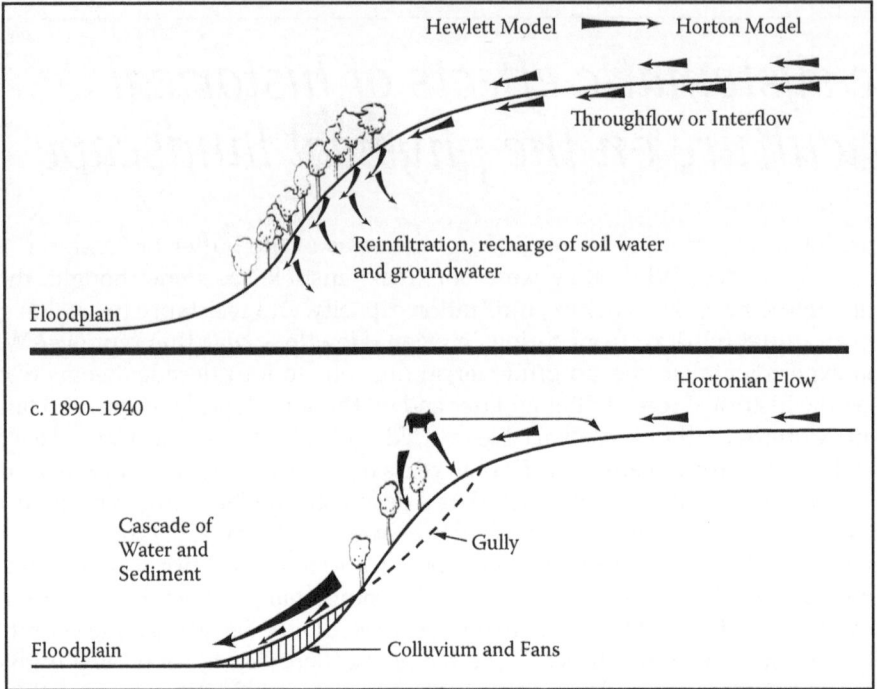

*Figure 3.1* Schematic of response of hillsides to agriculturally induced overland flow from upland areas. Top: Early; with increasing crops on the upland and deteriorating soil hydrologic conditions, runoff increasingly becomes surface (Horton Model) rather than subsurface (Hewlett Model), but hillside woodlands are able to reinfiltrate much of this. Bottom: With increasing overland flow from above and decreased infiltration on grazed hillsides, gullies form and create fans at the toe of the slope.

lowest level, so it flows along the lowest path down a slope. Greater depth of flowing water means greater velocity, and that means greater erosive power, as we will later see. As the force of the water exceeds the resistance of the soil, a rill forms. With this increased depth, erosion is enhanced so that the process accelerates. Untreated, a rill can become a gully. The difference is merely a matter of degree: farm machinery can pass over a rill but not a gully.

While some of the level fields in both Regions I and II may have eroded only slightly, most of the steeper Hill Country was highly affected by channel erosion. Most eroding agricultural fields had obvious rills (Figure 2.12). With time, larger channels formed, becoming hillside gullies where the overland flow from cultivated fields ran off the field and down a steep, grazed hillside (Figure 3.1). Another type of gully was a valley gully or "trench" (Happ et al., 1940). Some of these were just the enlargement of preexisting channels, while others were in valleys or draws where there had been no presettlement channel. Of these, the most notorious were the gullies across high Pleistocene terraces, and these as well as hillside gullies will be covered in some detail in the next chapter.

The erosional history of the Hill Country cannot be understood without understanding the full role of rills and gullies. They are serious in their own right because they damage the landscape and interrupt both economic and even social activity, first by erosion and then by deposition. But as already mentioned, their role as *connectors* can be just as important, especially in the Hill Country. That is, they very efficiently convey and thus speed the delivery of water and sediment to the downslope terrain and streams. Already discussed has been the role of poor agricultural land use and management in increasing the runoff and erosion that created the gullies in the first place. And likewise, improvements in land use bring a decrease in the density of rills and gullies. We can clearly see this decrease in Figure 2.12, which contrasts a portion of the 1934 and 1967 landscape in Coon Creek. A much larger view of the decrease of drainage density between 1938 and 1978, most of the decrease being due to gullies having become inactive, is presented by Fraczek (1987; Figure 3.2). We do not know the density of rills and gullies on the presettlement landscape, but, from the evidence presented already, it is reasonable to assume there were few or none. Certainly, the density would have been far less than at the present time. Thus, we can safely say that the density of rills and gullies went from extremely low on the aboriginal landscape to extremely high in the 1920s and 1930s, but has been greatly reduced since then.

But form and process are related, particularly in this respect. Rills and gullies help accelerate the upland erosional process *as well as* the downstream effects.

These channels formed because of the excess overland flow. And as those formed, water and sediment were more efficiently conveyed off the upland landscape, further enhancing erosion, a positive feedback process. An estimate of the increase of storm flow from a small basin, resulting from the formation of rill and gullies, suggests that the runoff peak for a large storm was almost doubled compared to what it would have been without the channels (Figure 3.3). This is an excellent example of connectivity; that is, the uplands become better hydrologically connected to the streams. While the numbers are only approximations, this demonstrates part of the vicious downward spiral that characterized the Hill Country landscape of the early 20th century. And more brightly, in the late 20th century, it gives an indication of the efficacy of soil conservation on the regional hydrology. Again, understanding these physical processes and effects is essential to understanding the dramatic landscape changes in the Hill Country that are about to be presented.

The growth of such hillside gullies was an accelerating process. That is, the channels conveyed water more efficiently and thus faster, accelerating the flow of water from uplands to valleys, increasing erosion in the gullies. Moreover, the rapid movement of water to small streams greatly increased their discharge for a given rainfall event, thus causing the bed and banks to erode. Coarse erosional debris, primarily gravel and cobbles from hillside gullies and eroding banks, filled channels and caused them to erode and widen even more. These accelerating processes are excellent examples of positive feedbacks in hydrology and geomorphology.

## *Increasing hydrologic change and soil erosion in the Hill Country*

Historical soil erosion probably began early in the lead country around Galena. It seems certain that spoil heaps from mining, some near streams (Chapter 2, Figure 2.2), would have eroded and furnished sediment to streams. Indeed, a visitor to the lead district in the early 1850s described what he termed the "forbidding and desolate hill country of the lead region:"

> All is poverty on the surface. . . . Storms have furrowed the hills in every direction, and the shovels of the mines have dotted the whole surface with unsightly pits, walled around with heaps of limestone and sand . . . (*Harpers New Monthly Magazine*, June, 1853)

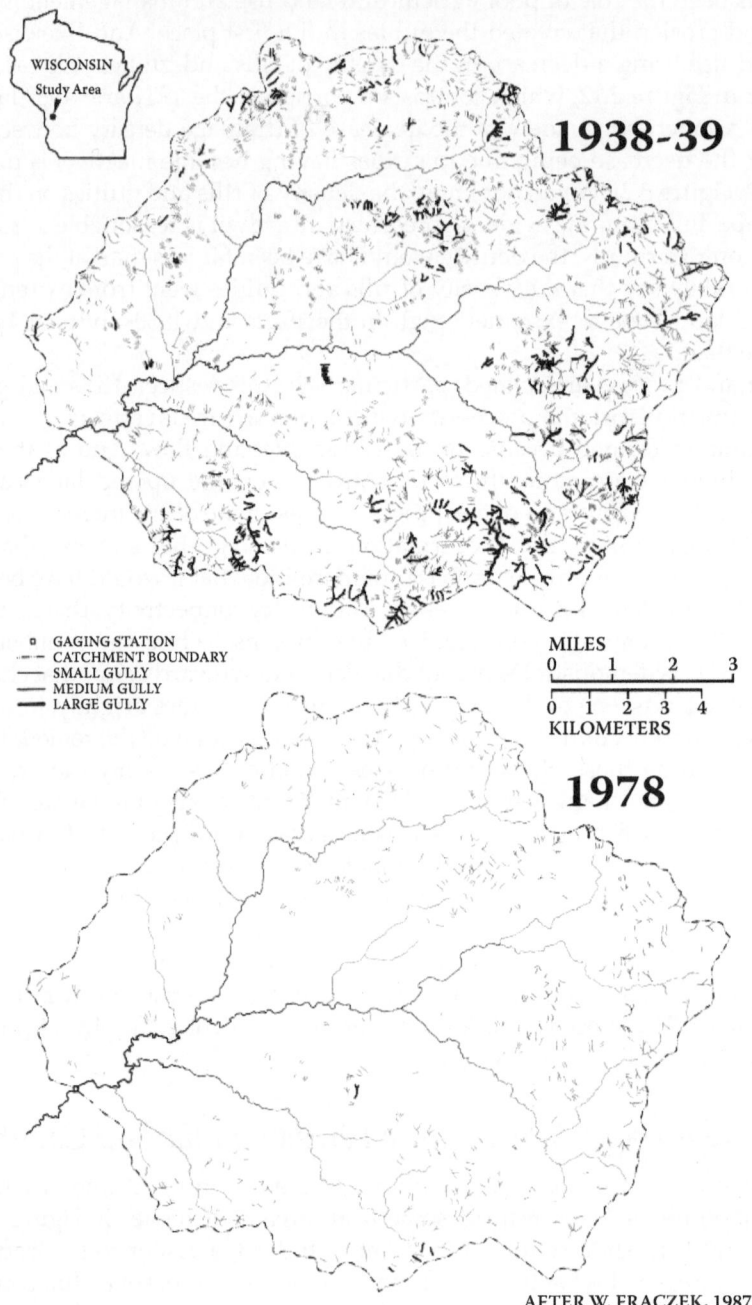

*Figure 3.2* Decrease of gullies in the upper Coon Creek basin as the result of improved land use and management, 1938–1978. This decrease of drainage density is strong evidence of decreased overland flow and, in turn, decreases the rapidity of runoff. (Redrawn from Fraczek, W., Unpublished MS thesis, University of Wisconsin, Madison, 1987.) See also Trimble and Lund, 1982, p. 6.

*Chapter three:   The systematic effects of historical agriculture on the physical landscape*     59

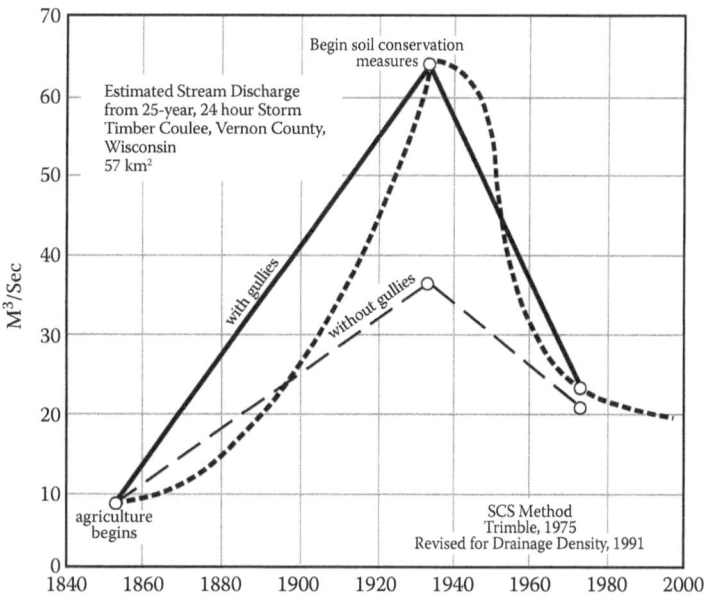

*Figure 3.3* Estimated change of stream discharge from a small tributary of Coon Creek for a moderately large storm, 1853–1995. The vertical axis is discharge in cubic meters per second (1 m³ = 35 ft³). Calculations suggest the significance of (a) land use and management and (b) the formation of gullies for increasing storm flow peaks. The short dashed line suggests the role of response lag discussed in the text. (Redrawn from Trimble, S., *Geomorphology* 108, 8–23, 2009.)

Despite this grim description of rill and gully erosion in the lead district, later studies suggest that mining and refining furnished only a fraction of the total measured sediment; most came from the agriculture, which sprang up shortly after mining started. In any case, by the 1850s, enough sediment had accumulated in the Galena River to interfere with navigation between the Mississippi River and the town of Galena (Hobbs, 1939; Adams, 1940, 1942).

Elsewhere in the Hill Country during the late 19th century, reports began to appear about the occurrence of increased runoff and soil erosion. As early as 1884, flooding in Grant County, Wisconsin, was said to be increasing and was blamed on cultivation (*Grant County Herald* [Wisconsin], February 21, 1884). This charge was repeated in 1892 (*Grant County Herald* [Wisconsin], June 30, 1892). Also in 1884, brook trout were reported to be disappearing from the streams of Winona County, Minnesota (Winchell, 1884). While no cause was given, too much sediment and more variable streamflow are logical suspects. By 1888, steep hillside fields in Buffalo County, Wisconsin, were eroding (Kessinger, 1888). On the west margin of the Hill Country in Dane County, Wisconsin, hillside erosion was reported in 1895 (*Hoard's Dairyman*, November 15, 1895). Much more serious was an 1895 geological report on Alamakee County, Iowa, which described hillside gullies with fans being formed on roads and sediment being swept into streams and onto floodplains. Also mentioned was more erratic stream flow as the result of cultivation. The report stated:

> We shall have, instead of clear rivers and springs and creeks, such as the older residents of the state well remember, flowing the year through, nothing but waterways, now flooded by destructive muddy torrents confined by no legitimate channel, now dry runs, now wide

reaches of sand; with dearth of water in all fields and pastures. (Iowa Geological Survey, 1895, p. 114)

In 1906, the same agency reported even more erosion problems in northeastern Iowa. Tributaries were beginning to trench from the excessive runoff, and roadside ditches were forming in some places (Wilder and Savage, 1906). The same report states that some floodplains were beginning to aggrade. During the summer of 1903, Elkport on the Turkey River in Clayton County had several floods, together leaving 1 to 6 in. (2.5–15 cm) of modern alluvium. Along smaller streams, fences were almost buried in fans, and new alluvium was almost a foot (30 cm) thick in some places (p. 289). However, the tone of these reports is that these phenomena are mostly exceptional. Later, they would become widespread if not ubiquitous.

Interestingly, some farmers welcomed the initial cascade of sediment because it mostly came from fertile soil. As Alex Siebenaler (1955), a lifelong resident, remembered it:

> By this time [c. 1900] the valley farmers experienced some flooding of their land, but at first the damage wasn't very great. It was said that the soil deposited on their land by these first floods was as good as an application of manure, in fact, some farmers welcomed it. The floods were carrying off some of the good topsoil from the farms above.

Siebenaler goes on to say that within a few years, the floods were severe and the sediment had become coarse and unfertile.

In 1907, scientists from the US Geological Survey examined the quality of streams in Minnesota. They noted that the Mississippi River received "considerable suspended matter from its tributaries" (Dole and Wesbrook, 1907, p. 96). However, the emphasis of this survey was on sanitary conditions in streams as the result of municipal sewage. As might be expected, the big polluter was Minneapolis–St. Paul, the sewage from which flowed down through Lake Pepin with little change (Scarpino, 1985).

In 1908, a survey of water power and mill sites in the region stated that cultivation had increased flooding and decreased the flow between storms (baseflow). This increased flooding, along with lower streamflows between floods, creating problems for the operation of mills (Smith, 1908).

By the second decade of the 20th century, general observations about increased runoff and soil erosion were becoming dire warnings. In 1916, the agricultural experiment station at the University of Wisconsin issued a bulletin admonishing farmers to "keep our hillsides from washing." It featured cases of severe erosion including gullies, and gave hints, some perhaps a bit naïve, on reducing erosion (Whitson and Dunnewald, 1916). The early soil survey of Buffalo County stated that soil erosion there was a problem and that large gullies half a mile (800 m) long had formed in one year (Whitson et al., 1917). In 1922, the University of Wisconsin extension assigned the agronomist Otto R. Zeasman to conduct erosion control, although at the time he, like most agriculturalists, knew very little about erosion (Zeasman and Hembre, 1963). After first visiting the Hill Country in 1922, Zeasman was astonished:

> My vocabulary is not adequate to describe my shock at the destruction observed. There was the dissection and destruction of level river terrace land by gullies . . . sheet erosion of the sloping to rolling

fields made conspicuous by rilling; stream bank cutting, the filling of bridges with sediment; ... and the burying of roads near the foot of bluffs. (Quoted in Johnson, 1991, p. 13)

In 1929, the Lakes States Experiment Station at the University of Minnesota was given a large grant to study erosion. In the same year, one of its directors issued a large study on the forests of the Hill Country, suggesting that heavy grazing and other poor forest practices were amplifying floods in the region (Zon, 1929).

However, the magnum opus for raising consciousness about soil erosion in the Hill Country was *Soil Erosion—A Local and National Problem* by Bates and Zeasman (1930). Carlos Bates was director of the Lakes States Experiment Station in Minneapolis and became a highly active soil erosion researcher in the region. A major point in this study was how severe the problem, especially gullying, had become in the previous decade or two and that the destructive processes seemed to be spreading exponentially (Figure 3.4). For example, Bates (1936, p. 961) describes an 11 ft (3.4 m) "wall of water" coming from a small basin just north of Winona, Minnesota, in 1932. Clearly understanding that they were witnessing an accelerating juggernaut, they presciently discounted any major role of "wet years," but assigned grazing a major role.

At the same time, USDA soil surveyors were also finding severe erosion. For example, by the late 1920s, Crooked Creek, a stream in Houston County, Minnesota, had apparently filled to a depth of about 15 ft with modern sediment from soil erosion (Gray et al., 1929). While most USDA soil surveys of the period clearly pointed out that erosion was a problem, there was generally little sense of urgency, and little was usually said about floodplains aggrading with modern sediment.

Another early worker calling public attention to the extreme erosion problem in the Hill Country was Melville Cohee, an agricultural economist of the SES, later SCS. He maintained that the then-raging Depression had forced farmers to push the land harder in order to make a living. Moreover, the droughts of the early 1930s led to planting and crop failures that left some fields more vulnerable to erosion and forced farmers to push their land even harder. In particular, he pointed out the folly of overgrazing pasture and grazing forested slopes. He maintained that nearly all forests were grazed and that larger, wealthier farmers were just as likely to do this as smaller, poorer farmers (Cohee, 1934). Cohee was later a key person in the Coon Creek Demonstration Area project.

Such a regional disaster elicited even the entrepreneurial spirit in the Hill Country. Locally, some contractors built stone steps to allow water to cascade safely downslopes, especially at the heads of gullies. Another device, made at least locally in central Region II by Gottlieb Muehleisen, an inventor from Alma, Wisconsin, was a metal flume to direct water around or over the head or overfall of a gully (McLeod, 1984). These were sometimes used by townships where the head of a gully threatened a road (Figure 3.5).

The hydrologic processes of the 1920s and 1930s had a disastrous effect on roads, bridges, railroads, and utility lines. Indeed, erosion control was begun in 1922 largely because of erosional debris blocking public roads (Johnson, 1991; Sartz, 1971). Bates and Zeasman (1930) found that some roads were being raised to keep them above floodplains. In some counties, about one-third or more of all road maintenance monies were being used to remove sediment and repair associated damages to roads and bridges (Bennett, 1939). In the Coon Creek basin, 40% of township road maintenance was clearing erosional debris (SCS, undated, a). On a larger scale, fairly precise data are available for the Whitewater River basin in Minnesota, c. 1940. Direct damages to roads and bridges from floods and sedimentation were estimated to average $16,150 annually, but this does not include very

*Figure 3.4* Alarming erosion and sedimentation in the Hill Country, 1929–1931. Top: Severe gullying in the Black River terrace, 3 mi (5 km) east of North Bend, Jackson Co., WI, June 15, 1929. See also book cover. Bottom: a 40-acre (16 ha) fan, deposited mainly 1923–1929, covers formerly good hay land on the Buffalo River floodplain, 10/24/31. (Photos credit: C.G. Bates.)

*Figure 3.5* Early private erosion control. Sheet metal flume used to convey water around the head of gullies, Proksch Coulee gully, Vernon Co., WI (see Chapter 4). (Photo credit: Nadine Kleinhenz.)

costly longer-term remediation such as raising road grades and bridges, and building dikes, revetments, and debris basins. For comparison, the total annual taxation on rural land in that basin was $189,053 (SCS, 1942). Thus, at least 9% of *total* tax monies was being used to deal with and correct short-term human-induced hydrologic problems relating to roads.

These messages, plus those from other affected regions of the United States, were reaching Washington, D.C., where there was increasing awareness of a national soil erosion crisis. This period was also the beginning of the severe wind erosion in the Dust Bowl, but many other areas were suffering severe water erosion. Many were interested in these problems, but the undisputed leader was Hugh Hammond Bennett, a former soil scientist with the USDA Bureau of Soils and Chemistry. His long experience surveying soils in the eastern United States had convinced him of the gravity of the problem, and he had become something of a crusader for soil conservation even though no one at the time knew exactly how to best do that. His clarion call, with W. Chapline, was "Soil Erosion: A National Menace," which appeared in 1928.

As noted in Chapter 2, the efforts of Bennett and others led to the formation of the federal Soil Erosion Service (SES) in 1933 as part of the Department of the Interior. One of the first tasks of the SES in 1933 was to establish the severity of soil erosion in the United States. This was done during a relatively short period in 1933 using all available personnel and was termed the *Reconnaissance Erosion Survey* (RES). Mapping was done in the field for each county and superimposed over old soils maps or USGS topographic maps usually at a scale of 1:62,500 (1 in. = 1 mi or 1 cm = 625 m). These were then reduced to published state maps at a scale of 1:500,000. More complete information on utility of the RES is found in Trimble (1975d).

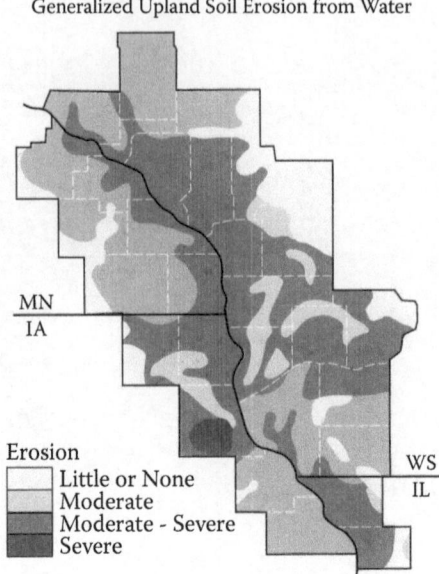

*Figure 3.6* Generalized upland soil erosion from water, 1933, Hill Country. Compiled from state maps of WI, IO, MN, and IL. USDA-SES *Reconnaissance Erosion Survey*, 1933, 1:500,000.

The state RES maps were used to make a highly generalized map of erosion for the Hill Country (Figure 3.6). It is important to note that this is a subjective appraisal primarily of impact on agriculture rather than just volume of soil lost. It was done rather hurriedly, and the categories do not necessarily correspond to quantitative measures such as depth or volume. It will be noted that there is a general correlation with relief or slope (or landform ruggedness) so that the western part of Region II shows moderate to severe erosion (note here that "severe" erosion would suggest that an area is essentially destroyed for agriculture). Given that slope appears to be a controlling factor, Region I had perhaps surprisingly severe erosion. However, that may be easily explained by erosive land use (ELU; Figure 2.16). And within Region I, it will be seen that severity of erosion does generally correlate with relief, although the erosion appraisal shown for the three northern counties of Iowa appears to be excessive.

The amount of historical erosional debris stored in streams and valleys is a reasonable surrogate for volume of eroded material (Happ et al., 1940). The earlier studies and data available plus my own measurements and observations across the Hill Country over 39 years suggest a general correlation and suggest that the map has some validity. However, the map probably does not fully consider the volumetric aspects of gully erosion, particularly in the areas with high, sandy, easily erodible terraces, and those alone account for much valley sediment in many cases. Moreover, much erosion occurred after the RES was done. One area for which I could find little correlation of indicated erosion was the three northern counties of Iowa, where the amount of valley sediment did not appear to be of the magnitude found further to the northeast in Region II. The general lack of high terraces may account for at least part of this disparity.

## Effects of soil conservation on the physical landscape

In Chapter 2, we looked at the vast improvement of land management starting in the 1930s. Now, we examine the physical effects of those improvements. First, how much was soil erosion reduced? Unfortunately, there is no direct way to measure soil erosion per se from fields at the scale we are considering here. Thus, the fallback approach is to estimate it using a model, in this case the Universal Soil Loss Equation (USLE) described in the Introduction. As useful as this model is, I emphasize that this approach gives only an *estimate* of the soil *moved* on a slope.

Using ten small sample basins for Coon Creek, the average soil erosion c. 1975 was calculated to have been only about one-fourth of that occurring in 1934 (Trimble and Lund, 1982). This was for a sample of the entire landscape, including forest. Looking more at cropland and also using the USLE, Argabright et al. (1996) estimated that cropland over several counties of the Hill Country in 1992 was eroding at about one-third of the rate in 1930. Given that the two studies used different dates, places, and assumptions, the two estimates are remarkably similar. Those wishing more details should consult those studies.

It should be kept in mind that calculated estimates as described above are of soil *moved* on a field from sheet and rill erosion and do not necessarily indicate how much soil leaves a field. And it certainly does not tell how much soil is accumulating in streams and valleys or leaving a stream basin as sediment yield. Keeping in mind that soil is matter, most of which is not soluble, any eroded soil particle must be accounted for somewhere downslope or downstream (Trimble, 1999; Trimble and Crosson, 2000). Once eroded and entrained by rainfall and overland flow, there is usually a tendency for particles to be deposited as sediment on more level areas or depressions on slopes, at the base of slopes, and along streams in channels and especially on floodplains and in ponds or reservoirs. As *Science* described it, if soil erosion is the murder of soil, the downstream particles of sediment are the corpses (Glanz, 1999). Thus, another approach is a *measured* estimate of erosion by locating and measuring the volume of downstream sediment.

With this concept in mind, Trimble and Lund (1982) took steps in an effort to verify and complement the estimates above. The first was to measure the sediment yields for small reservoirs for the old and new period. For the old period (1930s and early 1940s), we were able to locate five suitable reservoirs, but these samples probably exaggerated sediment yield because they were built to control badly eroding areas. For the more recent period, we used ten flood retention dams built in 1962 that we surveyed in 1975. Indeed, their ten sample stream basins were selected because the waters from each flowed into one of these ponds, allowing a sediment budget to be estimated. The differences from the old to the new period were astounding: sediment accumulation in the ten ponds for the period 1962–1975 was only about 1% of that for the old period. Note the disparity in measurements between erosion and sediment yield: the estimated (modeled) improvement ratio for upland soil erosion was 4:1, but the measured improvement ratio for reservoir sediment accumulation is 100:1! To a large degree, this disparity is explained by the fact that the earlier landscape featured far more deep rills, gullies (Chapter 2, Figure 2.12; Figure 3.2) and stream channel erosion than the new period. That is, much of the sediment yield was coming from channel erosion that is not estimated by the USLE (Trimble, 2009a). And with all the rills and gullies, connectivity was higher so any eroded sediment was more efficiently transported downstream

The next step in our attempt to establish a *measured* improvement from the implementation of soil conservation practices was to measure the rates of sediment accumulation in the streams and valleys allowing a sediment budget to be estimated, the approach created by Stafford Happ and discussed both in the preface and later in this chapter. This approach has the advantage of integrating all erosional debris not trapped on slopes on the one hand, or carried out of the basin as sediment yield on the other hand. But it also has the problem of including the debris from upstream channel erosion as well as soil erosion from slopes. Conversely, it does not include sediment deposited at the base of slopes as colluvium. By this measurement, vertical accretion rates in the main valley declined from about 6 in., or 15 cm/yr, in the 1920s to about 0.2 in, or 0.5 cm/yr, in 1975–1993, an improvement of 30:1. Moreover, the area of accretion was reduced.

A study similar to that done in Coon Creek was done by S.C. Happ in the Beaver Creek basin north of Galesburg, Wisconsin. Although no measurements were made of present rates and, indeed, no measurements were made after 1983, Happ estimated that sedimentation rates in the 1980s were less than 10% of those of the 1930s (Happ, 1985).

Much has been said already about the time lag or delay of erosion and sedimentation behind the onset of erosive land use (ELU). As indicated at the beginning of this chapter, the lag in erosion was largely a function of the ability of the soil to absorb abuse before serious erosion began, leading to rills and gullies with the positive feedbacks. The lag for sediment was even greater because it needed time to move downstream and be deposited. Moreover, as the water and sediment cascaded over slopes and down hillsides, it created the rills and gullies that not only added more sediment but more effectively delivered the water and sediment to the valleys. And as will be seen, much sediment was later derived from streams and floodplains and moved farther downstream. This lag function was investigated in detail for Coon Creek (Figure 3.7). The input or cause, ELU, has already been discussed. Valley sedimentation rates in Figure 3.7 were those made from careful field measurements, discussed above. Upland sheet and rill erosion was an estimate from the Universal Soil Loss Equation, discussed already.

Note that ELU increased rapidly from the time of settlement, peaking out about 1880 or so (see Figure 2.16), but there was apparently little commensurate effect on erosion during that period. But from c. 1880 to c. 1930 with ELU almost held constant, erosion was seemingly increasing on its own, the landscape being "out of control" as alluded to earlier. The lags in sedimentation are even greater with little accumulation of sediment until after about 1900. But by the 1920s and 1930s, even with ELU held relatively constant, sediment was accreting in the main valley at average rates of 6 in. (15 cm) per year.

Then, as amelioration started to take place in the late 1930s and ELU was reduced, sedimentation rates held constant for another decade. After that, improvement was rapid and has continued to the present time.

At a more theoretical plane, the graph represents the slow transition from the Hewlett concept of runoff to the Horton concept. Then, there is the partial transition back to the Hewlett concept, a transition that will never be complete as long as the land is used for agriculture.

If this lag function were consistent throughout the region, one would expect the onset of the critical period to be roughly a decade or so earlier south of the Wisconsin River and perhaps a bit later in Minnesota, the time corresponding to the onset of agriculture. And indeed, this appears to be the case. Knox (1987) has carefully dated the onset of massive accretion of sediment in a Grant County tributary, and it is roughly a decade earlier than in Coon Creek (Trimble, 1993). Likewise, evidence from the Whitewater River and elsewhere in Minnesota suggests that processes there lagged those at Coon Creek slightly. Thus, the time difference of peak erosion in the Hill Country could be as great as two decades. Case studies later will make the point more clearly. The systematic spatial differences in the peak rates of erosion as driven by ELU seem to preclude a major role for periods of heavy regional precipitation as suggested by Knox (2001).

Chapter three: The systematic effects of historical agriculture on the physical landscape 67

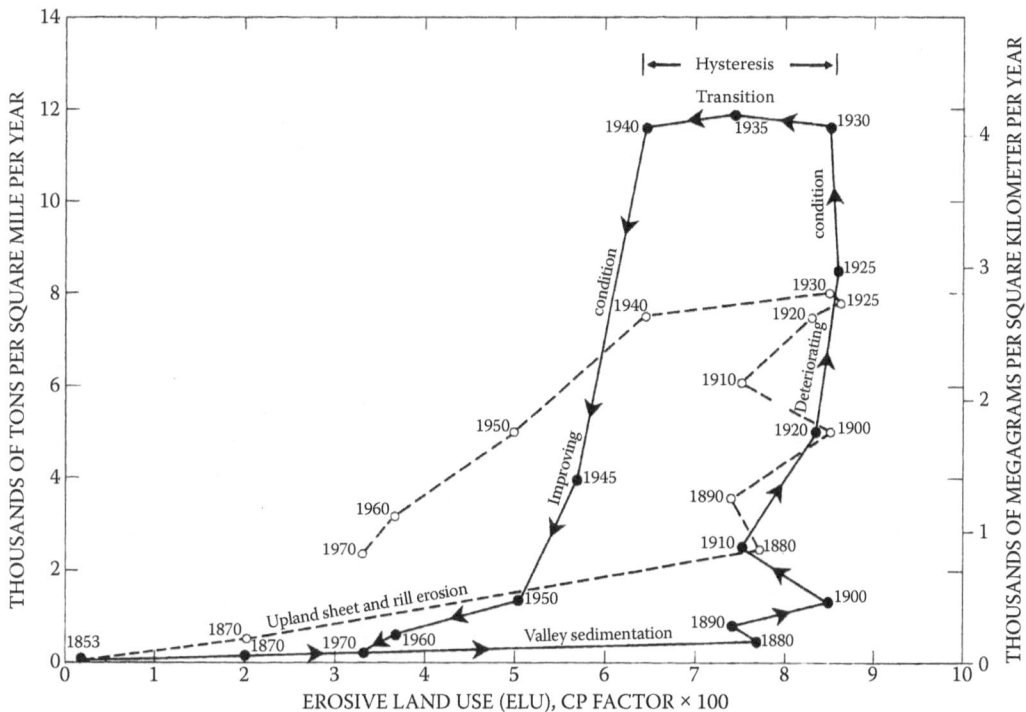

*Figure 3.7* Nonlinear and hysteretic relationship of erosion and sedimentation to erosive land use, 1853–1975, Coon Creek. Sedimentation rates were measured but erosion was modeled with the Universal Soil Loss Equation. The lags are largely functions of (a) soil condition and (b) connectivity between uplands and bottoms by rills and gullies. See text for more explanation. (From Trimble, S. and Lund, S., *US Geol. Surv. Prof. Paper* 1234, 1982.)

In 1935, SES was renamed the Soil Conservation Service (SCS) and moved into the USDA. Many of the duties and technical services of SES/SCS to farmers were described in the previous chapter but of note here was its research arm, the Climate and Physiographic Research Group. It included a group studying sedimentation in reservoirs, another studying severe hillside erosion, and yet another studying stream and valley sedimentation. Again, word was reaching Washington that erosion in the United States was more severe than anyone had imagined.

The head of the Stream and Valley Section of the SCS Climate and Physiographic Research Group, Stafford C. Happ, set out on reconnaissance to find the most severely eroding regions in the United States so they could be studied closely. Not only did this research group want to ascertain the damage already sustained so as to know how much remediation was needed, but they also wanted to be able to measure the effects of soil conservation after soil conservation practices were installed. Happ designed 16 of these stream and valley studies. These studies were pathbreaking: streams had never been so closely studied before (Trimble, 2008). An Easterner, Happ told me in 1974 that he had in the early 1930s heard rumors of roads, bridges, and farms being buried in the Hill Country, and his visits there confirmed the rumors. Of the 16 complete stream studies he designed for the entire United States, five were in the Hill Country, which may give a good insight into his perception of the problem there. These basins were (1) Coon Creek, Wisconsin, which he considered to be a "typical" basin in the region (McKelvey, 1939), (2) Beaver Creek, mostly in Trempealeau County (Kunsman, 1944), (3) Kickapoo River, just east of Coon Creek (Happ, 1944), (4) Whitewater River, Minnesota, mostly in Winona and Olmstead Counties (full study never published), and (5) the Galena River, in Grant County, Wisconsin, and Jo Daviess County, Illinois (Adams, 1940, 1942, 1944). In all of these basins, the investigators established cross-sectional profiles or "ranges" across the valleys and streams at intervals of about a mile. For every range, borings were made at close intervals to find the old floodplain soil and thus establish the depth of modern sediment. Then, precise instrument surveys were conducted over the ranges so that future changes might be later ascertained (see discussion in Preface). Details of the studies were mapped over air photos. All this allowed the calculation of the total amount of eroded soil deposited in the valleys. Using a similar research plan, a sedimentation survey was made of Lake Marinuka on Beaver Creek near Galesburg, Wisconsin. The measurements made in these studies more than substantiated the rumors about erosion in the Hill Country. These studies, especially the Coon Creek, Beaver Creek, and Whitewater River, were the inspiration, and much of the basic data sources, for my own 39-year study of the Hill Country.

In addition to the systematic sediment studies of selected stream basins, random spot observations, including detailed annotated photographs, were made as the researchers were traveling in the region during this period. Also, a seven-year systematic hydrologic study was done on Coon Creek and Little La Cross River, Wisconsin (SCS, 1942a).

With all this experience under their belts, Happ and two SCS colleagues in 1940 wrote what some consider the "bible" of human-caused stream and valley sedimentation (Trimble, 2008). It was appropriately but modestly titled "Some Principles of Accelerated Steam and Valley Sedimentation" (Happ et al., 1940). Drawing heavily on examples from the Hill Country as well as other areas in the United States, it gives 45 "principles" or concepts that have stood the test of time and still help guide us in understanding what happened and in dealing with the lingering effects. The study was far ahead of its time and is now considered to be a classic in geomorphology (Trimble, 2008).

Happ and his colleagues were perhaps most interested in measuring the storage and fluxes of sediment. The following work could not have been accomplished without the baseline data they provided.

## Sediment budgets over the historical period

We first look at the overall mass of sediment produced by erosion and its movement or flux over the historical period (Figure 3.8). These measurements are from the Coon Creek study mentioned earlier, with its continuation to the present time, and show the distribution of sediment sinks and sources over time. Note that every figure shown is an actual measurement except the "net upland sheet and rill erosion," which is a tare value obtained by subtracting sources from sinks plus sediment yield or efflux to the Mississippi River. Upland erosion is here termed "net" because it includes only the sediment making its way to the stream system and does not include colluvium, the eroded material deposited at the base of slopes and in concavities. Beach (1994) actually measured the colluvium in a small watershed in the Zumbro River basin and found it to be a significant proportion of eroded material.

There are several important aspects of this diagram (Figure 3.8). First, it shows how massive the fluxes were. Second, while it gives a general idea about the changes of sediment flux over time, the numbers are averages of often highly disparate values and can be deceiving. For example, the 1853–1938 values include both extremely low presettlement rates and the peak values of the 1930s. Likewise, the 1938–1975 values include the greatly improved conditions from conservation efforts by 1975. Third, the diagrams give some idea of the complexity of the sediment flux. Fourth, note that only a small proportion of the

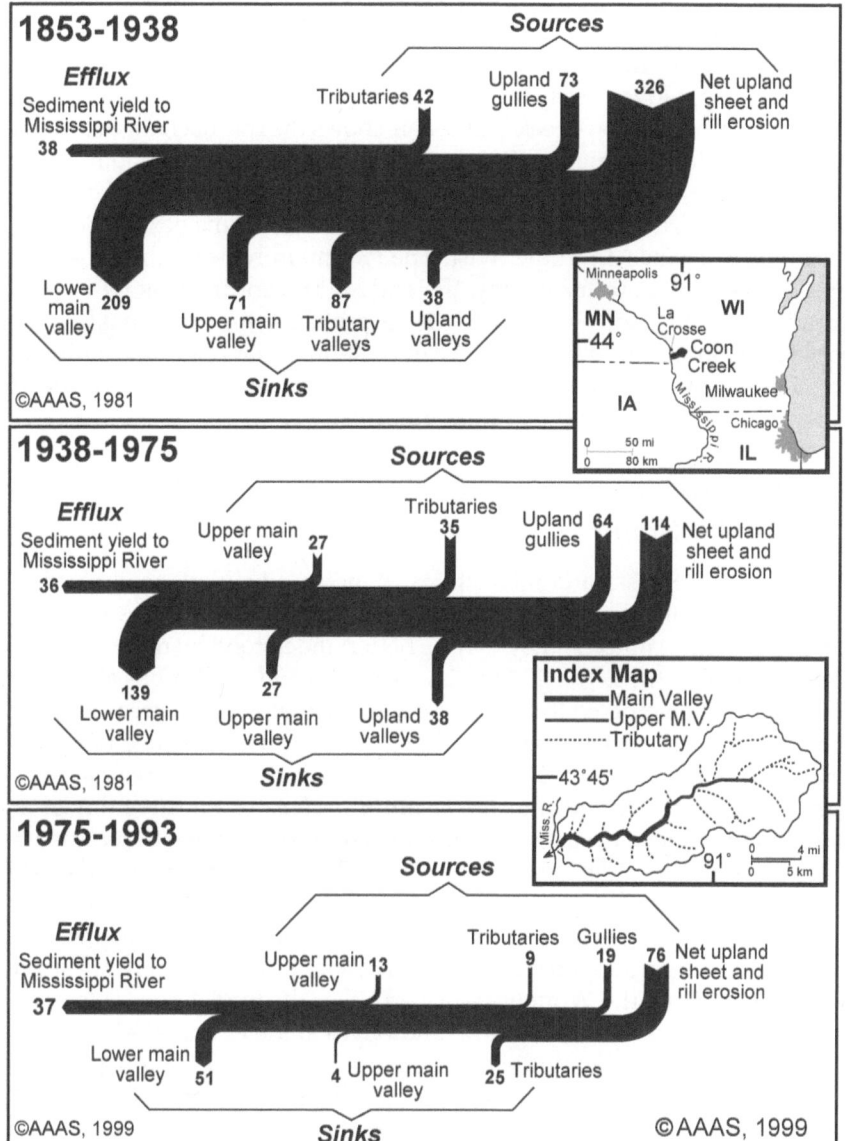

*Figure 3.8* Sediment budgets for Coon Creek, 1853–1993, showing the complexity of sources, sinks, fluxes, and efflux over time. Numbers are annual averages for the periods in tonnes/yr (1 tonne = 1.1 short tons). All values are direct measurements except net upland sheet and rill erosion, which is a tare: the sum of all sinks and the efflux minus the measured sources. The lower main valley and tributaries are sediment sinks, whereas the upper main valley has been a sediment source. Note that sediment yield to the Mississippi River has held relatively constant over historical time. (From Trimble, S., *Science* 285, 1244–46, 1999. With permission.)

eroded sediment has been moved as far as the Mississippi River and has apparently not changed over time. Expressed another way, most of the eroded material remains deposited in the basin not far from its source.

The last two points should be contrasted with the perception many still have that once a particle of soil is eroded from a field, it moves on directly to the sea. Instead, it can be millennia before the particle even leaves the small stream basin. Finally, it will be noted that sediment yield to the Mississippi River remained about the same no matter how much upland erosion was occurring within the Coon Creek basin. This shows the role of changes in sediment storage rates. When erosion was high, sediment went into storage at the bottom of slopes and in stream valleys. When erosion was curtailed, sediment came out of storage to augment the sediment yield. Indeed, these storage changes are the subject of the next section.

The last point also has an ironic twist. The two main reasons soil conservation work was begun in the Hill Country during the early 20th century was (1) to protect roads and railroads from erosional debris and (2) to protect the navigational channel of the Mississippi River (Johnson, 1991). While it can be shown that roads and railroads are well protected, there is little evidence to show that soil conservation has yet decreased the amount of sediment going into the Mississippi River.

## Zones of physical processes within stream basins of the Hill Country

While the physical processes and interactions of increased runoff, accelerated upland soil erosion, gullying, stream channel erosion, and stream and valley aggradation all within a stream basin are highly complex, they may be better understood by reducing them to a general model of space and time. The first fluvial zone, *rills and gullies*, has already been introduced. It should be remembered that a rill or gully is actually a new stream channel, created to transport surface runoff that had not previously been great enough to require a channel.

We now move down channel to what most consider to be "streams," and consider *tributaries, upper main valley*, and *lower main valley*. These three fluvial zones have been studied intensively in Coon Creek since 1938 so the movement of sediment in and out of the zones could be quantified and the operative fluvial processes identified (Figure 3.9). This is termed a *distributed* sediment budget (Trimble, 1993). Like any model, it is a generalization and there can be deviations, but the model is based on detailed measurements in Coon Creek, and it is also tempered by data and observations from other basins in the Hill Country, especially from the Whitewater River. What the model does is to show how the various processes, driven by both cultural and physical forces and effects, act and interact over historical time and space. While the numbers would vary from place to place in the Hill Country, I believe the general trends to be almost ubiquitous. As shown, the zones, or actually stream "reaches," are *tributaries, upper main valley*, and *lower main valley*.

### Tributaries

Presettlement tributary channels tended to be deep but narrow, so that a person might "jump across" them (McKelvey, 1939). They may be characterized as having early sediment deposition on floodplains as soil erosion and runoff increased. Depths varied, but 1–4 ft (0.3–1.2 m) of vertical accretion was normal, and the sediment varied greatly in texture.

With increasing runoff from upland fields, rapidly concentrated by expanding rills and gullies, the tributary channels began to erode and enlarge, sending copious sediment cascading downstream (Figure 3.10). This was especially true when violent bank erosion

Chapter three: The systematic effects of historical agriculture on the physical landscape 71

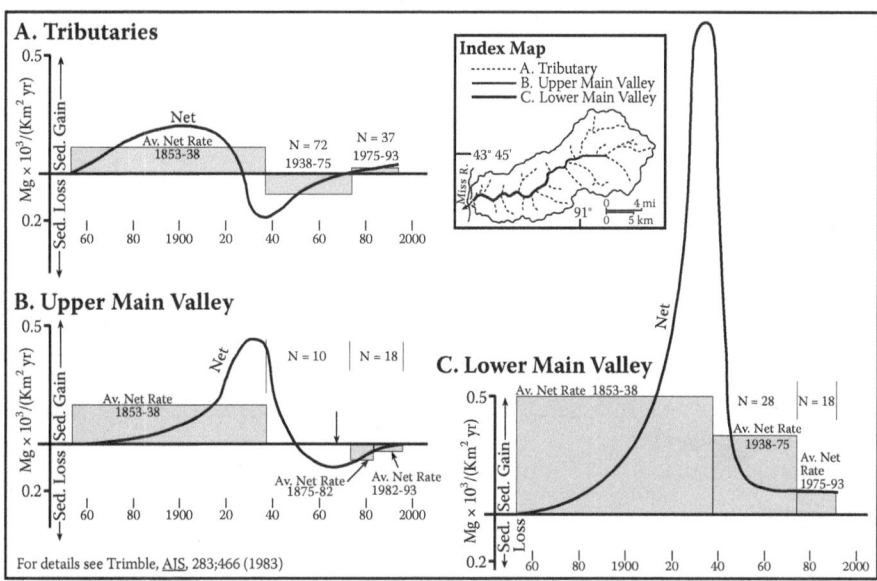

*Figure 3.9* Differential stream and valley sediment budgets, Coon Creek, WI, 1853–1993 (1 Mg = 1 tonne = 1.1 short tons). (Redrawn from Trimble, S. *Geomorphology* 108, 8–23, 2009.)

moved into adjacent high Pleistocene terraces (Figure 2.15). In other words, the tributaries had been transformed from a sediment *sink* to a sediment *source*.

By the early 1900s or so, tributaries looked like flumes in some places and gravel roads elsewhere, apparently depending on the supply of coarse sediment. Not only did the higher flows erode the channels but the coarse material they often transported from both upland and stream erosion widened the channels by erosion of the banks (Figure 3.11). This widening of tributary channels has been clearly demonstrated by Knox (1972, 1977) for southwest Wisconsin, and other work indicates this was typical for most tributaries in the Hill Country (McKelvey, 1939; Adams, 1940; Happ et al., 1940; Kunsman, 1944; Trimble, 1975a, 1976). Even after 1938, large sections of floodplain were eroded away until the ameliorative effects of conservation measures could take hold (Trimble, 1975a,b). Bank erosion often took large sections of agricultural fields and adjacent roads. The eroded channel was usually so large by c. 1940 that floods over the old floodplain were rare. In the detailed studies of Coon Creek, only *one* of the many tributary profiles had experienced sediment deposition on the historical floodplain after 1938 (Trimble, 1976 a,c,d). In other words, the old floodplain had become a terrace. These processes were abundantly clear even after my first field season in Coon Creek. In the fall of 1974, I was able to write:

> ... tributary channels enlarged greatly before stabilizing, with some having twice the 1938 bank full discharge capacity (to the level of the old flood plain). New, lower, floodplains have been created while the old floodplains have essentially become terraces. (Trimble,1975b, p. 24)

These preliminary conclusions were clearly demonstrated by detailed ground surveys in Coon Creek that indicated virtually no deposition on the old historical tributary floodplains for the period 1938–1975 (Trimble, 1976a).

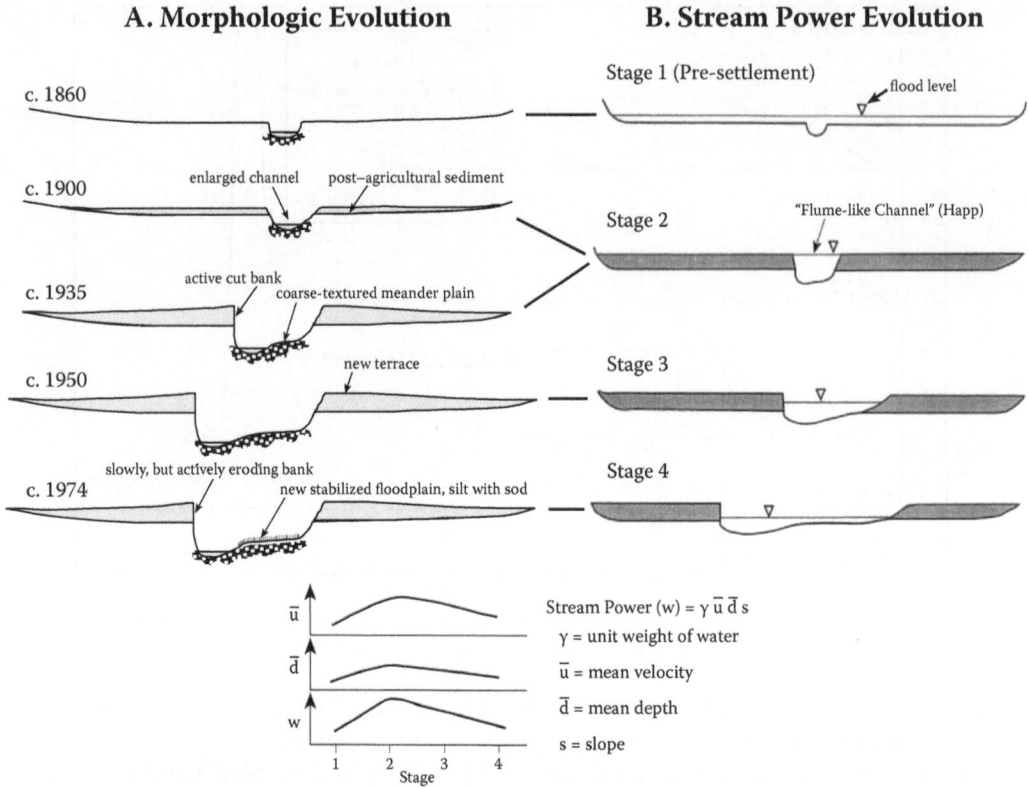

*Figure 3.10* (A) Schematic model of changes of historic stream and valley morphology for Coon Creek and other Driftless Area tributaries, 1860 to 1974 (from page 16 of a mimeographed pamphlet by S.W. Trimble for a field trip to the Driftless Area, April 1975, sponsored by the Association of American Geographers and led by G. Dury, J.C. Knox, W.C. Johnson and S.W. Trimble [Trimble, 1975b]). Lateral migration of the stream is not shown in this model. (B) Changes of stream power and the transformation of stream and valley morphology. This model assumes a constant discharge for each stage. With the small stream channel of stage 1, floods spread out over the floodplain, keeping depth, velocity, and stream power low. With accretion of the floodplain and stream banks with historic sediment in stage 2 (c. 1900), greater flows were restricted to the channel, thus increasing depth, velocity, and stream power so that the channel erosion shown in the 1900 stage (left) must have been very rapid. In stages 3 and 4, the channel erodes laterally, so that floods are spread, with decreases of depth, velocity, and stream power. By the latter stage, fine sediment covered the old gravel meander plains and new floodplains are formed as shown to the left (from presentation to the Association of American Geographers, San Diego, CA., April 20, 1992. Ron Shreve of UCLA made important suggestions for preparation of this diagram in 1991, from Ward, A. and Trimble, S., *Environmental Hydrology*, CRC, Boca Raton, 2004.

With soil conservation measures and amelioration of runoff and erosion, the channels began to heal, although cut banks are still present in some places. But within the eroded high banks of the old channel, new, lower floodplains of fine sediment developed, diagnostic of a milder flood regime (Trimble, 1975a; Figure 3.10). A comparison of bankfull discharge capacities, as calculated to floodplain level, of the 1930s and the 1970s shows an extraordinary decrease (Figure 3.11). Even so, the channel widths c. 1970 were still significantly larger than those at the time of settlement as measured by the original land surveys (Knox, 1972). Even in the 1970s and later, channels still eroded along cut banks

Chapter three: The systematic effects of historical agriculture on the physical landscape    73

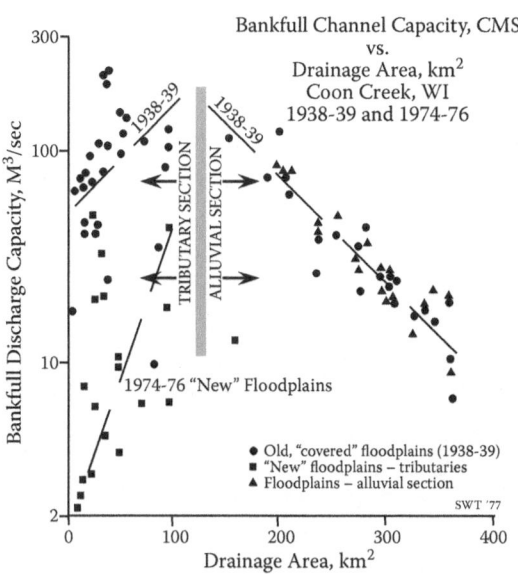

*Figure 3.11* Changes of bankfull discharge capacity (BFQ) for tributaries, 1930s vs. 1970s, Coon Creek (1 m³ = 35 ft³, 1 km² = 0.37 mi²). Alluvial reaches downstream of tributaries (upper and lower main valley) were not calculated for the 1970s, but this analysis suggests a different set of channel forming conditions from the tributaries, at least for the 1930s. The overall relationship between BFQ and drainage area (DA) in the 1930s is a third-degree polynomial curve (Gatwood, 1989). For selected streams elsewhere in the Hill Country, Knox and others found a log-linear relationship for bank widths and DA (e.g., Knox, 1987). The relationship of stream size in the Hill Country, however measured, is a function of many factors, including land use and basin morphology, particularly drainage network.

because less-laden flows (Kondolf, 1997 used the term "hungry water") had the power to erode and transport sediment. Fish shelters installed since c. 1975 by governmental agencies along tributaries have marginally helped stabilize channels, but such fish shelters could never have worked had the stream regime not been moderated already (Trimble, 1997, 2009). Presently, tributaries are generally stable and are mild sediment sinks. Indeed, the transformation of tributaries from the disheveled conditions of the 1930s to the stable conditions of today is sometimes hard to believe without much visual reinforcement (Figures 3.12, 3.13; see also Trimble and Crosson, 2000, p. 249).

In some parts of the Hill Country, the evolution of tributaries may have been somewhat different. Faulkner (1998), for example, found that some tributaries of the Buffalo River adjusted to the reduced sediment loads of the post-soil conservation landscape by vertical channel erosion rather than the lateral or bank erosion as described above.

Another important aspect of the tributaries, and indeed, the entire stream system, is the change of soil and groundwater hydrology. With poor land use and water rushing off hillslopes, there was less recharge of groundwater so that springs began to dry up, and there was less flow of water between storms. With improvements in land use over the past 70 years, more water infiltrated, and there is now more baseflow in the tributaries (Potter, 1991; Gebert and Krug, 1996; Krug, 1996; Kent, 1999; Shilling and Libra, 2003), an important factor in the great improvement of fish habitat and the reintroduction of brook trout (Thorn et al., 1997; Vondracek et al., 2005). Not only does the increased baseflow help facilitate fishing, it also helps facilitate other recreation such as canoeing and rafting, also a growing regional industry.

*Figure 3.12* Evolution of tributary channels as explained in the text and in Figures 3.10 and 3.11. North Branch, Whitewater River, MN, 1 mi (1.6 km) west of Elba, MN, looking upstream (west) (NE1/4, Sec9, T107NR10W, Winona Co.). For reference, note barn, bridge, and road intersection. 1905, Note narrow, apparently deep stream lined with trees and dense riparian vegetation. (Photo credit: 1905 and 1940 photos by Soil Conservation Service [SCS].) (Continued)

## Upper main valley (UMV)

Upper main valleys acted much like the tributaries, only later. Accretion was generally of finer material, and depths of vertical accretion ranged up to more than 10 ft (3 m), often burying old terraces on which roads, farms, and even villages were located, or at the very least making them more liable to flooding. With soil conservation and consequently reduced stream sediment loads, streams in these reaches began to widen the channel by lateral erosion of banks. But as the highcut bank of the old historical floodplain retreated from erosion, a new, lower floodplain was created on the opposite bank (Figure 3.14). Just as the high floodplain of the 1930s was diagnostic of the flood flows of that period, the new low floodplain is diagnostic of more recent flows. Over time, a stream creates a channel adjusted to its flood regime, so the change of bank heights is a good indicator of changes of mid-valley flood regime for the historical period. As this process continues, the cross section of the floodway between the two high banks of the historic floodplain expands. This larger floodway has greatly reduced flooding on the old historic floodplain so that there are positive aspects to these processes.

There has until recently been a huge net removal of material, making this zone a prolific source of downstream sediment. Hence, governmental agencies have wanted to "fix" this problem by regrading the high banks to a low slope and armoring them with rip-rap. As the new floodplain widens, however, there is increasing space for deposition.

Chapter three: The systematic effects of historical agriculture on the physical landscape 75

*Figure 3.12 (Continued)* Evolution of tributary channels as explained in the text and in Figures 3.10 and 3.11. North Branch, Whitewater River, MN, 1 mi (1.6 km) west of Elba, MN, looking upstream (west) (NE1/4, Sec9, T107NR10W, Winona Co.). For reference, note barn, bridge, and road intersection. By 1940, by lateral erosion, the stream has widened 2 to 5 times in response to increased storm flow and the abundant supply of bedload from hillside gullies and the eroded stream itself. Indeed, a hillside gully and rock fan is barely visible in the top right quadrant. (Photo credit: 1905 and 1940 photos by Soil Conservation Service [SCS].) (Continued)

And the amount being deposited, at least for the period 1975–1993, has become almost as great as the amount being eroded. Therefore, these reaches appear to be no longer major sediment sources as they were c. 1950–1980 (Trimble, 1999, 2009a). Hence, the long-term prognosis for this reach is uncertain. While this reach may not be presently a large net source of sediment, it has an effect on sediment texture. As material is eroded from the high bank, the fine material is carried downstream to the floodplain or even out of the valley as sediment yield. The sand, however, is more likely to be retained in the channel, where it migrates downstream, resulting in effects described in the next section. As the hydrologic regime continues to moderate, the processes in this region move downstream (Figure 3.14).

*Lower main valley (LMV)*

This zone has been continuously accreting at an accelerated rate since some time after settlement. The process was very slow at first, but by the 1920s and 1930s, Coon Creek was aggrading at an average rate of about 6 in. (15 cm) per year, and the maximum depth of historical sediment there is now about 15 ft or 4.6 m (Figure 3.15). The graphs dramatically demonstrate the extremely rapid increase of erosion and the increase of runoff that

*Figure 3.12 (Continued)* Evolution of tributary channels as explained in the text and in Figures 3.10 and 3.11. North Branch, Whitewater River, MN, 1 mi (1.6 km) west of Elba, MN, looking upstream (west) (NE1/4, Sec9, T107NR10W, Winona Co.). For reference, note barn, bridge, and road intersection. 1976: The stream has narrowed again to similar dimensions as seen in 1905 with dense riparian vegetation. (Photo credit: 1905 and 1940 photos by Soil Conservation Service [SCS].)

would have been necessary to erode both upland and upstream channels, bringing sediment into the main valley at such a rapid rate. This rapid increase has been mentioned by earlier investigators (Sartz, 1961) and will be continually demonstrated in this study. As to total accumulation in the entire main valley, the greatest depths are mid-valley, 10–15 mi (16–25 km) upstream near the center of the old pre-European longitudinal profile (Figure 3.16). Note that the newer profile is straighter and even convex in some places. It has been argued by Knox (2006) that raising the base level of streams by the navigation dams on the Mississippi has induced much of the aggradation in the extreme lower portions of lower main valleys, and that may indeed be true for some streams (Hendrickson, 1990; Knox, 1993; Knox and Faulkner, 1994). But Knox then goes further and states that much of the sediment deposition in Coon Creek is induced by the higher base level of Pool 8 created by the dam at Genoa (Knox, 2006, p. 294). However, Figure 3.16 shows that unsupported contention to be inapplicable to Coon Creek. It will be seen that Range 36 is only a half mile (0.8 km) from the mouth of Coon Creek but the 1850–1975 aggradation there is the least in the entire longitudinal profile shown. In other words, the historical sediment thins out as Coon Creek approaches the Mississippi River, a condition clearly shown in Figure 3.16. Moreover, the deposition rate 1938–75 was less at the mouth than in mid-valley. In an unpublished report to USGS, Happ (1977) discussed this thinning of sediments in the lower reach of Coon Creek and suggested that it means that little sediment from Coon Creek has reached the Mississippi River. Indeed, Happ's point is clearly demonstrated in Figure 5.8. Another very significant point is that, while the accretion rate has been low near the mouth, the rates have stayed more or less constant over historical time. Again, this is clearly shown in Chapter 5, Figure 5.8.

Chapter three: The systematic effects of historical agriculture on the physical landscape 77

*Figure 3.12 (Continued)* Evolution of tributary channels as explained in the text and in Figures 3.10 and 3.11. North Branch, Whitewater River, MN, 1 mi (1.6 km) west of Elba, MN, looking downstream west (NE1/4, Sec9, T107NR10W, Winona Co.). For reference, note barn, bridge, and road intersection in the first three frames. 1905, close-up of the reach just downstream of the barn. 1940, close-up of the reach upstream of the barn looking east (downstream). (Photo credit: 1905 and 1940 photos by Soil Conservation Service [SCS].)

*Figure 3.13* Lewiston Fork of Garvin Branch, 2 mi (3 km) SW of Stockton, MN, on US Hwy 14, looking upstream. Top: 11/2/39. This is the typical "gravel road" tributary of that period, which Stafford Happ described here as a "bouldery torrent plain formed by bank erosion." Bottom: July 1977; rocks have been covered with silt, and the floodplain is a fine pasture. Note the eroded silt banks where cows have trampled them. (Continued)

*Figure 3.13 (Continued)* Lewiston Fork of Garvin Branch, 2 mi (3 km) SW of Stockton, MN. on US Hwy 14, looking upstream, July 2009; With cessation of grazing, floodplain has now grown up with woody plants.

Locally, the depth of historical sediment is usually great where tributaries flow into the main valley, creating alluvial fans. In some cases, a "hump" or convexity was formed in the longitudinal profile. This may be seen between profiles 33 and 33b, where several tributaries were bringing in massive amounts of sand from the erosion of high terraces. In a few cases, such tributary fans forced the main stream to move some distance laterally across a valley. An example is the mouth of Wing Creek, which, between 1924 and 1983, pushed Coon Creek more than 200 ft (60 m) to the west (SE ¼ Sec 25, T14N, R7W, Vernon County, WS). See Chapter 4 for examples of gully fans displacing master streams.

Many valleys in the Hill Country have less sediment than Coon Creek, but some may have more. In many places, roads, bridges, farms, and even villages have been buried, or so affected by flooding and/or wetness that part or all of the place or function had to be abandoned.

Aggradation of main valleys has slowed greatly since the 1940s, and measured accretion rates in the main valley of Coon Creek 1975–1993 were only about 3% of the 1930s rates. Although average accretion rates are low, the problem is still serious because the migrating sediment is becoming coarser, as explained under UMV, and is increasingly likely to be deposited in and along channels in these low-gradient areas. Measurements show the channel and banks accreting faster than the distal floodplains (Trimble, 2009a). Whereas aggradation with finer material over the previous century tended to accrete the entire width of the floodplain, recent aggradation with coarser material tends to accrete the channel and adjacent bank Much has happened on the Southern Piedmont, an area of sandy soil (Trimble, 1969; 1974). The disproportional

*Figure 3.14* Downstream end of the Upper Main Valley of Coon Creek, as defined by the process of a stream eroding a high bank and depositing a lower bank on the other side of the stream adjusted to a lower flood regime. 1940; looking downstream from bridge on Lietke Lane, just off WI Hwy 162, stream is fairly straight at this point. (Photo credit: S. C. Happ.) (Continued)

accretion along the stream is termed a natural levee, and it places the distal floodplain at a relatively lower elevation. This selective aggradation of streams and their banks is a problem because it increases the incidence of floods; that is, as the stream itself aggrades higher and higher above the distal floodplain, any overflow of the banks can fill the distal floodplain to a greater depth. Also, the higher natural levees do not allow the water from adjacent tributaries to drain easily into the main stream, resulting in the distal floodplain becoming wetter. In many cases, the distal floodplain, likely at one time cropland or pasture, has become a swamp and in a few cases, a lake. Both because of wetter conditions on floodplains and because of present agricultural economic conditions, large floodplains and streambanks are not grazed as they once were. The result is the replacement of grass with riparian brush and trees, the net effect of which is to cause channel erosion. Thus, reaches of the LMV reverting to forest may well undergo erosion with net enlargement of the stream channel (Trimble, 1997a, 2004; Lyons, et al, 2000).

## Stream erosion of high terraces and high banks

A recurring question is the role of streamside high terraces in furnishing sediment to streams. This occurs when a stream flows by the base of a high terrace and actively cuts into it (Figure 2.16). Because the stream may be cutting away a friable bank perhaps 20 ft or

*Chapter three: The systematic effects of historical agriculture on the physical landscape* 81

*Figure 3.14 (Continued)* Downstream end of the Upper Main Valley of Coon Creek, as defined by the process of a stream eroding a high bank and depositing a lower bank on the other side of the stream adjusted to a lower flood regime. Late 1980s. Note that the stream has cut away the bank to the left foreground and is depositing a slightly lower bank in the right foreground. The same pattern exists for two more meander loops in the distance (looking downstream from bridge over Coon Creek, NW1/4, Sec 24,T14N, R6W).

more (6 m) high, a small amount of lateral migration into the terrace face releases a large amount of sediment into the stream. While not common, such eroding streamside terraces may be found in all three of the fluvial zones, but most are in the lower main valleys. Because they are uncommon, often of short length, and often not easily accessible or even observed, they are not usually included in most sediment budgets. Yet, they may locally furnish large amounts of sediment directly into the stream.

Is the erosion of these terraces natural or is it the result of human use of the land? A balanced answer is that this type of erosion existed before European settlement, but it was uncommon. Owen (1852, p. 55) described and sketched a long cut bank in a high terrace on the Chippewa River, calling it a "theater of sand." He also terms it "conspicuous," perhaps meaning that it is singular since he does not mention any other instance. Because the Chippewa is heavily loaded with sand from glacial outwash, it might be expected to erode its banks and widen. And although it seems likely that other instances existed in the primeval landscape, none have been found in the literature.

All the evidence that I have examined suggests that very few such high cut banks existed under natural conditions. Given the mild hydrologic regimes of primeval streams, the apparent lack of bedload, and the luxuriant riparian vegetation, stability would be

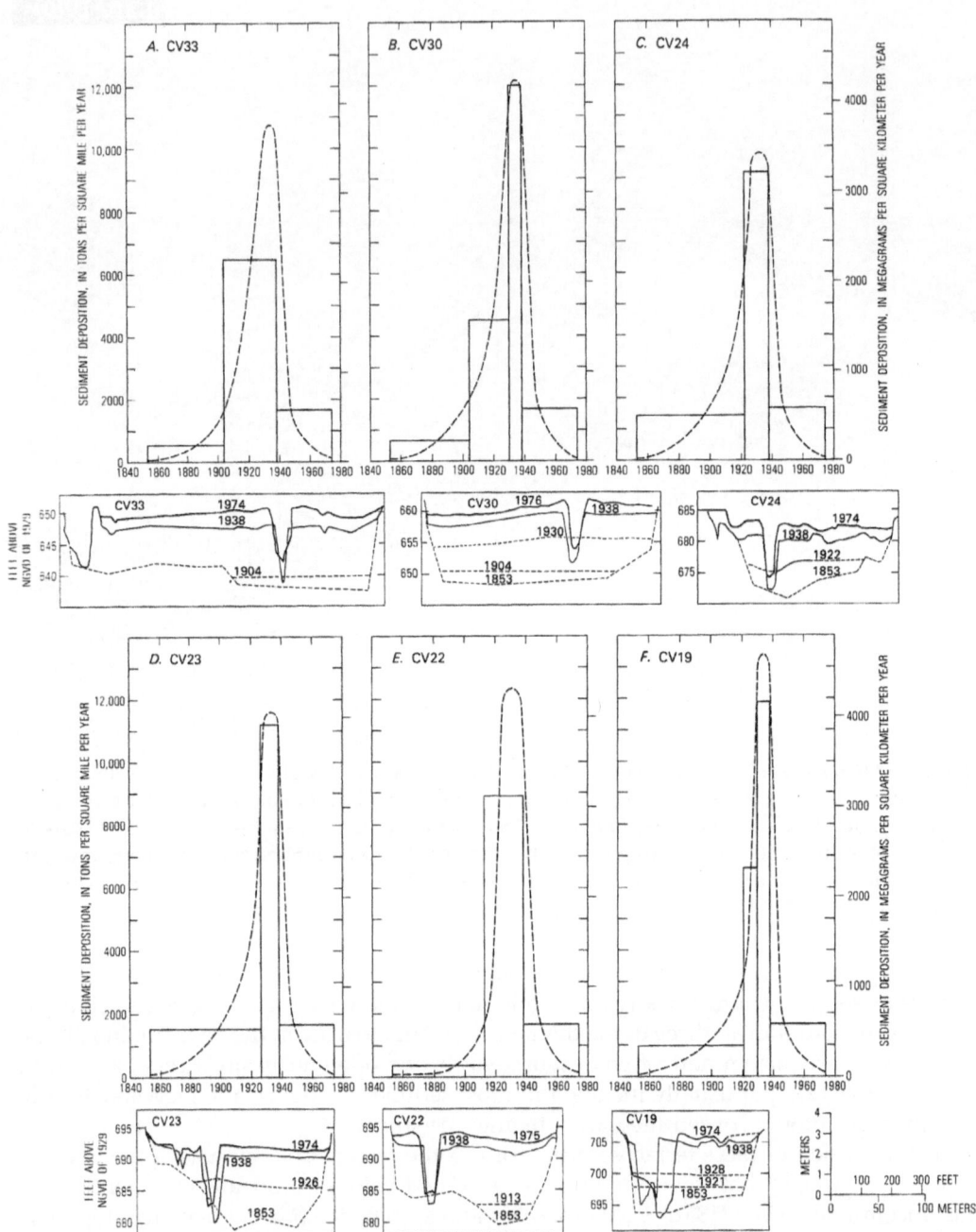

*Figure 3.15* Deposition rates at selected Lower Main Valley sites in Coon Creek, 1853–1977. These are derived from cross-sectional reconstructions of deposition rates based on surveys, excavations of soils, bridges, roadways, and other dating methods. (From Trimble, S. and Lund, S., *US Geol. Surv. Prof. Paper* 1234, 1982).

Chapter three: The systematic effects of historical agriculture on the physical landscape 83

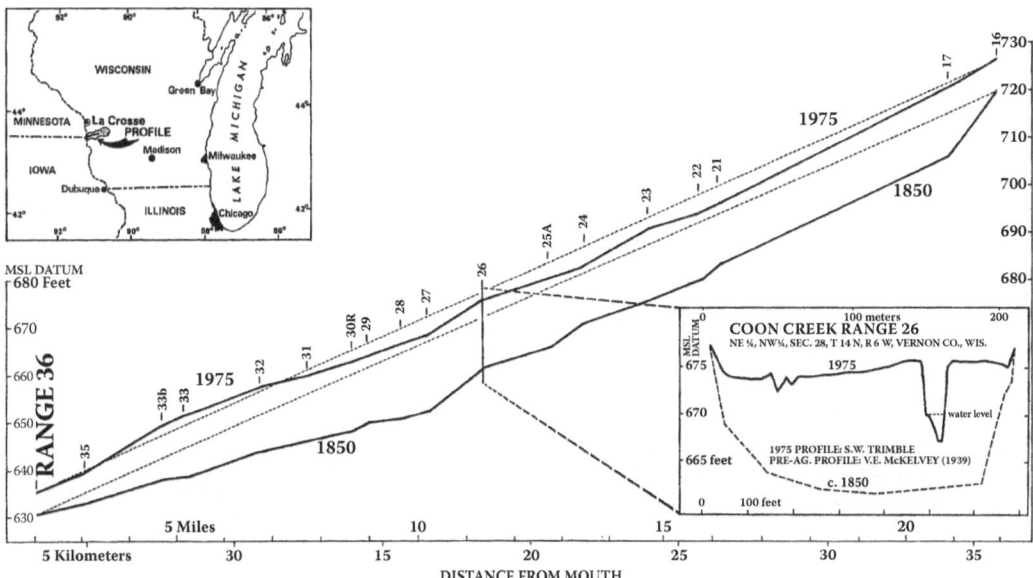

*Figure 3.16* Longitudinal profile of Coon Creek stream banks showing depth of historical sediment accretion for the entire main valley, 1850 and 1975. Dashed lines in the longitudinal profile are straight lines for reference. Profile 16 is 2 mi (3 km) upstream from Coon Valley and Range 36 is less than a mile (1.6 km) from the Mississippi River. Note that the longitudinal profile in 1850 was concave upward, typical for a humid area stream in equilibrium. By 1975, however, the profile has become almost straight, indicating a stream overloaded with sediment. Indeed, from Range 32 to 35, where erosion from high Pleistocene terraces in Wing Hollow was especially severe, the profile became convex upward. Note also that the raising of the base level of about 8 ft (2.5 m) by navigation pool 8 in 1938 seems to have had negligible effects on sediment depths in the lower 5 mi of the stream. In fact, the accretion rate near the mouth 1938–75 was less than in mid-valley.

expected. With the increase of flood regime, greater supply of coarse bedload to streams from gullies, and destabilization of streambanks by domesticated animals, it is logical that streams would more actively cut into these terraces. A case in point at Fairwater, Minnesota, is discussed in Chapter 5 where it can be clearly demonstrated that the cutting of a high terrace began only in the early 20th century and then furnished tens of acre-feet (thousands of cubic meters) of sediment downstream to the Whitewater River.

However, recalling the discussion of causative factors in the Introduction and Figure 0.3, we can still see a strong interplay between natural and cultural actors. An example demonstrating the complexities is found in the LMV of Coon Creek (Figure 3.17). There, the stream ran close to a terrace about 20 ft (6 m) high. While we do not know the primeval state of this location, we do know that by the 1930s, the stream was cutting mildly into the high terrace, but the reach did not appear unstable. During the very wet year of 1993, a combination of moderately high but prolonged stream flows, excessive rainfall, and cattle grazing caused the high bank to collapse in a series of slumps (Figure 3.17). This pushed the stream toward the opposite bank, causing the bank to be undercut with the collapse of riparian trees into the stream. In turn, the fallen trees diverted the stream into the opposite bank, causing it to be eroded. Thus, more than 10 m (30 ft) of channel banks had been destabilized.

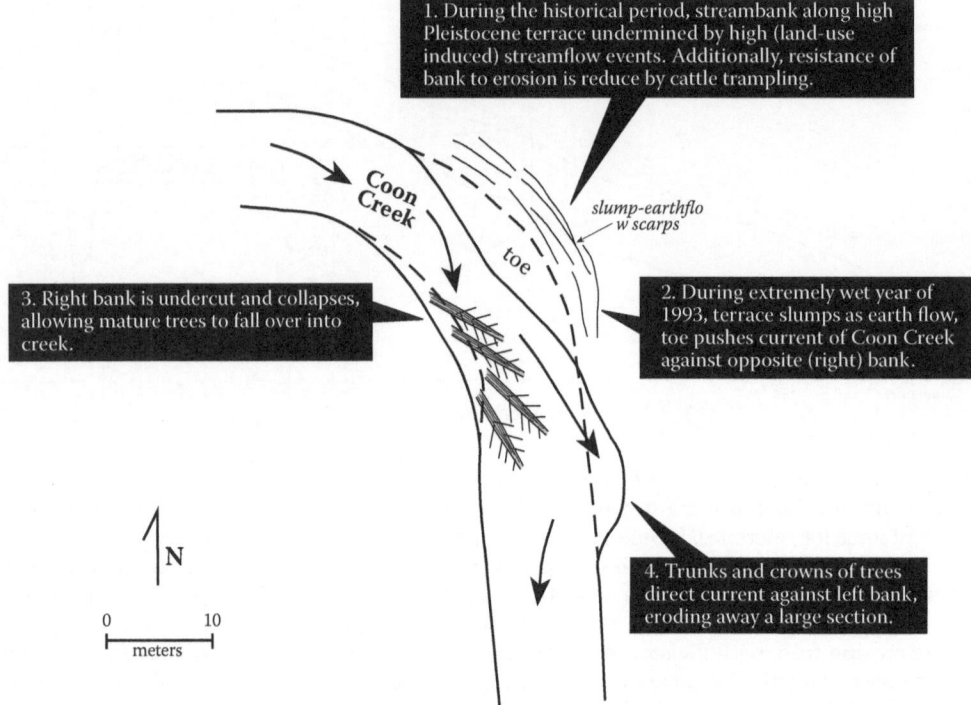

*Figure 3.17* Destabilization of stream reach by collapse of undercut high terrace, showing natural and human factors. This reach may be observed by looking upstream (south) from WI State Highway 162 bridge over Coon Creek (see Fig. 8.14).

At first blush, the cause of the destabilization would appear to be the very wet year and thus might appear to be natural. But it is unlikely that all this would have happened in the absence of preexisting conditions brought on by humans. Indeed, these conditions appear synergistic.

With these processes and zonal models of physical changes for the Hill Country in mind, it is now appropriate to discuss some effects of this hydrologic rupture and recovery on several aspects of settlement, including villages, isolated farms, mills, roads, bridges, and farmland. Starting in the next chapter, these effects are considered for the four fluvial zones just described: **gullies, tributaries, upper main valleys**, and **lower main valleys**. Because the water and sediment were passing down the watershed in a cascade, the four zones often demonstrated some unexpected relationships and feedbacks.

*chapter four*

# Upland gully erosion and its effects

"Upland" is here defined as being at a higher elevation than local stream valleys and includes farms on high terraces. Many farms located entirely on level to very mild upland slopes, especially in Region I, perhaps saw few major problems during this destructive period. Of course, the runoff from their fields cascaded down slopes and valleys to farms below, where it did create problems in many cases.

Gully erosion was first noted in the Hill Country in the 1880s, was widespread by 1900, and was of disastrous proportions in many areas by the 1920s (Chapter 3, Figure 3.4). It was highly associated with the steep slopes and high relief of Region II. A 1917 soil survey of Buffalo County, Wisconsin, noted that gullies one-half mile (0.8 km) long could form in 1 year (Whitson et al., 1917). As the result of improved land use and gully amelioration engineering, most gullies have become generally inactive since the late 1940s. Indeed, the decrease of gullies in the Hill Country was dramatic (Chapter 3, Figure 3.2).

Gully erosion includes deep channel incision where none existed before and also enlargement of ephemeral channels, the latter sometimes termed *channel trenching* (Happ et al., 1940), but the distinction between these two types is often blurred. In the Hill Country, gully erosion occurred more or less simultaneously in two environments that are different enough to consider them separately. These are (a) *hillside gullies* and (b) *high terrace gullies*. While hillside gullies were more widespread in the Hill Country, the gullies in the high Pleistocene terraces, as seen in the cover photo of this book, were much more spectacular and destructive. These sometimes were as deep as 30 ft (9 m) and covered tens of acres, greatly disrupting human activity.

Because gullies sometimes cut deeply, they normally transported coarse material ranging from sand to boulders. The fans often debouched onto roads and railroads so that traffic was disrupted with every storm. As already noted, much of the township, county, and state road budgets were to repair bridges and roads, and to clear debris after storms. Perhaps just as important economically were fans extending out over the surface of floodplains and terraces, so thick they often buried fences (Chapter 3, Figure 3.4). The coarse material was unfertile and, with fresh deposits from every rainstorm, it was difficult to grow crops there. Another sad instance repeated many times in the Hill Country was where a house had been sited at the foot of a hill where a small stream or spring run issued. In many cases, the small valley was occupied by a gully that brought down debris and often buried the farmstead.

Just as gullies were widespread in the Hill Country, so were there fans of erosional debris. In the Whitewater River watershed alone, 209 sand and boulder fans were large enough to be mapped on air photos (Happ, 1940). The sand fans averaged 5 acres (2 ha) in size, while the boulder fans averaged 0.9 acre (0.36 ha). One gully in a high terrace was more than 700 ft (215 m) long, and more than 100,000 yd$^3$ (7,650 m$^3$) of sandy material had been lost. The fan was more than 12 ft (3.7 m) deep, and its volume was 65,000 yd$^3$ or 5,000 m$^3$ (SCS, 1940). The total mapped area of fans was 568 acres (227 ha), so almost 1 mi$^2$ (2.6 km$^2$) of valuable valley land was covered by coarse, mostly infertile debris. Long reaches of road were covered and, in one case, a large sand fan laterally displaced a length of the main channel of the Whitewater River (SCS, 1940). Of fans not large enough to map, there were "several times" as many as those mapped (Happ, 1940). The area of fans was expected to increase at the rate of 11 acres per year.

## Hillside gullies and their fans

Hillside gullies were sometimes incised into loess and/or residual material, but Sartz (1961) maintains that many historic hillside gullies occupied prehistoric gullies that had filled with debris, presumably by mass movement, over geologic time. This accounts for the often large rocks, usually limestone, which came from these gullies. Hillside gullies include those incised into existing swales or down cattle trails or roads, zones where channels were not usually preexisting. It also includes the enlargement of small preexisting ephemeral channels in draws and smaller valleys that would often flow directly into tributaries, which were in many cases also trenching. In both instances, excess water was coming from pastures and cultivated fields upslope, creating the gully.

The rapid formation of hillside gullies as well as their rapid disappearance is shown by a series of three photographs of the hillside just west of Elba, Minnesota (Trimble and Lund, 1982, Figure 16). In 1905, the hillside was forested and there were no apparent gullies. By 1940, the hillside had been cleared for agriculture and gullies were well formed. Then, by 1978, the hillside was in grass with no visible gullies. The rapid disappearance of small gullies and rills is also shown by the transformation of the landscape from eroding rectangular fields to contour strips (Chapter 2, Figure 2.11).

A classic example of a hillside gully with a rock fan on the valley floor is seen on the middle branch of the Whitewater River, Minnesota (Figure 4.1). In this case, the fan is covering farmland, taking it out of production, and making farming more difficult. In some cases such gullies advanced headward (upslope) and threatened farmsteads and roads (Figure 4.2). By the 1970s, nearly all such gullies were inactive (Figure 4.1).

### The Appleby farm

A remarkable example of the interaction of the landscape processes described in this study is the Appleby farm in the Whitewater Valley (NW1/4, Sec 10, T108N, R10W, Winona, County, Minnesota). Located at the base of a hillside in the lower main valley, all the floodplain pastures had become wet by the 1930s (Figure 4.3). Appleby then had no choice except to graze his animals on the steep slopes behind the house. Within about a decade, sediment from the gullies had buried the backside of the house. When the state bought up land in the area for a state park in 1944, the house and the 270 acre (108 ha) farm were sold for only $4400 (H. Johnson, 1976). The Appleby house was large and at one time elegant. By the time it sold, it was derelict. A photo made at the time of sale is found in H. Johnson (1976, p. 210).

### The Zink farm

While Appleby's experience was perhaps unusual, a fairly common situation was a farmstead being located in a small valley, often near a spring since water was so essential. Over time as the landscape deteriorated, stormflows down the valley increased, eroding the channel and bringing down cascades of sand and rock ranging from cobbles to boulders. Thus, what had been originally an idyllic house site became a nightmare of flooding and rocks. Since moving the farmstead was usually out of the question, farmers used the rock

*Figure 4.1* Stabilization of hillside and gullies, North Fork Whitewater River, looking north just east of Fairwater, MN (SE1/4, Sec 5, T107N, R10W, pan of two photos made 10/23/40). Top: 1940, note treeless "goat prairie" and two hillside gullies with fans at base of slope. Bottom: July 2009, note the transformation from "goat prairie" to mostly deciduous forest. Gullies and fans are relatively inactive. Compare to Chapter 5, Figure 5.1 of the same slopes. (See also Trimble, 2009a, p. 18.)

*Figure 4.2* Gully heads (overfalls) threatening economic activity in the late 1930s. Top: A threatened farm possibly in Trempealeau Co., WI. Bottom: A threatened roadway in Wabasha Co, MN, 2 mi (3 km) SW of Weaver.

Chapter four: Upland gully erosion and its effects

*Figure 4.3* The Appleby farm, Whitewater Valley looking north (NW1/4, Sec 10, T108N, R10W, Winona County, MN). Top: The farm seen across the floodplain, grown too wet for grazing. Thus, Appleby was forced to turn his animals loose on the hillside behind the house, and gullies are visible. Bottom: Close-up of gullied hillside behind house with partially buried stone wall visible. Corner of house visible to left. Both photos 8/27/37. A present remake of the top photo would show the lake formed in the foreground as the main channel of the Whitewater River aggraded to a level higher than the distal floodplain here (see Chapter 7).

debris to make levees, creating a large flood channel or flume to bypass the farmstead. An example is the Zink farm, located at the confluence of a small tributary (which also trenched) with Coon Creek (NE 1/4, NE ¼, Sec 33, R10W, T14N, Vernon County, Wisconsin, Figures 4.4, 4.5, and 4.6). Zink's first nemesis came in the form of a side gully that formed in a small draw north of his house and flowed between the house and barn (Figure 4.4 and 4.5). Carrying water and sediment from the upland with almost every rain during the early decades of the 20th century, the gully built a large boulder fan in his farmyard. At the same time, the larger dry run from the southeast was also now bringing torrents of water and sediment. Indeed, this torrent was so powerful and massive that, between 1900 and 1924, the coarse alluvial fan it was forming deflected Coon Creek about 1500 ft to the west across the floodplain.

Zink then used the fan materials recently deposited in his farmyard to build dikes to protect his house and farm buildings from the water and debris (Figures 4.4 and 4.5). While this did not solve the problem, it at least allowed him to survive the bad years. Even with the protection of the dikes, Zink would often evacuate his family to a safer location at the onset of a storm (interview with Gene Zink, great-great grandson of the original owner, June 2011).

With the implementation of the soil conservation program and the resulting improvements in land use and land treatment, the storm flows subsided. Presently, the side channel and dikes are largely vegetated and carry water only during major rainstorms (Figure 4.5; see Chapter 8 for effects of an extreme storm on the Zink farm).

Streamflow trends at the Zink farm are similar to those found elsewhere in the Hill Country. The presence of spring here was a determining reason for locating the house but, by the early 20th century, the spring was failing so that the local stream was dry most of the time. But by the 1980s, as the result of improved land use and increased infiltration of rain, the spring was rejuvenated and has increased in strength ever since (Figure 4.6).

I have identified perhaps three dozen valley farmsteads in the Hill Country with a history similar to that of the Zink farm (see Chapter 8, Figure 8.7). However, the total number was doubtlessly much larger, numbering perhaps in the hundreds. The irony is that the problem was rarely the fault of the victim. Rather, the water was coming from the land of neighbors on the uplands, sometimes at some distance.

Some hillside gullies could transport stunning amounts of water and sediment out onto a floodplain. The small dry tributary running near Zink's house trenched in its upper reach but deposited material into a huge alluvial fan as it reached the lower gradient by the south side of the farmstead and moved onto the floodplain of Coon Creek. Thus, as already noted, Zink had to build another dike on the south side of his farmstead to protect his property from flooding (Figures 4.4 and 4.6). With the continuing deluge of coarse erosional debris, the alluvial fan eventually pushed Coon Creek across the valley. When the Zink family acquired the property in the 19th century, Coon Creek flowed just to the west of the present bridge on County Road O. That had been the position of the creek at the time of original survey in 1847 and it remained as so until at least 1900 (Mississippi River Commission, 1900). By 1924, a new USGS topographic sheet *Stoddard, Wisconsin* (1:62,500), showed a large fan to the east of the bridge with the creek apparently driven to the other side of the valley, where it has remained. Thus, Zink's dairy cows lost easy access to the main stream of Coon Creek. Presently, a line of tree stumps just southwest of the bridge mark the old channel of Coon Creek. Not only did Zink lose easy access to the creek, but the fan there of generally coarse texture is much less fertile than the land that was covered. Being higher and porous, it is dryer than the original floodplain surface and supports a sparse meadow. It is still affected by occasional storm flows that still bring more sediment.

Chapter four: Upland gully erosion and its effects

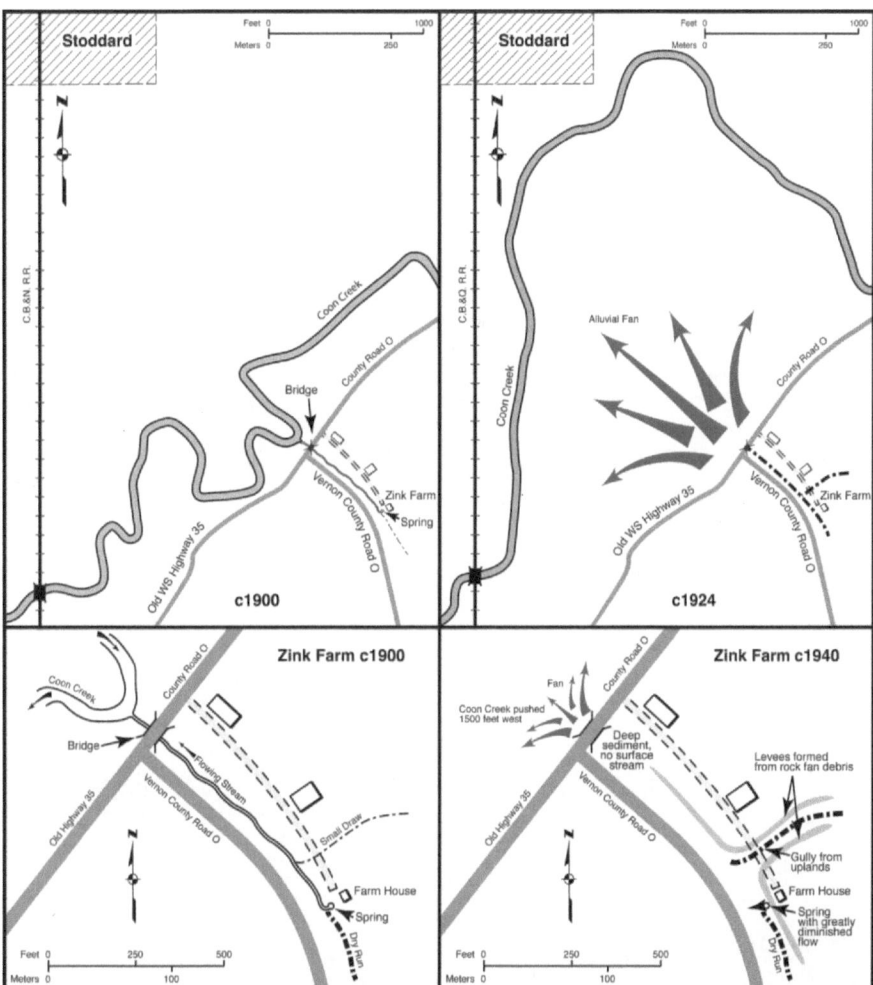

*Figure 4.4* Changes at the Zink farm about 1 mi (1.6 km) southeast of Stoddard, WI, c. 1900–1940 (NE1/4, Sec33, T14N, R7W). In 1900, Coon Creek still ran on the east side of the valley close to the Zink farmstead. By 1924, the small tributary from the southeast had formed an alluvial fan that has pushed the creek across to the west side of the valley. This process clearly demonstrates the rapid onset of disastrous erosion in the early 20th century.

A large cottonwood stands just about 100 ft (30 m) west of the bridge, perhaps marking the earlier route of Coon Creek. In the early 1970s, a large limb grew from the tree at ground level (see photo in Costa and Baker, 1981, p. 188). Oscar Zink (1913–1982), who then owned the farm, stated in 1973 that in his younger days he had been able to drive a loaded hay wagon beneath the limb. Accretion from the fan has since covered the stub of that limb. This information alone suggests well over 10 ft (3 m) of accretion since the 1920s and 30s.

The present bridge over the fan on County Road O (old Wisconsin Highway 35) at the Zink farm is the latest in a long line of bridges there. I excavated the wingwalls of an earlier bridge in 1975, but those were since buried by the fan (Figure 4.6). The small but permanent stream now flowing under the bridge and off to the southwest in a ditch deserves

*Figure 4.5* The Zink farm, Coon Creek basin. Top: Looking southeast (upstream) from bridge on County Road O. Note dike in middle background to protect farm from flooding (arrow). The former gully and fan, now a stream, flows toward viewer on the bridge. The side channel flows in from the left (north) between the barn and the house. (Photo credit: Nadine Kleinhenz, 2006.) Bottom: Looking upstream (northeast). Mostly dormant rock fan between barn and house reshaped into flume with dikes to either side. The fan formed in the early 20th century and carried water and rock with most rainfall events in the early 20th century. Material from the fan was used to construct the dikes. Flume is shown after 4 in. (10 cm) rainfall event, c. 1990.

*Chapter four:   Upland gully erosion and its effects* 93

*Figure 4.6* The Zink farm. Standing on fan looking east (upstream) at bridge on County Road O. Top: 1975. Wingwalls of buried earlier bridge are still visible. An excavation 5 ft (1.5 m) deep here in 1975 and 1976 did not reach the base of the wingwalls nor did it fill with water, indicating the lack of permanent flow and significant underflow at that time. Bottom: 1999. The wingwalls have been buried in the aggrading fan. Most notable to observe is the permanent streamflow, which began in 1988. For scale, the person to the right is 77 in. (193 cm) tall.

special mention. From the first time I saw the channel in 1973 until 1988, it was always dry. The wingwalls of the buried bridge shown in Figure 4.6 were excavated to depths of about 5 ft (1.5 m) in both 1975 and 1976, but no water even seeped into the hole, thus showing that there was no underflow at that level. But by 1988, a trickle of surface water was passing under the bridge. Coming from the rejuvenated Zink spring, the flow has grown stronger with each passing year. This pattern has been repeated countless times in the Hill Country and accounts for the higher base flow of streams (Gebert and Krug, 1996; Potter, 1991; Kent, 1999; Shilling and Libra, 2003).

The restoration of the Hill Country springs, such as the one at the Zink farm, and the increasing baseflow of streams mark the partial restoration of the presettlement hydrologic system. It also has importance for hydrologic theory, which had long stated that such springs could be restored by land use changes that increase the infiltration of water. Sartz (1961) states that, while often pronounced, this concept had never been proved in practice.

There may be many similar instances of large stream channels being laterally displaced by tributary or gully fans such as those seen at Zink's, but only three others are documented. One was in Iowa on the Upper Iowa River (Happ et al., 1940, pp. 73–74, Plate 9), where 5 acres of farmland were said to have been covered by coarse, infertile sediment in the process. Happ pronounced the 5 acres as "destroyed" and suggested that he had seen other instances in the Hill Country but unfortunately did not name them. The second instance was a huge sand fan, in this case from a high Pleistocene terrace on the Whitewater River. An approximately 0.5 mi (0.8 km) length of river, about 1 mi (1.6 km) northwest of Elba, Minnesota (E 1/2, Sec 2 R10W T107N), was laterally displaced as much as 800 ft or 250 m (SCS, 1940). A third example is Wing Creek, a tributary of Coon Creek (see Chapter 5).

Many hillside gullies debouched onto farms, farmsteads, roads, and railroads (Figure 4.7). In at least one incident, cars were actually buried in the fan of a gully while under way on a federal highway (Figure 4.8). The event happened so fast that the occupants were trapped in their cars by the sediment that was 7 ft (2.1 m) thick in some places, and they had to be rescued. The nearest rain gauge at Beaver, 8 mi (13 km) east, showed only 2.91 in. (7.5 cm) in 2 h (*Winona* [Minnesota] *Republican-Herald*, June 8, 1940).

The cost of cleaning and controlling this sediment to state, county, and township road departments, as well as to railroads, was extraordinary. In the late 1930s, annual costs of cleaning roads of sediment was about $6,000 in the Zumbro watershed and about $12,500 in the Whitewater, amounts which constituted a large proportion of tax revenues at that time (SCS, c. 1940). All over the steeper parts of the Hill Country, elaborate structures were built to control or contain the sediment (Figure 4.9). What must be understood is that such blockages were not just the result of large floods. Rather, they were frequent, often occurring even with small rain events. Thus, there was the need for expensive controls along highways at the mouth of gullies.

Most highway departments built debris dams to both trap sediment and partially control such gullies by raising the base level. Three of these were surveyed and found to have been accumulating sediment at rates ranging from 7,000 to 40,000 tons per year per mi$^2$ (16,000–95,000 t/ha$^2$) of drainage area (Trimble and Lund, 1982). The worst of these was delivering about 3000 t/yr (2700 t/yr) onto Wisconsin Highway 35 directly across the Mississippi River from Winona, Minnesota. Another, Middle Michell debris dam (NE ¼, Sec 35 T101N R7W, Vernon County), was built in 1941 to stop debris from constantly stopping traffic on Wisconsin Highway 56 near Romance, Wisconsin. The dam filled in only 2 years but, in the meanwhile, soil conservation measures were being taken on the contributing area. Grazing was reduced, the steeper slopes were taken out of cultivation and

Chapter four: Upland gully erosion and its effects

*Figure 4.7* Examples of gully fans covering roads and railroads. Top: Sand fan over MN highway 74, 2 mi (3 km) west of Weaver, 9/7/38, looking south. (Photo credit: Vince McKelvey). Farm in background was already abandoned. Bottom: sand fan over railroad, location unknown. (Photo credit: C.G.Bates, 8/2/30.)

*Figure 4.8* Top: cars halted and partially buried by sand fan on US Highway 61 about 2.5 mi (4 km) north of Weaver, Wabasha Co., MN, 6/7/40. The fan covered 0.2 acre (0.08 ha), was 7 ft (2.1 m) thick at the apex, and contained 4.6 acre-feet (5,700 m3). Bottom: The gully that created the fan, 6/9/40. Note the almost vertical walls. (Photo credit Soil Conservation Service [SCS].)

*Figure 4.9* Examples of elaborate drop structures to control hillside gullies creating rock fans on transportation routes. Top: Near Genoa, WI, to protect railroad, 6/27/30. Bottom: Between Wabasha and Lake City, MN, 1932. (Photos credit: C.G. Bates.)

reforested, and the more level upland areas were planted to alfalfa. When we surveyed it in 1976 and 1977, the gully from the upland had become relict from the infrequent flows and had grown up in forest (Trimble and Lund, 1982). The effects of an extreme storm on this debris dam are discussed in Chapter 8 (see Fig. 8.10).

## High terrace gullies and their fans

The second type of gully was found on the high Pleistocene terraces, which were usually underlain with sand and were very erodible. Here, as the noted Wisconsin agronomist Leonard Johnson put it, "disaster lurked" (Johnson, 1991, p. 12). Ephemeral streams on such terraces usually had not formed distinct channels under primeval conditions, but rather flowed through a U-shaped valley a few tens of feet across (Kunsman, 1944, Figure 4.10). In many cases, flow originated on residual slopes above and flowed across the terrace. But once water breached the silty soil of the valley floor and cut into the cohesionless parent material of the terrace, often sand, the gullies grew at breathtaking rates. Some likened it to "melting sugar": One gully advanced 1000 ft (300 m) during a single rainstorm (Johnson, 1991)! There were hundreds, perhaps thousands, of such gullies, some small and some very large arroyo-like channels, some tens of feet deep and thousands of feet long, and highly destructive both onsite and downstream.

Fans from the high terraces were normally formed from sand and finer materials. While seeming more benign than the often coarse materials from hillside gullies, the sand could spread out over a larger area. Because the sand was often infertile, such fans often ruined tens of acres of agricultural land.

### The Buffalo and Black River terraces

The most famous, or perhaps infamous, of the high terrace gully systems were those along the Buffalo River in Buffalo County and the Black River in Trempealeau and Jackson Counties, Wisconsin (Chapter 3, Figure 3.4; see cover of this book). Knox (1993) suggests that most gullying occurred there between 1910 and 1940. According to Bates and Zeasman, the Black River gullies were initiated in the early 1920s. By 1929, there were 170 deep gullies with a total length of 18 mi (30 km) and having lost over 3 million cubic yards (2.3 million m$^3$) of material (Bates and Zeasman, 1930). They found that perhaps 75%–90% of the eroded material was deposited in fans on the floodplain of the Black River. They also noted similar gullies in Jackson County, Wisconsin, which had become "very active" in 1916.

### Proksch Coulee

While not as expansive as the Black River systems, a spectacular gully formed in Proksch Coulee, a tributary of Coon Creek near Stoddard, Wisconsin. Although the coulee drains hundreds of acres, there was no stream channel at the time of European settlement, again suggesting the mild presettlement hydrologic regime. As already indicated, there was a wide, roughly elliptical vale that kept water depths, velocities, and thus stream power, low (Figures 4.10 and 4.11). This vale was originally most likely vegetated with grass before agriculture started because a deep, black soil had developed there. In the 1970s, older residents remembered being able to drive a wagon across at most any point. With valley sides being cultivated, however, the sediment from local soil erosion was conveyed into the vale. As erosion increased in the early part of the 20th century, the bottom often became a mudhole, making it not only economically useless but also a nuisance and difficult to cross.

Chapter four:   Upland gully erosion and its effects

*Figure 4.10* Untrenched drainageways in Pleistocene terraces showing how such valleys appeared before trenching. Such untrenched valleys are now rare. Historical sediment usually covers the valley floor to a depth of several feet. Top: Legue Coulee, 6 mi (10 km) NE of Ettrick, Trempealeau Co., WI. (Photo credit: Stafford Happ, 9/26/40.) Bottom: Small valley leading into Coon Creek. (Photo made from Cedar Valley Drive, looking downstream (south) about 1 mi [1.6 km] southeast of Proksch Coulee gully [SE1/4, Sec 26, T14N, R 7W], c. 1977.)

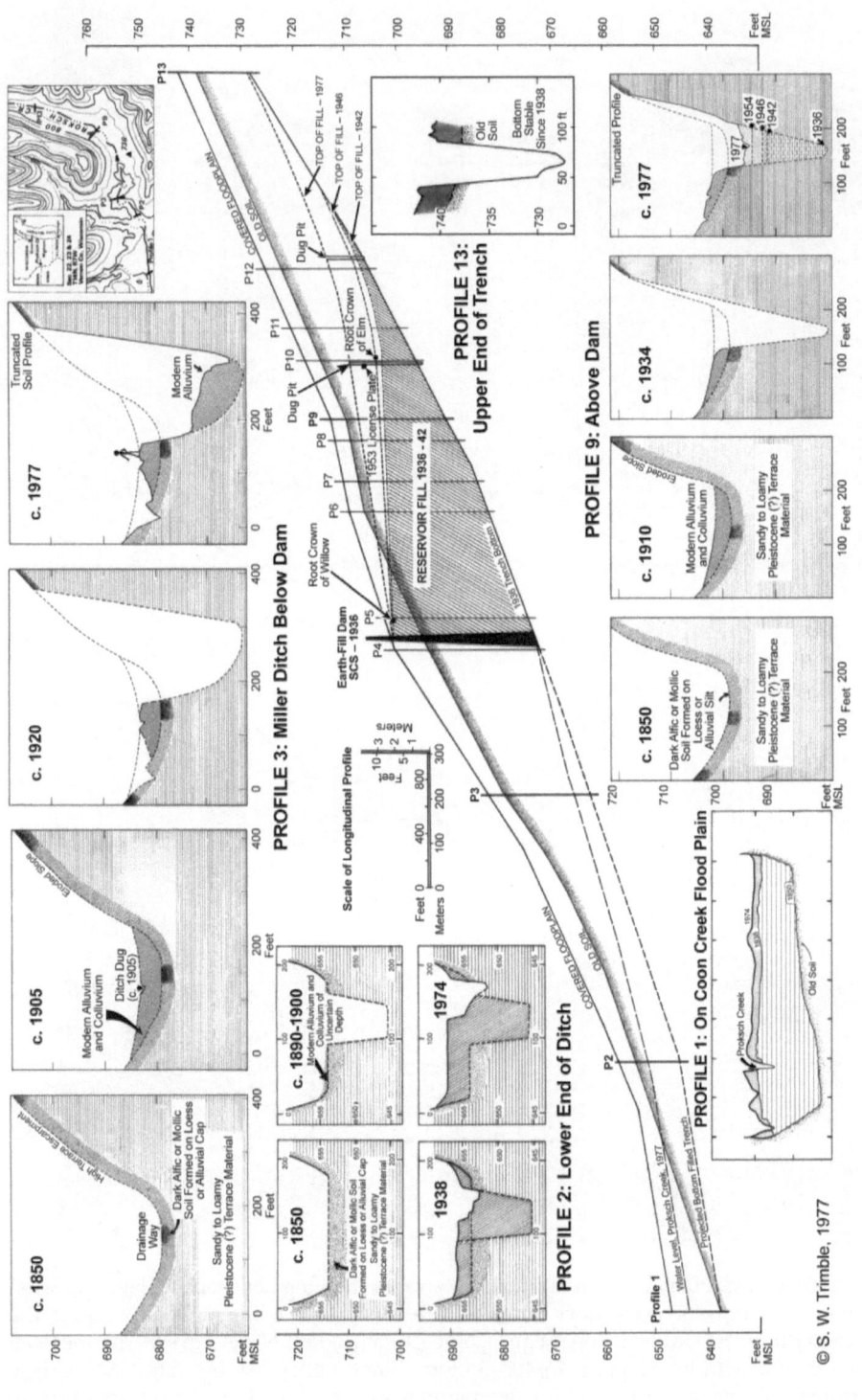

*Figure 4.11* Explanatory diagram of Proksch Coulee Trench (Gully) development based on intensive field research and precise surveys done 1974–1977.

Chapter four:   Upland gully erosion and its effects

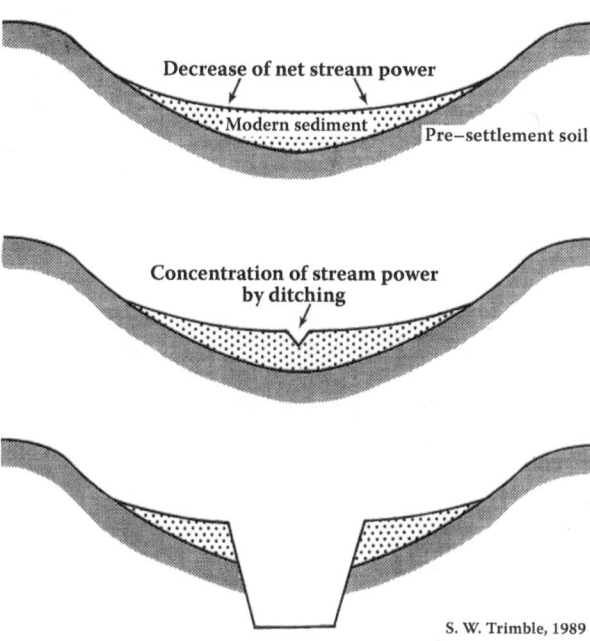

*Figure 4.12* Schematic explanation of the trenching of Proksch Coulee. Concentrating the flow and increasing the depth increased the power of the stream to erode (see Chapter 3, Figure 3.10). This diagram may apply to many if not most of the trenched Pleistocene terraces of the Hill Country.

The "solution" was to cut a drainage ditch down the vale with a plow. This concentrated the flow, increased stream power, and caused incision (Figure 4.12, Profile 3). Since there was little vegetation in the vale, the water cut rapidly through the mud, further concentrating flows and stream power (Figures 4.11 and 4.12) so that the buried native turf was easily breached. Soon, there was a huge gully looking much like an arroyo in the western United States. Indeed, some of the western arroyos were started in much the same way as in Proksch Coulee (Cooke and Reeves, 1976; Sir Ron Cooke, personal oral communication, April 1985).

In the case of Proksch Coulee, at least part of the ditch was dug in 1905 by the grandfather of Elmer Miller (1901–1994), shown in the diagram pointing into the ditch in 1977 (Figure 4.13). By a fluke, the eroding gully bypassed at least two parts of the old channel and left remnants of the original ditch superposed as shown in the photo. By the early 1930s, the arroyo was huge and about 25 ft deep, effectively dividing farms into two parts, not mutually accessible (Figures 4.13 and 4.14). The rawness and recentness of the trenching is evident from Figure 4.14, a photo made at the Clement farmstead, just upstream from the Miller farmstead, in the late 1930s. In 1936, the SCS built a huge drop-inlet structure, a "soil saving dam," to stabilize the worst part. The pool created by the dam, about 3000 ft (900 m) long, filled with sediment within about 5 to 6 years and continued to aggrade near the head for many years, creating a graded valley upstream (Figure 4.11, Profile 9). Then, as upstream soil conservation measures took hold and the runoff from the upstream valley became clearer and "hungry," the fill in the reservoir began to erode. Thus, we have

*Figure 4.13* Top: Elmer Miller in 1977 points to a portion of the ditch his grandfather dug in 1905 to drain the bottom of the channel. Subsequent trenching bypassed this reach of the ditch and cut around the left side (to the viewer here looking upstream) leaving this part isolated and elevated. Bottom, 1977: Looking downstream at the terrace floor remnant on which Miller stands in top photo. Dr. Stafford Happ stands in the gully remnant. For some reason, perhaps a blocked channel, the gully cut around to the right and bypassed this site. The two white flags on slope to right (arrow) mark the buried dark soil of the original (pre-trench) valley bottom, covered with about 6 ft (2 m) of historical sediment. The gully floor in the foreground to the right is now covered with about 6 ft (2 m) of alluviation adjusted to the aggraded floodplain of Coon Creek 1000 ft (300 m) downstream.

*Figure 4.14* Proksch Coulee Trench. Top: air photo, Feb 1934. North is to the left and the photo covers about 4000 ft (1,220 m) of the trench. Bottom: photo made in the Clement farmyard (arrow in top photo) looking up the valley, late 1930s. Note the bare, almost lunar-like, grazed landscape and the recent, vertical quality of the gully walls in the background.

*Figure 4.15* Alluvial fan from Proksch Trench on Coon Creek floodplain looking northeast up the channel. (Photo credit: S.C. Happ, 11/11/38 (NW1/4, Sec27, T14N, R7W, Vernon Co.) The extremely straight channel appears to have been dug and is completely occluded (filled up) in the foreground. Note sandy wash on floodplain and obviously wet conditions. However, this huge influx of sand has created an elevated fan out across the floodplain quite visible on the most recent USGS *Stoddard* sheet (1:24,000, 1983).

another example of stream equilibrium in operation—first, an overloaded stream deposits, then an underladen stream attempts to erode.

During the 1920s and 1930s, before the dam was built, much of the sandy debris from this gully was deposited in a huge fan on the Coon Creek floodplain, only a few thousand feet downstream (Figure 4.11, Profile 1), accreting it to depths of more than 10 ft (3 m). Crops were destroyed several times during that period, and the coarse debris and wetness lowered the agricultural value of this land (Figure 4.15).

A postscript to the Proksch Coulee Trench was that the "soil-saving" dam that helped stabilize the valley survived for 71 years. It was destroyed by the flood of August 2007 when about 12 in. (30 cm) of rain fell in 24 h (Chapter 8). Woody debris blocked the drop inlet and the cascade of water over the dam eroded it away, bursting the dam. With the base level thus lowered, a trench was immediately formed in the fill behind the dam, cutting upstream. This demonstrates the futility of building small dams without emergency spillways in the eastern Unites States.

Chapter four: Upland gully erosion and its effects

## Ratz gully

One of the most spectacular gullies in the Hill Country was the Ratz gully located 2 mi (3 km) north of Beaver, Minnesota, in Glendale Hollow, a tributary valley to the Whitewater River coming in from the west. The time and immediate cause of the initiation is not known exactly, but the gully was obviously recent and out of control by the mid-1930s. Photographs made in 1939 show a line fence, rebuilt in 1933 or 1934, suspended about 10 ft (3 m) above the abyss of the gully, suggesting how rapidly the gully was eroding (Figure 4.16; for another time-lapse photo of this gully, see Trimble, 1998, p. 292–293). The gully separated the Ratz farmstead from the public road so that it was necessary for the Ratz family to traverse it on a regular basis. A bridge was erected across a narrow part of the new gully, but the rapid growth meant that arrangement would not last long. A tragic yet humorous story was told about a member of the Ratz family returning to the farm late on a very dark night. It had rained during the day and, unbeknownst to him, the gully had enlarged significantly. Trudging in the gloom along his familiar route on his way to cross the bridge, he fell into the newly enlarged section where just that morning, his road had been (Figure 4.16). While he was reportedly unhurt, the tragedy was that access to the farm had been cut off again, but the point of the story is the extreme rapidity of erosion. Using several photos of the fence crossing and later detailed surveys, a cross section of the gully was reconstructed (Figure 4.17).

While the Ratz gully itself affected only a few farms in Glendale Valley, the fan of its debris spread out to the east over the already aggrading floodplain of the Whitewater River and the State Highway 74, which ran along the west side of the river. A 1939 photo of the fan shows the almost barren sandy landscape with a partially buried fence in the foreground. In the background is the terrace face with a small active gully visible (Figure 4.18). With almost every rain, copious amounts of sand were washed out onto and over the road to the rear of the photo.

## The Peterson event

As with the hillside gullies, there were hundreds if not thousands of terrace gullies active in the Hill Country during the 1930s. Three large ones have been recorded here, but most of the rest, especially the small ones, are undocumented. However, one in Peterson, Minnesota, was recorded on film simply because it was public, and it was so dramatic. So dramatic, in fact, that the frontispiece of this book depicts that event. In September 1938, the entire yard of a house was buried in a fan from a gully behind the house (Figure 4.19). Although no further details are known, the circumstances suggest that this event was a surprise, that neither the gully nor the fan were particularly active beforehand. For example, an active fan would have probably elicited a diversion wall of some kind in the yard to keep the water and sediment away from the house, but none appears in the photo. This demonstrates the fragility of these Pleistocene terraces and how out of control the Hill Country environment was by the 1930s. A visit to the site in 1977 revealed that the photo had been made from the top of the terrace, which was 30–35 ft (9–10 m) high. Several relict gullies, completely stable in 1977, were evident on the terrace surface and face, and these had presumably furnished the sediment.

*Figure 4.16* Ratz gully. Top left: 11/27/39, looking south (between Sec 3 and 4, T108N, R10W, Winona County, MN). Line fence was rebuilt 1933–34, but just 6 years later was undercut and suspended about 10 ft (3 m) above the valley floor. Distance to far bank was about 50–60 ft (16–19 m). Top right: repeat of photo, made July 1979. Distance to far bank was about 275 ft (84 m). Bottom: c. 1940, looking downstream (east) toward remains of Ratz bridge, constructed in the 1930s after the gully became too deep to cross. Gully has cut around right side, isolating the Ratz farm to the left (north).

Chapter four:   Upland gully erosion and its effects    107

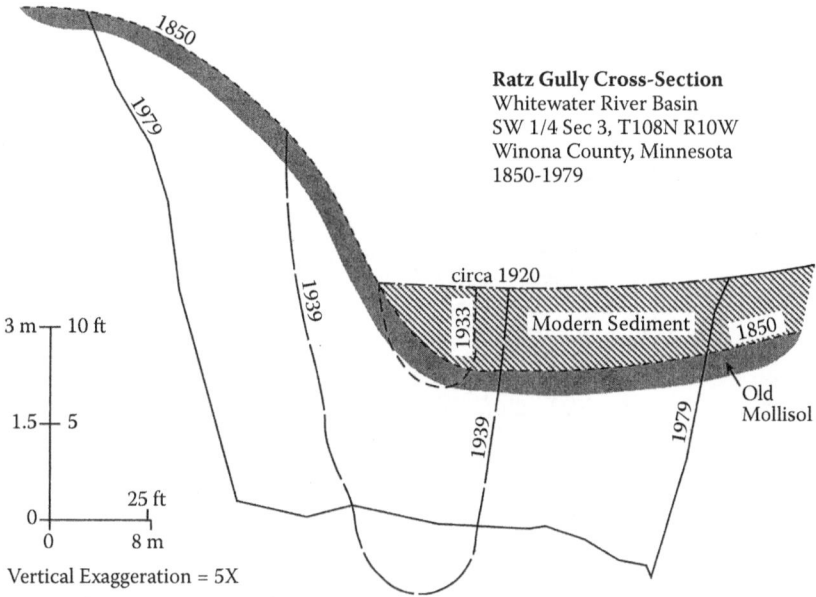

*Figure 4.17* Cross-sectional reconstruction, growth, and change of Ratz gully based on photos, surveys, and soils. (Redrawn from Trimble, S., *Catena* 32, 1998, 283–304.)

*Figure 4.18* Mouth of Glendale hollow and fan of Ratz gully on Whitewater River floodplain looking northwest, 1939 (NW1/4, Sec 10, T108N, R10W). Note sandy deposits in the foreground and partially buried fence in middle background. In far left background is the face of the Pleistocene terrace with a new gully actively incising.

*Figure 4.19* Fan from gully near Peterson, MN, 9/12/38. Top: standing on Pleistocene terrace face looking south toward the village of Peterson on the Root River. Methodist church in village in left background for reference. Gully to rear of viewer. Bottom: detail of car and house. (SCS Photos.)

*chapter five*

# The tributaries
## Zone of early, complex changes of process and form

Just downstream from many gullies, the tributaries may have been the most tumultuous zone of all since so much varied action—gully fans, flooding, floodplain aggradation, and extreme streambank erosion—was packed into such a relatively small area (see Chapter 3 for details on form and process). The processes affected villages, farms, roads, railroads, bridges, and communication lines, often disastrously.

## Villages

### Fairwater, Minnesota

The "poster child" of tributary zone devastation, at least that can be documented, is Fairwater, Minnesota, on the North Fork of the Whitewater River about 3 mi west of Elba. Fairwater, settled c. 1860, was hit by every destructive process considered in this study and was essentially made almost unlivable. Built around a mill site and based on productive local agriculture, the village as seen about 1905 presented an attractive, serene scene of a small, rural American settlement of the very early 20th century (Figures 5.1–5.4). What this remarkable set of four time-lapse photographic views shows is that the stream was stable and mostly tree-lined well into the 20th century (c. 1905, photos made 1902–1908), but, even then, a sand fan was beginning to infringe on the west side of the village. Undoubtedly, the floodplain had already aggraded somewhat by this time but that does not show in the photos. Flows were modulated enough and sediment yield low enough to permit a mill to operate in the village.

By about 1940, the cumulative effects of poor land use were manifest and the stream had eroded its banks severely, widening to 2–3 times its earlier width, destroying the milldam and both bridges. Just downstream of the village, the stream had cut into a high Pleistocene terrace, destroying several acres of excellent farmland and sending vast amounts of sediment downstream to the upper main valley of the Whitewater River, just upstream of Elba. The gully active on the west side of the village had deepened and cut through its own fan all the way to the creek. It, along with many other gullies unseen in this photo, was also sending more coarse material (cobbles, small boulders) into the creek, which was helping to further destabilize the stream by widening it. The millrace appears to be almost filled with sediment. Only scattered buildings remain in the once-flourishing village. By 1975, modern soil conservation measures had taken effect, and the 1975 photos show a much more stable, tree-lined stream with no other visible signs of erosion or sedimentation.

A bucolic close-up of the upstream bridge crossing c. 1905 shows the stability of the stream (Figure 5.5), but the repeat in 1940 suggests the violence and power of the flows at that time. Coarse sediment, which adds greatly to the bank erosion, is clearly evident.

*Figure 5.1* Fairwater, MN. Top: c. 1905; photo made from bluff west of town looking northeast (downstream). Note tree-lined stream (North fork, Whitewater River) and millrace (arrow). Bottom: 1940. Stream has widened by 2–5 times. Bridge in left foreground has been destroyed and a new one placed about 500 ft (150 m) upstream. Millrace is filled with sediment. Note bare "goat prairies" in background. (Continued)

Chapter five:   The tributaries         111

*Figure 5.1 (Continued)* Fairwater, MN. 1975: stream channel is again lined with trees and is no longer visible from this vantage point. Goat prairies are becoming revegetated.

The west pier is still to be seen in Fairwater, partially buried in the present bank. The downstream bridge suffered the same fate. About a mile downstream, a large section of the public road was washed out on several occasions increasing the isolation of the village (Figure 5.6).

An interesting and significant landscape feature seen best in Figure 5.3 is the disappearance of the "goat prairie," the south- and west-facing hillslopes that were largely grass at the time of settlement, environments once thought to be treeless because of xeric edaphic conditions. But the absence of trees was related to fire and grazing. Once those practices were curtailed, trees appeared. In Figures 5.1 and 5.3, one can, c. 1905, still see a long stretch of grassy hills facing south. By 1940, there is scattered woody vegetation, much of it juniper. Then by 1975, the slopes are covered in trees. When I last visited the hilltop photo sites in 2001, the trees had grown too tall to allow a repeat of the 1975 photos. The forest, now largely deciduous, has responded to the absence of fire and grazing. There are two important implications of these changes to this study. First, the absence of grazing improves the infiltration capacity of the slopes. Second, the trees transpire more water than the grass, providing more potential water storage in the soil. The result is less overland flow and erosion. The ridge seen in Figure 5.2 had several active hillside gullies in the 1930s, but none are active now (see the same hillside in Chapter 4, Figure 4.1). To see more forest growth on previous goat prairies, one should climb the old fire tower just east of Elba. Photos of the main valley c. 1900 are posted there, showing most slopes in goat prairie, so that one may contrast those with the mostly forested slopes at present.

*Figure 5.2* Fairwater, MN. Photos made from bluff on N. side of town (visible in Figure 5.1) looking upstream (west). c. 1905. Stream and millrace (arrow) lined with trees. Note small sand fan on right from hillside gully from valley to the right (north). (Continued)

## Similar villages

Other villages met a similar fate to that of Fairwater, but none are as well documented. One is Newton (Sec 23T13N, R6W, Vernon County, WI), which in 1896 was a flourishing village on the north fork of the Bad Axe River in Vernon County, Wisconsin (Figure 5.7). As late as 1924, there were still 13 houses, 2 schools, and 1 church there (USGS topographic sheet *Stoddard*, 1:62,500, 1924). Presently, the milldam and most traces of the village are buried, with only one house and the cemetery remaining.

Another village is Millville, Clayton County, Iowa, on Little Turkey River near its confluence with Turkey River. In 1886, it was already an established village built around a mill site (*Plat Book of Clayton County, Iowa*, 1886). By 1989, only one building was left, a once very substantial and beautiful Federal-style brick building crouched below a very high flood dike (Figure 5.8).

While flooding and sediment were obvious problems here and elsewhere, it is also probable that some similar small, relatively isolated villages have largely disappeared for social and economic reasons. Thus, it is sometimes difficult to blame the demise of a village entirely on physical factors. This can only be done on a detailed case-by-case basis, but these villages were clearly impacted by flooding and sediment.

Chapter five: The tributaries    113

*Figure 5.2 (Continued)* Fairwater, MN. Photos made from bluff on N. side of town (visible in Figure 5.1) looking upstream (west). October 22, 1940. Stream has widened by 2–5 times, millrace (arrow) has filled with sediment, and gully from the right (north) has enlarged and deepened, and has created a boulder fan into the creek. The coarse bedload from this and other sources in this reach probably have played an important role in the severe bank erosion and widening of the stream. (Continued)

Some villages and towns were able to adapt. Good examples are McGregor and Clayton, Iowa, across the Mississippi River and downstream from Prairie du Chien. Established in narrow valleys directly tributary to the Mississippi River in the 1840s, the upland areas above them went into agriculture. The first significant flood hit McGregor in 1866. An early but undersized storm sewer system ameliorated the problem until 1890 when a series of floods from the cultivated ridges above the towns sent torrents of water and sediment through town (Perfect and Sheetz, 1942). In 1916, an 8 ft (2.4 m) wall of water destroyed much of the town (Price, 1916). Since the mid-1930s, a combination of gigantic storm sewers and improvement of land management on the upland has largely ameliorated the problems.

## Other locations and functions

### Farms and farmsteads

Few villages were located along tributaries, but many farms were. Settlers of the Hill Country built not only their villages on low terraces near streams but also located many of their farmsteads there, even when higher terraces were close by. Evidently, the proximity to running water was an important locational factor to many. Indeed, Hamlin

*Figure 5.2 (Continued)* Fairwater, MN. Photos made from bluff on N. side of town (visible in Figure 5.1) looking upstream (west). 1975: channel is again narrow, lined with trees and all is stable. 2001: old gully from the north now stable (arrow on 1940 and 1975). Note trees growing on old gully bottom. This is presently typical of most old gullies in the Hill Country.

Chapter five: The tributaries 115

*Figure 5.3* Fairwater, MN looking downstream (east). Top: c. 1905. Pleistocene terrace in fore and middle background. Stream in middle background is hidden by riparian trees. Goat prairies on hillsides in background to left on south side of hills. 1940: Stream has widened and eroded deeply into terrace. As the result of floods and the coarse sediment, the stream is now braided. Goat prairies in background now have more trees (see Figure 4.1 for a close-up of the same hillsides). The same farmhouse is marked by an arrow in all photos for reference. (Continued)

***Figure 5.3 (Continued)*** Fairwater, MN looking downstream (east). 1975: stream had cut further into very high Pleistocene terrace. The arrow to the right points to the same spot on all three photos and indicates the extent of lateral bank erosion by 1940 and 1975. Thus, it can be seen that the stream had cut several hundred feet into the high terrace. Note that the high cut bank of the terrace in 1975 is partially obscured by a tree branch. See Figure 5.4. (Continued)

Garland (1917. p. 26) points out that many early farmsteads were intentionally located near streams. A good example is the Hass farm (NE 1/4 Sec 19, T10N, R6W) Vernon County, Wisconsin. Built by New Englanders along Deadman Creek near its confluence with the lower main valley of the Bad Axe River, the New England style house was located on a low terrace. By 1915, the farmstead was under fluvial siege as evidenced by the stone levee near the house (Figure 5.9). The rocky streambed in the foreground is indicative of the flashy stream flow of that period, with the large size of the boulders reflecting the occasional intense power of the peak flows. No water can be seen in the stream channel even though the creek has a watershed of about 1.5 mi$^2$ (4 km$^2$), and the years 1914–1916 were wetter than normal in the region (Trimble and Lund, 1982). Many springs and small tributaries were drying up by the early 20th century, and that result would be characteristic of the poor hydrologic conditions existing at that time. However, it is also true that the early 1900s had less precipitation than the average for 1850–1975 (Trimble and Lund, 1982; Knox, 1987). In any case, the stream now appears to be perennial and the location of the farmstead near the channel suggests that the stream had permanent flow at the time of settlement.

The house and buildings were dismantled shortly after the photograph was made in 1915, and a new house was constructed on a high terrace surface, about 600 ft to the east. I excavated the foundation of the old site in 1977 finding that it was buried in about 20 in. (50 cm) of sediment (Figure 5.9). But had the house been located only 600–700 ft (200 m) farther downstream where a fan was being constructed by Deadman Creek on the Bad Axe floodplain, the burial would have been much deeper. At that location, a barn was

Chapter five:   The tributaries                                                                                            117

*Figure 5.4* Fairwater, destruction of the high Pleistocene terrace, looking north/northwest (upstream) from hill on southeast side of the village. c. 1905: stream is narrow and tree-lined. House (arrow) is the same as marked in 5.3. Large, dark building just upstream from house is the mill. 1940: same view but taken farther back. Using the house as a marker, it will be seen that the stream had eroded laterally perhaps 150–200 ft to the left (west) and a comparable distance toward the viewer (south, see Figure 5.3). Not only is the high terrace eroding from stream action but is also eroding from agricultural runoff as evidenced by the lateral gullies. As a result of flooding and coarse sediment, the stream is now braided. The visible bridge is a replacement for a bridge similar to the one seen in Figure 5.5. A comparison of this area c. 1940 (Brown and Nygard, 1940) with the 1972 USGS topographic sheet *Elba* (1:24,000) suggests that about 4 acres (1.6 ha) of terrace were eroded away in those 32 years. Conservatively, assuming an average net removal depth of 10 ft (3 m), about 40 acre-feet (49, 400 m$^3$) of material were lost here with much being moved to the main valley of the Whitewater River. Note that goat prairie in background is now covered with cedar (same as seen in Figure 5.1). (*Continued*)

*Figure 5.4 (Continued)* Fairwater, destruction of the high Pleistocene terrace, looking north/northwest (upstream) from hill on southeast side of the village. 2001: view now shifted to north/northeast because old camera position was eroded away. For reference, note different perspective on house. Stream has cut several hundred additional feet to the west and south and is now cutting deeply into the high terrace in the immediate foreground and on which the viewer is standing.

built c. 1915 on an intermediate terrace approximately 6-ft high (1.9 m) above the floodplain. By 2009, the 6 ft (1.9 m)-high escarpment had disappeared, the floodplain having been aggraded above the level of the ground floor.

In most parts of the Hill Country, many tributaries heavily laden with sediment have built huge alluvial fans (or deltas) out onto the floodplain as seen at the Hass barn and at the Zink farm (Figure 4.4). These fans are very noticeable because of their higher elevation and often their dryness (from the coarseness of the sediment), which results in a vegetation cover quite different from the lower parts of the floodplain, which are wetter. An example is the fan or delta of Wing Creek, a tributary of Coon Creek (NE1/4, SE1/4, Section 25, &14N, R7W).

Tributaries over geologic time often created a wide embayment or "estuary-like" tributary valley where they flow into the main valley, all of which makes the main valley wider at that point. The additional depth of the fans then makes the average depth of historic sediment deeper across that wide part of the main valley.

## Roads, railroads, bridges, and communication lines

While flooding was obviously later perceived as a problem of living in Fairwater, another concern was communication. Roads close to eroding tributaries were very vulnerable and were often severely undermined and even washed away. Already seen is the effect on the bridges in town. Farther out, the two roads into the village were interrupted on several occasions. Indeed, much of the road eastward to Elba was taken out by a flood in

Chapter five: The tributaries

*Figure 5.5* Fairwater, bridge crossing visible in 5.1 and 5.2 looking upstream. c. 1905: the structure is probably slightly later than the one visible in 5.1 and 5.2 but is at the same crossing. Note stability of stream and banks. 1940: stream widened by about three times here so that right pier is near the center of the widened stream. The left pier is still visible (2006) in the village, but the recovering (narrowing) stream has buried the right pier in the new bank.

*Figure 5.6* Fairwater, road from Elba to Fairwater about 1 mi east of Fairwater, 9/38, looking west. Several hundred feet of roadway eroded away.

1938 (Figure 5.6). The annual damage to roads from streambank erosion alone during the late 1930s was about $7000 in the Whitewater River watershed and about $11,000 for the Zumbro River watershed (SCS, 1940).

Another example of a tributary undermining a public roadway is found in Trimble, 1975b (Plates 2A and 2B). Between 1938 and about 1951, Bohemian Creek, a tributary of Coon Creek (Region I), removed large areas of floodplain and terrace, eroding the base of La Crosse County Highway G (NW1/4, Sec27, T15N, R5W). In the late 1940s, a heavy timber and wire revetment was built to protect the highway, but the creek continued eroding right around it so, in desperation, the county channelized the stream away from eroding banks in 1951. The revetment still stands isolated at the edge of a pasture, one of many failed engineering devices still relict on the landscape. In this case, on a heavily traveled road, there was justification to keep the route open, but other routes in more remote areas did not always receive the same maintenance and resources, resulting in uncertainty and often the closure of some roads.

Eroding streams can undermine bridges, causing them to fail or even be washed away, as shown in Fairwater. McKelvey (1939) discusses this at length and includes a photo of an undermined tributary bridge in Coon Creek in 1938 (Figure 5.10). In some cases, substantial bridges were buried or washed away altogether (Figure 5.10). All of this makes it clear that maintaining bridges and roads on tributaries during that period was difficult. Kunsman (1944) states that most bridges in the Beaver Creek basin had been replaced several times, some from being destroyed in floods and others from being partially buried. In addition, telephone and, later, power poles were sometimes toppled by floods. The social

Chapter five: The tributaries 121

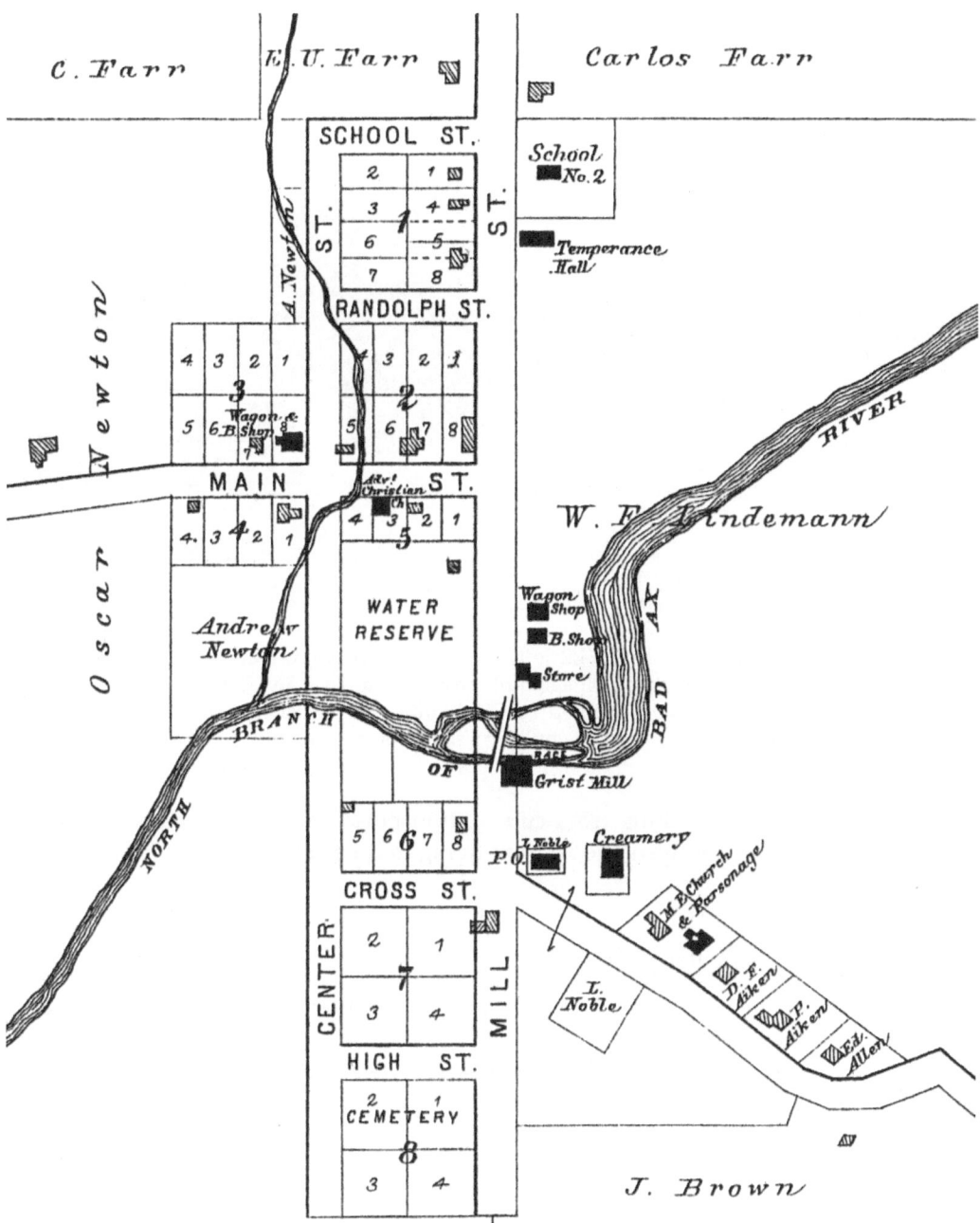

*Figure 5.7* Newton, WI, on the north branch of Bad Axe River (Sec 23, T13N, R6W, Vernon Co.), 1896 (from Hood, E., *Plat Book of Vernon County*, 1896). The mill and dam are now buried and only one original building remains. The river flows westward (right to left).

*Figure 5.8* Millville, Clayton Co., IA, on Little Turkey River near the confluence with Turkey River. In 1886, it was already an established village built around a millsite. In 1989 (top), only these two buildings were left, huddled inside a flood dike. The highway in the background is on the far dike. The nearest, a once-beautiful Federal-style building suggests the one time wealth of Millville and how long it has existed.

and economic isolation resulting from cutting transportation, power, and communication lines may have discouraged people from living in the tributary zone as much as did flooding.

## Mills, reservoirs, and water power

Most dams in tributaries typically had short lives. As seen at Fairwater, small dams and reservoirs could be swept away by the violent floods common in such channels a few decades after European settlement. There were many such dams early on, but few traces exist of them now. Indeed, at Fairwater, there is no trace of the dam.

However, one clever miller in Vernon County, Wisconsin, was able to keep his mill operating right through the critical period and even operated until 1968. Instead of creating an in-channel reservoir, he located his mill on a tributary having an adjacent stream terrace. The reservoir was then excavated from the terrace, and the mill was located on the terrace alongside the stream but above most floods. A low diversion dam was built upstream to shunt water from the stream through a canal to the reservoir (Wisconsin Dam Commission, 1913). The dam itself was expendable, being merely a low, linear pile of the plentiful rock swept down by each flood, allowing the floodwater and sediment to bypass the pond and mill. The accumulation of rock in the channel was a fortunate circumstance because the dam had to be rebuilt after each flood (Figure 5.11).

*Figure 5.9* Hass house, Deadman's Creek (NE 1/4 Sec 19, T1ON, R6W, Vernon County, WI). 1915: Creek in the foreground flowing to the left (east). House sits behind flood dike made of boulders brought down by creek. 1977: stable, grassy floodplain, stream out of view in foreground. House foundation is buried in floodplain in middle ground.

*Figure 5.10* Bridges over tributaries destroyed by flooding and sediment. Top: Undermined bridge in the Coon Creek basin, exact location unknown, 1938 (from McKelvey, V.E. 1939. Stream and valley sedimentation in the Coon Creek drainage basin, Wisconsin: M.A. thesis, University of Wisconsin-Madison). Bottom: Abutment of destroyed bridge over Trout Brook near Dumfries, MN. Zumbro River Valley, 1939 (Sec 10, T110N, R11W, Wabasha County) (SCS Photos).

Chapter five: The tributaries

*Figure 5.11* Oium Mill dam site, Timber Coulee, Coon Creek basin (N1/2, Sec8, T14N, R4W, Vernon County, WI), looking upstream from bridge on Lars Hill Road, 1919. Note typical "gravel road" appearance of stream and the almost lunar, bare quality of the heavily grazed landscape behind it. The expendable rock dam, rebuilt after every flood, diverted water into a race (to the right, not visible) that led to a reservoir excavated into a terrace downstream. This was above most floods thus saving it from destruction. The mill was about ¼ mi (400 m) downstream at the edge of the terrace (Files of Wisconsin Railroad Commission).

At least two relatively large reservoirs were built on Hill Country tributaries. One of these, Lake Como at Hokah, Minnesota (Houston County), experienced dramatic physical changes. Originally built in 1858 near the confluence with Root River, the lake covered 90 acres (36 ha) and was 20 ft (6 m) deep. Because of the lake's clarity and attractive natural setting, it soon became a regional resort with excursion trains bringing bathers and boaters from surrounding towns (Bissen, 1975). A flood in 1909 broke the dam and allowed the lake to drain. The dam was eventually rebuilt and the reservoir refilled in 1922 (Figure 5.12). But hydrologic matters were now far different: the massive erosion and sedimentation already described was under way. Thus, the lake was extremely turbid after storms, resulting in fish kills and water unsuitable for recreation. In a matter of just a decade or so, the reservoir was completely filled with sediment, thus ending Lake Como (Figure 5.12). Today, the filled lake serves as a playground for the village youth and as garden plots.

Another large reservoir was at Lanesboro, Minnesota (Fillmore County). Built on the Root River in the 1870s, it was completely filled with sediment by 1937 (Figure 5.13).

## Fish habitat

As discussed in Chapter 2, the presettlement tributaries were full of brook trout, *Salvelinus fontinalis*, which require dependable flows of clear, cool water. It will be apparent from the postsettlement changes seen above that the quality of water in tributaries was anything other than high quality by the 1920s and 1930s and thus, except in a few

*Figure 5.12* Lake Como, Hokah, MN, looking south 1926: lake full and 20 ft. deep. Note barriers in water (left middle ground) to prevent swimmers from venturing too far into deep lake. 1936: lake completely filled with sediment. Old lake bed now used by villages for recreation and gardens. (Photo credit Soil Conservation Service [SCS].)

*Figure 5.13* Lanesboro Lake, Lanesboro, MN, on the upper Root River. C. 1890: large expanse of water. 1938: completely filled with sediment. Arrow marks top of dam in both photos. (Photo credit Soil Conservation Service [SCS].)

protected locations, brook trout were extirpated (Thorn et al., 1997). With the wish to retain trout fishing, exotic brown trout, *Salmo trutto*, were then annually stocked into creeks. By that time, neither brook nor brown trout could reproduce in that environment, but brown trout could at least survive. As soil conservation measures took hold, some tributaries in the 1970s became suitable enough for brook trout to be stocked. Later, conditions improved to the point in some locations that both brown and brook trout could reproduce. Starting in the 1970s, fish structures (in-bank shelters) were installed along some reaches of tributaries, which helped stabilize streams but only slightly (Trimble, 1997). However, the installation and maintenance of such structures was contingent on the ameliorative effects of soil conservation. That is, these structures could not have survived the stream environment of the 1930s, so the improving streams and structures were synergistic in their effects. By the late 1990s, fish managers were reporting additional miles of streams where brook trout were reproducing and available for fishing. Thorn et al. (1997) state that reproducing or "wild" brook trout, almost extinct by 1900, were by 1997 found in 54% of southeastern Minnesota streams. In southwestern Wisconsin, the length of streams with "wild" brook trout increased by a factor of almost three between 1973 and 1999 (David Vetrano, Wisconsin Department of Natural Resources Fish Manager, written personal communication, March 10, 2000). The environmental reasons for this increase are (1) less sediment in streams, (2) more equitable stream flow, and (3) cooler water in streams as the result of enhanced spring flows and more shade along streams (Thorn et al., 1997). More detailed reasons are given by Vandracek et al. (2005).

One old fisherman in Elba, Minnesota, summed it up well in 1989. He said that any rainstorm in the 1930s made the streams unfishable for several days, but by the late 1980s, whenever it rained, he "reached for his fishing rod." Presently, trout fishing is a major tourist draw in the Hill Country, a generally unappreciated bonus of soil conservation.

*chapter six*

# The upper main valleys
## Zone of later complex changes of process and form

Physical changes in the upper main valleys were similar to those in tributaries, the main differences being that they came later, were less violent, and the sediment texture was generally finer. As explained in Chapter 3, these reaches were prolific net sediment sources from about 1940 until recently. As in the tributaries, the current prognosis is general stability with perhaps a mild accumulation of sediment in the future. Again, both villages and farms were severely affected.

## Villages

### Elba, Minnesota

Only about 3 mi downstream from Fairwater (Chapter 5) on the Whitewater River, the mid-valley town of Elba has had a somewhat different history and one that well demonstrates the processes in the upper main valleys. Settled about 1860, Elba was built on a stream terrace about 10 ft (3 m) above the floodplain of the Whitewater River, which probably placed it above all but the most extreme flooding for the first 60 years or so. A dam about 1 mi (1.6 km) upstream diverted water though a race to a mill in the village. With the increasing flooding and floodplain aggradation of the early 20th century, cellars filled with the rising groundwater and flooding of the village had become common by the late 1930s. Indeed, the floodplain of the river had aggraded almost to the level of the town site, and SCS (Soil Conservation Service) investigators thought that Elba would be become untenable with the 1-year flood projected to move into the village area by 1990 (SCS, 1940; Brown, 1941; see surveyed profiles of Whitewater Valley for 1940 and 1964 at Elba in Happ, 1975). Given the sediment supply from just the Fairwater area alone, that conclusion is easily understood (Chapter 5). No pictures of flooding in Elba itself were found, but the severity is shown by the conditions at the Siebenaler farm on the terrace just upstream of Elba (Figure 6.1).

The reaction of the village was to build a flood dike, which gave some relief, but the worst floods still invaded. By the 1970s, however, the stream channel, as defined by the historic floodplain, was eroding and enlarging. High banks that were 25 m (83 ft) apart in 1940 were 55 m (180 ft) apart by 1990, the process in this one reach sending about 16,500 m$^3$ (13.4 acre-feet) of sediment downstream (Trimble, 1993; Figure 6.2). Meanwhile, the old historical floodplain had aggraded about another 1 ft (30 cm) or so, probably before most of the channel erosion occurred, and a new, lower, floodplain was created within the bounds of the high banks. Thus, it will be seen that the floodway is now much larger and the danger of flooding has probably decreased. However, the biggest floods still get into the village, as happened in 2007. There is no place for the village to move since it is backed by bluffs.

*Figure 6.1* Nick Siebenaler's house just upstream from Elba, MN, on a terrace above the Whitewater River. Flood 9/8/38. Top: Floodwater was up halfway on screen door. The men are pumping out the basement. Bottom: Siebenaler's chicken house was moved off the foundation by the flood. Note sediment deposits in foreground. (Photo credit Soil Conservation Service [SCS].)

Chapter six: The upper main valleys

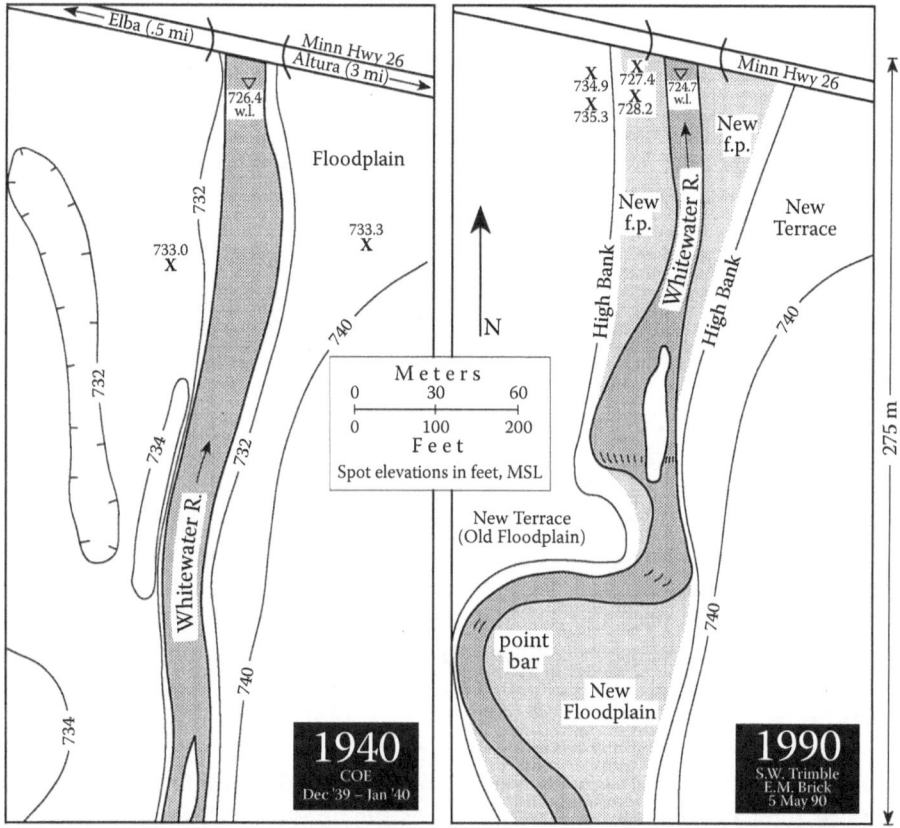

*Figure 6.2* Bank erosion and channel enlargement of Whitewater River at Elba, MN, 1940–1990. In 1940, the river flowed through a flume-like reach with high banks (the aggraded historical floodplain) to either side. Between 1940 and 1990, the floodplain aggraded an additional 1 ft (30 cm) or so and the channel was later greatly enlarged by bank erosion. This process, creating a greater floodway but sending copious amounts of sediment downstream, has been typical of the Upper Main Valley Zone since soil conservation measures took full effect. (From Ward, A., and Trimble, S., *Environmental Hydrology*, CRC, Boca Raton, 2004. With permission.)

## Coon Valley, Wisconsin

The setting of Coon Valley, on Coon Creek, is similar to that of Elba except that it was settled slightly earlier. Even though the village was built on a higher terrace, it seemed by the 1930s that it was just a matter of time before Coon Valley met the same fate as Elba. That scenario was perceived by the SCS scientists investigating Coon Creek, and they carefully mapped the village to see which areas would flood first. But something profound happened: soil conservation methods were installed and took effect in time to slow down and finally stop the aggradation of the historic floodplain. The village of Coon Valley was literally saved by soil conservation!

The situation for the lower part of Coon Valley is clearly shown by Figure 6.3. The 1912 photo of the west end of the village is remarkable because it shows the relationship of the floodplain and terrace so clearly and because it has such high resolution. Moreover, the cheese factory on the terrace surface in the right middle background was like a Rosetta Stone in allowing the measurement of heights by oblique photogrammetry. Another contemporary photo showed the same corner of the building along with standard milk cans. This allowed the determination of a precise height of the building, which in turn allowed approximating the height of the terrace face and the floodplain. The floodplain was still only 6 ft (1.9 m) higher than normal stream level although it would have already aggraded somewhat by 1912. The terrace surface in the background where the cheese factory is located was about 10 ft (3 m) higher than the floodplain.

By the late 1930s, the streambanks had aggraded to stand about 12–13 ft (4 m) above the stream level so that the floodplain was only a few feet lower than the terrace in the background, giving grounds to the fears that the village would be destroyed by floods. But by 1976, as is characteristic of this UMV zone, channel erosion was well under way (Figure 6.3). The high bank on the right was retreating from erosion, but the new, lower floodplain on the left was being built at a height of about 6 ft (1.9 m). In general terms, this morphological change suggests the stream here had by 1976 returned to the flood regime of about 1912. The presettlement bank height is not known but presumed to be considerably lower than 6 ft (1.9 m) because the flood regime was milder than in 1912.

As this process of eroding the high bank and building a low bank continued, the size of the floodway increased and the flood probability decreased at Coon Valley. Thus, within the lifetime of some of the older residents, Coon Valley had been direly threatened by the stream that probably inspired its location in the first place. But the danger is certainly less now than in the past. Indeed, a 500-year flood in 2007 flooded only the lower parts of the village on the north side, but little of the business district was threatened (Chapter 8). But all may not be well. In the 1990s, the eroding high banks were stabilized with reshaping (beveling), vegetation, and riprap. Because the high bank is now fixed in place and can no longer retreat, it means the floodway, as explained in Chapter 3, cannot expand further. Indeed, the new, lower floodplains have been aggrading, in part because their height was lowered by excavation to allow handicapped access to the water (Figure 6.4). With the high banks constrained and the new floodplains aggrading, the flood danger may now be increasing at Coon Valley. Of course, as long as the village continues to remove the new deposition from the floodplain, the flood danger will remain lower.

The recent experience at Coon Valley has theoretical as well as practical implications. As stated earlier, a stream creates a channel adjusted to its flood regime, normally capable of conveying the 1-year flood, and the bank height reflects that channel size. The present natural bank height in this reach is about 6 ft (1.9 m), but the banks at Coon Valley were cut down to about 3 ft (1 m), the purpose being to allow fishing for the handicapped from the edge of the bank. At first, this increased the floodway capacity up to the level of the historic floodplain, and Coon Valley is still enjoying some of that benefit. However, a stream should theoretically rebuild its floodplain and banks to the "correct" level and that is exactly what is happening, an almost experiment-quality verification of theory.

The practical application is obvious. If a stream is transporting sediment, especially legacy or historical sediment as here, a very neat control measure would be to lower floodplains by mechanical excavation on outside meander loops and just allow the stream to rebuild the floodplains over time, thereby reducing the stream's sediment load. All this

*Figure 6.3* Coon Valley, Coon Creek, looking upstream Vernon Co., WI. Top, 1912: photo by Lloyd Thrune. Streambanks are about 6 ft (2 m) high, and terrace face in background with cheese factory (white building) is about 10 ft (3 m) high. 1976: bank to right is about 12 to 13 ft (4 m) high and is eroding, but new bank (new floodplain) accreting to left is only about 6 ft (2 m) high, diagnostic of the newer, milder flood regime of the stream. (Continued)

*Figure 6.3 (Continued)* Coon Valley, Coon Creek, looking upstream Vernon Co., WI. Mid-1980s: photo made farther back and standing about 12 ft (4 m) higher. From this view, one can more clearly see the meandering stream with the retreating high bank to the right and the advancing low bank to the left. In the background, it will also be seen that the floodplain has aggraded to the level of the old cheese factory. As at Elba, a large floodway was being created here that would make Coon Valley safer from flooding, but that process has been artificially curtailed at Coon Valley (Figure 6.4).

could be done without having machinery actually get into the water. Sediment could be extracted from the stream with very little impact on aquatic life.

## Freeburg, Minnesota

Back across the Mississippi River in Houston County, Minnesota, is Crooked Creek, a direct tributary of the Mississippi. There are found the remnants of the village of Freeburg in a situation roughly comparable to Elba and Coon Valley. Located on a high terrace, the village was reportedly about 25 ft (8 m) above the floodplain in the mid-1880s (Minnesota Geological Survey, 1885). By 1929, Gray et al. reported that an average of 15 ft (4.6 m) of modern sediment covered the floodplain, suggesting that the village was then only about 10 ft (3 m) higher. Presently, the old terrace is hardly distinguishable from the aggraded floodplain and the 25 ft (8m)-high escarpment has disappeared (Figure 6.5). Not surprisingly, Freeburg has had many serious floods in recent decades, so its situation is actually more similar to that of Elba than Coon Valley. Only three or four buildings remain.

*Figure 6.4* Aggrading "new" floodplain in Coon Valley Village Park. Floodplains were cut down to facilitate handicapped access ramps (arrow) to Coon Creek (seen on far right). Subsequently, the stream has attempted to rebuild its floodplains to the height dictated by the present flood regime, that height being about 6 ft (2 m). The elevation of the lowered floodplain is shown by the paved walk on which the person stands, and the recent deposits are up to 2 ft (60 cm) thick, However, the formerly retreating high banks have been stabilized with rock so that the previously expanding floodway is now stable and is decreasing in size. Photo made July 2008.

## Farms and farmsteads

Most farmers in the UMV were able to find high places such as terraces on which to place their houses and farmsteads and thus kept them above flooding. But not all were that lucky, at least not with the passage of time. The Daffinrud farm about 2 mi (3 km) east of Coon Valley is a case in point. Built on a terrace about 7 to 8 ft (2.5 m) above the original floodplain of Spring Creek just as it enters the upper main valley of Coon Creek, it was safe from flooding for many years (Figure 6.6). But as the floodplain aggraded, the house came under siege and a dike was built around the house. Nevertheless, the house is still flooded on occasion. Much like the situation at Coon Valley, the stream has been artificially stabilized so that it cannot enlarge its channel, and the new floodplain continues to aggrade. Thus, the natural process that might have eventually saved the house from flooding has not been allowed to happen here.

Just across the valley from Daffinrud on the main channel of Coon Creek lies a long reach of UMV that has eroded greatly during the soil conservation period. Since the bank erosion process appears to move downstream, it is much more developed here than even at Coon Valley, only about 2 mi (3 km) downstream. Most of the reach has been "fixed" since c. 1985 by public agencies by chamfering the high banks to a lower slope and treating the bank toes with riprap and wooden fish structures. Only one "natural" reach remains at this writing, the "Marshall reach" just downstream of county roads P and G (NE1/4, Sec 5, T14N, R5W, Vernon County). I've had this reach under intense study since 1978 and as reflected in the declining elevation of the floodplain, it perhaps best demonstrates stream processes in the UMV (Figure 6.7).

Similar to most of the UMV, this reach was "flume like" with banks 10–12 ft (3.5 m) high in the late 1930s. With the reduction of sediment load, the stream used its excess energy to meander and carve away at banks on the outside of meander curves. At the same time, the stream was creating a "new," lower floodplain on the bank opposite the cut bank, the height of which was presumably adjusted to the flood regime at that time. It will be noticed that the level of the new floodplain next to the old bank, created earlier in perhaps the 1940s or 1950s, is higher. This higher elevation suggests the higher floods still extant at that time, that is, floods have been becoming milder over the past 70 years as reflected in the declining elevations of the new floodplains. This has been shown by the analysis of flood series in the Pecatonica River in southeastern Region II (Potter, 1990).

Net erosion from such reaches as this has, over the past 70 years, furnished huge amounts of sediment to be moved downstream, the amounts approximated by the depth (difference between the old and new floodplain, here about 7 to 8 ft or about 2.5 m) multiplied by the width of lateral migration times the length of the reach. It will be seen that the Marshall reach alone has furnished much sediment downstream, perhaps as much as 25–35 acre-feet (31,000–43,000 $m^3$). While it is true that such reaches of the UMV have been significant net sources of sediment in the past, that is not necessarily true at the present. As the new floodplains become wider, there is more space for deposition. So while the high banks are still eroding, the amount of sediment being simultaneously deposited over the period 1982–1993 approximated that being eroded, so there was no net loss, a negative feedback from morphological change (Trimble, 1993). In the longer run, however, the continuing difference of height between the high and low banks means that there would have to be continuing loss of material to downstream reaches (Figure 6.8). Also to be noted is the fact that the processes and form of the UMV are slowly moving downstream. The downstream-most reach of the Coon Creek UMV is shown in Chapter 3, Figure 3.14.

What are the effects of these processes on the farmer? There are both debits and benefits. The high banks being eroded away are largely fertile silt, while the new floodplains are largely less-fertile sand, a net qualitative loss. On the other hand, the newly enlarged floodway between the high banks of the old floodplain can convey all but the greatest floods. This makes the old floodplain, now a stream terrace, less likely to flood and thus now reasonably safe for cash crops such as corn and soybeans as well as for grazing livestock (Trimble, 1975a,b, 1993; Woltemade, 1994)

From the standpoint of watershed management, the effects are also mixed. While the enlarged floodway will reduce flooding in this reach, the rapid conveyance of water downstream may increase flooding downstream. "Fixing" reaches like this by public agencies as described before also has mixed results. Curtailment of bank erosion will probably reduce the sediment load moving downstream, but the resulting "hungry stream" might destabilize downstream channels already adjusted to prevailing sediment loads. And as we have noted at Coon Valley, a decrease of the cross-sectional floodway will make flooding more likely. Hence, stabilizing such reaches has questionable results. Aside from the expenditure of public funds, it might be well to remember that what is happening here is a quite natural stream process.

*Chapter six: The upper main valleys* 137

*Figure 6.5* Freeburg, Crooked Creek, Houston Co., MN. Crooked Creek Township Hall. Top: Turn of century, photo made from road. Viewer's eyes are at a level about 4 ft (1.2 m) below the bottom of wooden siding on the building. Bottom: Again standing on the road, 1980s. Viewer's eyes are level with bottom of first board below windows. This suggests that road is now about 8 ft (2.5 m) higher.

*Figure 6.6* Daffinrud house, near confluence of Spring Creek and Coon Creek about 2 mi west of Coon Valley (SE1/4, SE1/4, Sec 5, T14N, R5W, Vernon Co.). House was originally built on a terrace about 7 to 8 ft (2.5 m) above the floodplain. Top: Wedding at Daffinrud house in 1890. The crowd in front of the house makes it difficult to perform oblique photogrammetry, but the eye of the viewer appears to be below the base of the house. Bottom: Same view, 2010. The eye level is now slightly above the bottom of the windows. Note that there is a dike about 2½ ft (75 cm) high around the house on which Carl Daffinrud stands and which hides the bottom of the house. The floodplain is now almost level with the terrace on which the house sits.

Chapter six:   The upper main valleys

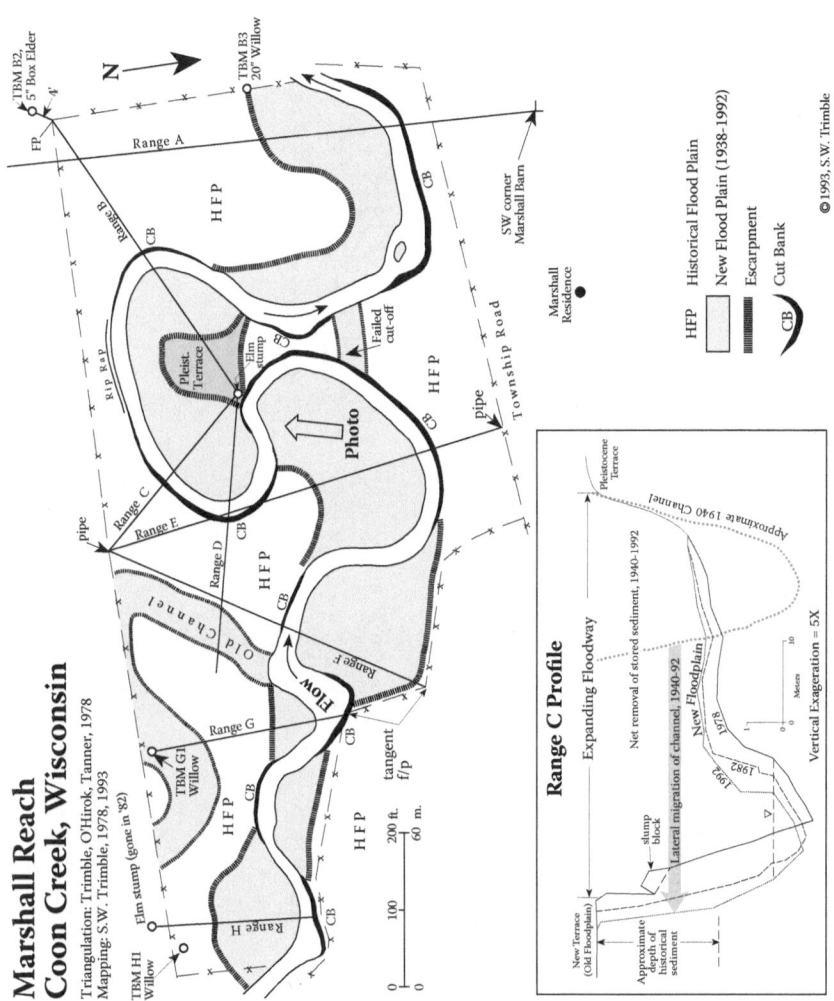

*Figure 6.7* An active agricultural reach of Upper Main Valley showing typical UMV processes and morphology, virtually untouched and unhindered by artificial modification. In 1940, the stream channel was like a flume with high banks on either side. Since then, it has cut away the high banks and left low ones as seen in Range C (inset) and the process continues. Just this short reach has furnished perhaps 20–30 acre-feet (25,000–38,000 m³) of sediment to downstream locations since c. 1940. This morphological transformation has both benefits and drawbacks to the landowner. The new, low floodplains, being sandy, are less fertile than the old floodplain, and they flood much more frequently. However, the increased size of the floodway between the high banks means that flooding on the historical floodplain is less probable. The old floodplain is now a new fluvial terrace and is now more valuable. Stream flows from left to right (westward).

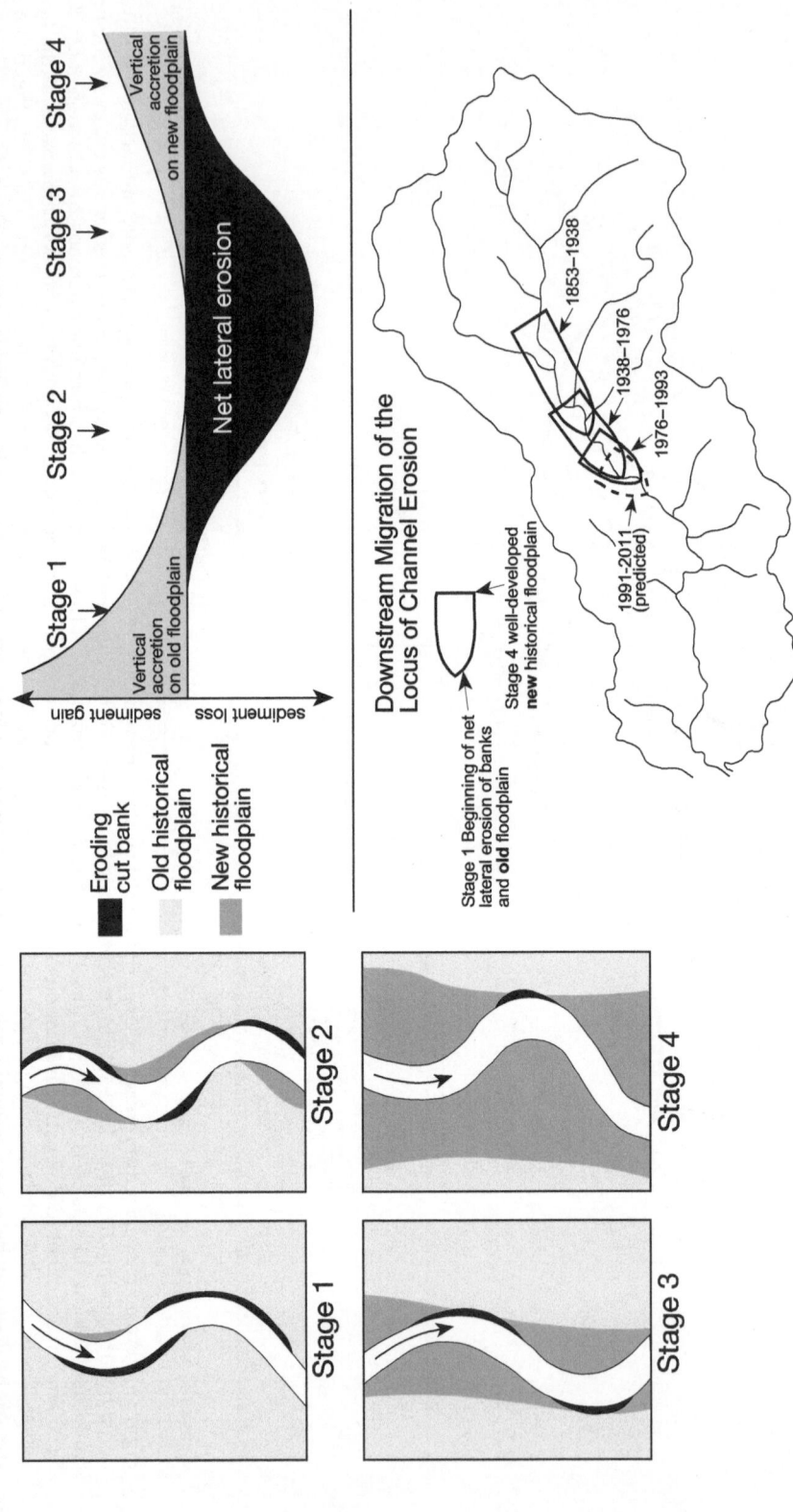

*Figure 6.8* Diagram showing how UMV reaches have developed since c. 1940 and why they might not be a long-term net sediment source: the expanding "new" floodplains may eventually accumulate more sediment than the high-cut banks lose. The inset shows how these processes move downstream. (From Ward, A., and Trimble, S., *Environmental Hydrology*, CRC, Boca Raton, 2004. With permission.)

## chapter seven

# The lower main valleys
## Zone of perennial sedimentation

While the natural history of the gully, tributary, and upper main valley zones was quite complex with alternate filling and cutting, the process in the lower main valley has been somewhat simpler: continuous accretion since shortly after settlement. To be sure, however, those rates varied greatly, but systematically. Using soil profiles, old surveys, excavations of bridges, roads, railroads, and buildings, along with archaeological methods, valley accretion rates were reconstructed for six locations in the lower main valley of Coon Creek (Chapter 3, Figure 3.15). Rates of vertical accretion increased from virtually zero at the time of settlement to about 6 in. (15 cm) per year in the 1920s and 1930s. My investigations since 1973 suggest that this chronology of deposition held over much of the Hill Country, although maximum rates differed locally. Recalling the settlement chronology from Chapter 2, it should be kept in mind that sedimentation peaks may have started a decade or so earlier to the south and perhaps a bit later to the northwest in Minnesota. While Figure 3.15 extends to only 1975, measurements for the period 1975–1993 showed that these floodplains were accreting at only about 3% of the highest rates, a tribute to soil conservation (Trimble, 1999, 2009a; Figure 3.9). Although the present average accretion rate may be low, it is important to realize that these rates are continuing, with some of the sediment migrating from storage in upstream reaches as seen in Chapter 6. The peril remains. It is a slow but sure process. With these rates as a conceptual model, we shall now see the effect on various elements of human settlement. It is in this zone that more human activity took place, and it was where much of the historical sediment accumulated, so it is appropriate to devote more attention to it. And for the same reason, much more documentation is available. Many settlements of the Hill Country were located in valleys, generally along streams on low to high alluvial terraces. Valleys provided excellent agricultural land, dependable water supplies, mill sites, and transportation routes. That the collective intelligence of villagers perceived their settlement sites to be safe from floods (and, indeed, they were safe for over a half century) gives us some prima facie evidence about the generally benign hydrologic regime existing during the second half of the 19th century.

### Villages and towns

The lower valleys are long and wide, and that is where much of the economic activity of the Hill Country was located. Thus, many villages and even towns, farms, roads, bridges, mills, and man-made lakes were subject to flooding, continuing sediment accretion, and rising groundwater. It is again to be noted that in most locations, the problems did not begin until the 20th century. These are the locations that could be identified and documented in a very large region. Evidence in some places was limited. For example, I could find no sign of a mill or mill site in many villages once noted for milling, the mill sites and shoals having been buried by historical sediment. In the town of Zumbro Falls, Minnesota, there are presently no falls or rapids to be seen. Many settlements not mentioned here certainly had severe flooding and sediment problems, but finding and documenting those is left to others.

## Chaseburg, Wisconsin

Chaseburg, Wisconsin (Vernon County, Region II), is a village that was forced to move uphill. If Fairwater, Minnesota, is the "poster child" village for tributaries, Chaseburg is the one for the LMV. It was established about 1855 on the low and intermediate alluvial terraces of an island on Coon Creek and was a flourishing town by the late 19th century (Figure 7.1). Bedrock outcropped in the streambed, creating shoals on either side of the island, and a saw mill and grist mill were built there in 1865. The magnitude and chronology of sedimentation was established from historical data on Chaseburg, especially dam inspections by the Wisconsin Dam Commission, together with two years of excavations and surveys (1976–1977) and later surveys in 1993. This chronology is approximated in Figure 3.15, and the local magnitude is shown by an archaeological cross section of the village (Figure 7.2). Perspectives on the village and the cross section are given in photographs made in the early 20th century (Figure 7.3).

There were two dams at Chaseburg creating a single millpond. The first was an auxiliary dam above a rock shoal at the head of the island on the east side of the island. It diverted most flow toward the west side of the island but had gates to sluice floodwater (and, later, sediment) around the east side. In the 1914 photograph made by the Wisconsin Dam Commission, (Figure 7.4) the sills of the bridge are about 13 ft (4 m) above the water at the foot of the shoals. The newer bridge there in 1977 sat about 1 ft (30 cm) higher than its predecessor but had only 4 ft (1.2 m) of clearance at that time. Thus, there was about 10 ft (3 m) of aggradation there between 1914 and 1977.

The main dam on the west side of the island was at the head of bedrock shoals and 6 ft (1.9 m) high, the probable height of the banks at that time. The race extended 140 ft (43 m) to the turbine at the mill, gaining another 3 ft (0.9 m) of hydraulic head or fall of the water for power (Figure 7.5). In 1932, the wooden dam was replaced by one of concrete but now relocated downstream at the mill. The hope was that sluice gates built into the dam would allow sediment to be flushed through but that failed and the dam was dynamited in 1946. By 1977, the top of that dam was 5 ft (1.5 m) below the surface of the floodplain (Figure 7.5). Presently, the only visible trace of the old mill is about 2 ft (0.6 m) of the vertical turbine shaft protruding from the floodplain right behind the Hideaway Restaurant and Inn.

Developments at the Martiny House (Figure 7.2) give a good idea of what happened to the village itself. The house was built on a low terrace in 1903 by Ole Martiny, an older, long-term resident of Chaseburg. One presumes that a half-century of observing the stream behavior would have instructed the village residents as to where flood damage would be probable, especially considering that other buildings existed, so that frequent flood damage would have been obvious and common knowledge in a small place such as Chaseburg. It thus appears that with this collective half-century of experience, Mr. Martiny believed it appropriate to locate his house on the low terrace.

Only 4 years later in 1907, a disastrous and unprecedented flood hit Chaseburg, submerging the Martiny house about 4 ft or 1.2 m (see Figure 0.2). Nothing had been experienced like this before, so Mr. Martiny decided to raise his house to the level of the 1907 flood. His reasoning was that being above the highest flood on record would preclude future flood damage. He was wrong: the flood plain aggraded, and floods became progressively higher so that the house had to be abandoned in 1925, only about two decades after initial construction

Chapter seven: The lower main valleys

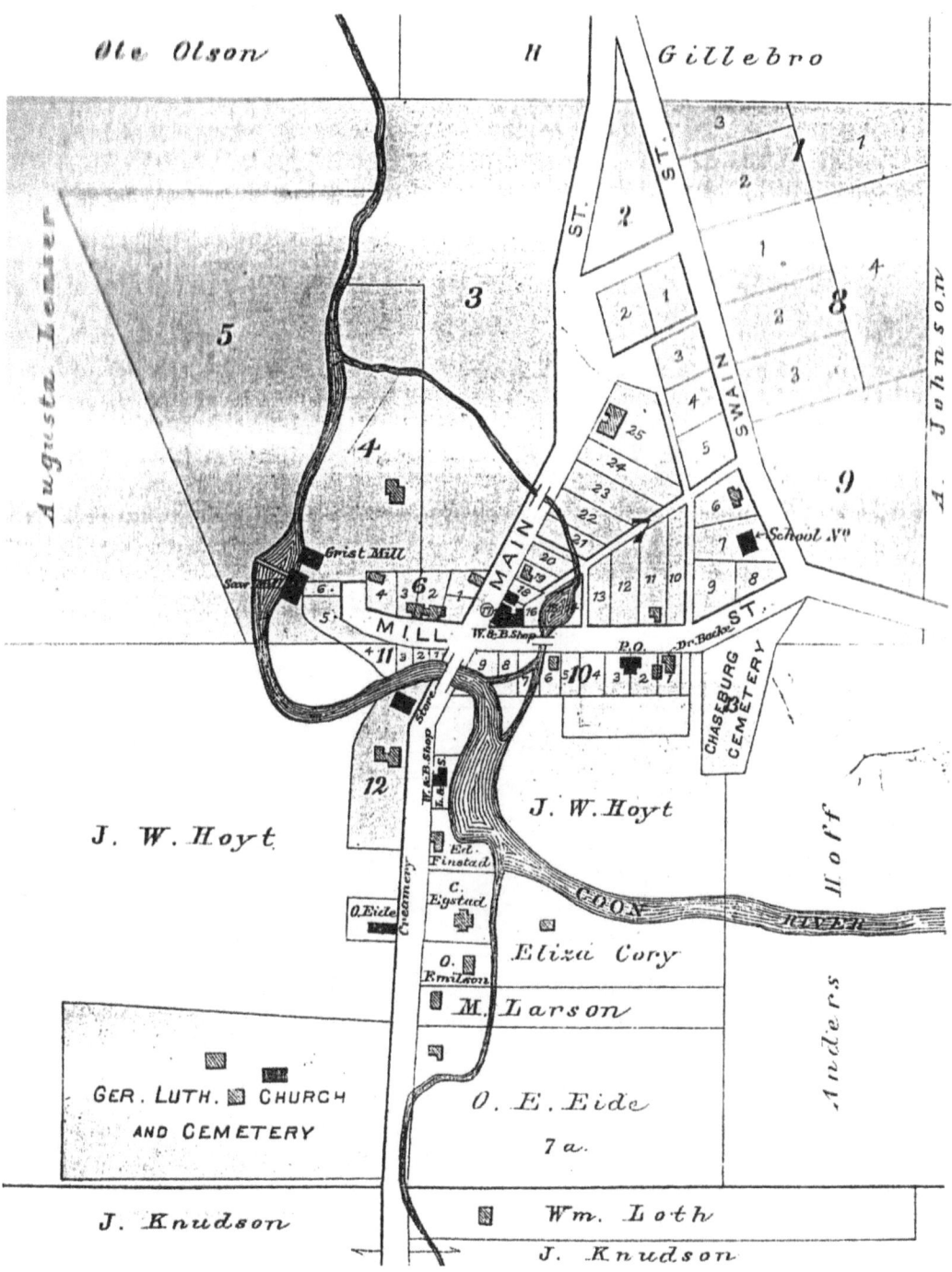

*Figure 7.1* Chaseburg, WI, on Coon Creek in 1896 (Hood, E. *Plat book of Vernon County*, 1896). Note that all buildings except two are in lower Chaseburg (on or near the island). By 2010, only one building was left in the lower part of the village. Stream flows right to left (northwestward).

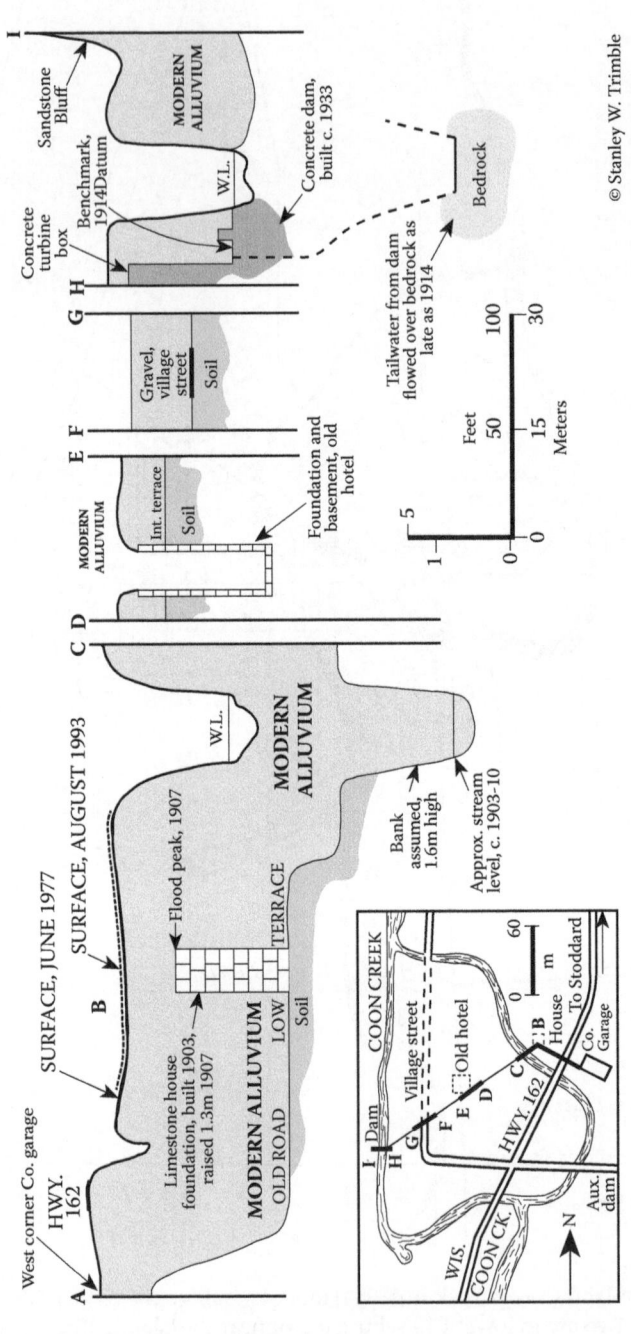

*Figure 7.2* Cross section of Chaseburg from field research and excavations, 1976–1977. The limestone foundation to the left is that of the Martiny house, built 1903. A new surface survey in 1993 indicated about 3–6 in. (7–15 cm) of additional sediment accumulation since 1977.

Chapter seven: The lower main valleys 145

*Figure 7.3* Chaseburg, refer to Figures 7.1 and 7.2. View southward across the village, c. 1905. The Martiny house west of Main Street is marked with an arrow. (Continued)

(Figure 7.6). Note that the flood stage of 1907 was about 3 ft (0.9 m) *below* the 1993 *floodplain* (Figure 7.2).

Even by the early 20th century, flooding and sedimentation problems were becoming manifest in Chaseburg (Figure 7.7). On the south side of the village, the floodplain of the small tributary from the south had already aggraded to the level of the road and buried fence posts. Only a quarter mile away to the north, the Main Street bridge over the millpond had to be raised in 1915 (Figure 7.7).

By the time we conducted our detailed surveys and excavations in 1977, much of the lower village was buried, and most of the remaining buildings had been raised to a higher position. The village hotel, a fine Italianate building, had been located on an intermediate terrace (Figures 7.1, 7.2, 7.3). The structure was abandoned in the 1930s and demolished in the 1940s. Although almost filled with rubble, water, and sediment and grown up bushes and small trees, the basement and foundation could still be discerned in the 1970s. The fact that buildings on the island had basements tells much about the lower stream levels at the time of construction and the low expectation of floods.

One older building which remained in slightly modified form to 2009 was the county road maintenance garage. Originally built about 1915 as Gardner Garage, it later became the Chevrolet dealership. There was a steep terrace face leading up to the garage door (Figure 7.8), and older residents of Chaseburg in the mid-1970s could remember this well: many of their period automobiles had been incapable of climbing the steep slope in low forward gear, so reverse gear had to be used to back up the slope. Both photos were made from the road in front of the garage, but in 1975 the elevation of the road was at least 10 ft higher than in 1915

*Figure 7.3 (Continued)* Chaseburg, refer to Figure 7.1 and 7.2. View to the northeast. Martiny house west of Main Street marked with arrow. Dark building with white trim to far left is the mill. The white Italianate building to the right of, and beyond, the mill is the village hotel. In the background is the site of the "new" town. Photo date unknown but after 1903. (Continued)

(Figure 7.8). Although the area immediately next to the building has been built up about 2 ft from the road, one then looked *down* into the garage door. The front facade had been changed, but the rest of the building remained the same. The house in the left background was built on another terrace surface about 3 ft (1 m) higher than that of the garage, but the house site was by 2008 almost level with the road and the terrace face had almost disappeared. Frequent flooding caused the county to abandon the garage, and the 2008 photo was made just after the June 2008 flood. Along with the rest of the buildings remaining in lower Chaseburg, the garage was demolished in 2009. Only the Hideaway restaurant remains, where one may see an excellent collection of historical photographs of Chaseburg and the surrounding area.

By about 1920, when it became clear that flooding would continue to be a problem in Chaseburg, most new buildings were erected in the "new town" on the high terrace surface about 1000 ft (300 m) east of the old town. Where one or two buildings stood about 1896 (Figure 7.3) now exists most of the residences and the present commercial buildings of Chaseburg. The new, higher area was platted in two successive sections known as Johnson's First Addition and Gilbertson's Addition, respectively (*Standard Atlas of Vernon County*, 1930). Most of the remaining buildings in lower Chaseburg were then raised to a higher elevation and set on modified foundations in the vain hope that would be adequate.

While flooding has generally decreased in the region, it has remained problematic in Chaseburg for several reasons. The floodplain and channel are aggrading slowly, probably

*Figure 7.3 (Continued)* Chaseburg, refer to Figures 7.1 and 7.2. View across the millpond and village looking northwest. Martiny house west of Main Street marked with arrow. Note recent sediment deposits on banks of millpond. Date: between 1903 and 1915.

about 20 in. or 50 cm/100 years (Trimble, 1999, 2009a), and, as explained earlier, some of that is coming from upstream channel erosion (Trimble, 1983). This trend and rate seems to ensure that matters will slowly but continually worsen in Chaseburg, much of the deterioration thus being the legacy from the earlier destructive period. But there are two problems specific to Chaseburg that are making floods worse. The first is that the valley narrows greatly here, making the floodwaters deeper, or perhaps faster. Second, the stream channel on the southwest side of the island was allowed to fill with sediment and become unusable by the stream. This reduced the capacity to conduct floodwaters through the village by more than half, significantly raising flood peak elevations. Why this particulary egregious condition was never emeliorated by governmental agencies remains a mystery. And on top of these problems, the increasingly wet climate of the region has brought several large storms with destructive floods in Chaseburg. So in 2009, FEMA and other agencies purchased most of the remaining buildings in lower Chaseburg and demolished them. In time, Lower Chaseburg will be simply a faded memory with perhaps a roadside historical marker.

## *Beaver and Whitewater Falls, Minnesota*

Beaver, Minnesota, perhaps the most well-known example of a buried village in the region, is already a mere historical marker. Established in 1854 by New Englanders and having a village green, Beaver was located on low-to-intermediate terraces at the confluence of Beaver Creek and the lower main valley of the Whitewater River. There were two mills within the village, suggesting considerable drop in elevation of the creek within a

*Figure 7.4* Bridge over Coon Creek on Mill Street in Chaseburg just below auxiliary dam (seen in background), looking southwest (upstream), June 1914. Note rock shoals below dam. The sills of the bridge were about 13 ft (4 m) above the water. For locational reference, note Lutheran church in background. (Continued)

short distance before flowing into the Whitewater River. The village of about 125 residents prospered for 60 years (Figure 7.9), but floods were becoming a nuisance by the second decade of the new century. With any large storm, the village was flooded first by Beaver Creek and then, a few hours later, the Whitewater itself would crest, flooding the village again. So rapid was the sedimentation of the valley and the increase in flood discharges that flooding was becoming disastrous by the 1920s. In 1938 alone, Beaver was flooded 27 times (Saari, 1956; Horn, 1972; Sillman, 1976). In one of those floods, the living room of one home was filled with sediment to the top of the keys on the piano (Siebenaler, 1955). The magnitude of aggradation at Beaver is suggested by the fact that several houses were originally built with basements, indicating the water table was at least 10 ft (3 m) deep. But by the late 1930s, some houses were buried almost to the second story floor, and the site was becoming wet. In one case, the second story floor had become the ground floor. By 1940, the village had been almost abandoned (Figures 7.9 and 7.10).

A particularly poignant portrait of Beaver's demise is found in the collections of the Winona County (Minnesota) Historical Society. In addition to photos and maps of the village, there are photographs of village 19th-century social life such as the baseball team and women's drill team. The diaspora of the refugees after c. 1940 was widespread, and I obtained some of my information in the mid-1970s from former residents then living as far

Chapter seven: The lower main valleys

*Figure 7.4 Continued)* 1975: Replacement bridge at the same site as above, looking southeast (upstream). The newer bridge sat about 1 ft (30 cm) higher than the old bridge but had only about 4 ft (1.2 m) of clearance above the water in 1977. The 2011 water level here is 10–12 ft (3–4 m) higher than in 1914.

away as California. The village site, now a swampy wilderness, was purchased by the state of Minnesota for a wildlife refuge (see H. Johnson, 1976, pp. 212–213).

About two miles up (south) the Whitewater Valley lay the hamlet of Whitewater Falls, laid out in 1856 and named for the rapids at that site (No author, 1883) (SE 1/4, Sec 27, T R Winona County, MN). The village was reported to be in decline by 1888, but 30 people remained (no author, 1883). An 1894 plat in a county atlas shows several buildings, including a post office and school. There was also a public square labeled "Franklin Square" (Foote and Henton, 1894). One of the houses, a gingerbread Gothic Revival, was reportedly moved several miles to the plateau upland (Winona County Historical Society, "Whitewater Falls" folder). There is presently no trace of either the village or the falls, both having been buried by sediment.

## Soldiers Grove and Gays Mills, Wisconsin

East of Coon Creek but flowing southward is the Kickapoo River, a tributary of the Wisconsin River. Early residents stated that the Kickapoo rarely overflowed its banks and then only during the spring thaw (Sartz, 1961). The floodplain of the main valley has been aggraded, although not so deeply as in previously discussed instances. By the early 1940s, an average of 3 ft (0.9 m) was common, but deposits were often more than 6 ft (1.9 m) thick. (Happ, 1944). Although it was uncertain as to just how much modern sedimentation had increased flood elevations, the 1935 flood covered the low terraces, on which were situated

*Figure 7.5* The main dam at Chaseburg looking upstream; refer to Figure 7.2. Top: wooden dam in 1914 at head of shoals. The race on the left extends 140 ft (44 m) to the mill, which originally gave 3 ft (1 m) more head or fall. This much fall meant that the dam was built on bedrock and that the bedrock extended at least to the mill. Bottom: Concrete dam built downstream at mill in 1932. Note sluice gate (center). (Continued)

*Figure 7.5 (Continued)* The main dam at Chaseburg looking upstream; refer to Figure 7.2. 1946. Dam has been demolished and is rapidly being buried. Note rectangular concrete pier on dam to hold flash boards, probably a delaying tactic to keep the mill going a few more months. (Continued).

the fifteen or so villages of the valley. Assuming that sedimentation had held at the 1930s rates, the village areas damaged by a 4-year return-frequency event were predicted to increase by 146% between 1938 and 1968. The number of flooded village houses was predicted to increase from 10 to 52; commercial buildings, 22 to 59; and subsidiary buildings, 28 to 89 (Happ, 1944). Channel improvements were later carried out, but flooding has continued. An upstream floodwater retention structure was designed in the 1970s to mitigate the problem, but it was cancelled halfway through construction and thus the problems remained. Many proposals to alleviate flooding have been considered, some rather heroic.

Soldiers Grove, a village of about 600 people, was built along the Kickapoo and by the late 1920s sprawled over 150 or so acres (60 ha) (*Standard Atlas of Crawford County, Wisconsin*, 1930). After more than a half century of flooding, much of the town was just moved to a higher location between 1979 and 1983.

About 8 mi (13 km) downstream from Soldiers Grove is Gays Mills (pop. 625), founded 1847. Major floods occurred there in 1912, 1935, 1951, 1978, 2007, 2008, and 2010. The 2008 flood covered 90% of the original townsite, and plans are now being made to demolish many buildings there, raise them to a higher elevation, or move them to higher ground. All such steps are expensive and will require state and federal subsidies (FEMA, 2008).

## *Galena, Illinois and Potosi, Wisconsin*

Galena, Illinois (Region I), was mentioned in Chapter 3 as being one of the first locations to be affected by historic sediment. It became a shipping point for lead about 1816 (H. Johnson, 1976; Brown, 1948) and was an important steamboat port by the 1840s (Adams, 1940, 1942).

At Galena itself, the Galena River in the "early days" was reported to be "400 feet [122m] wide and 15 feet [4.6m] deep" (Hobbs, 1939). According to a 1937 WPA guide, the river at the time of settlement was "340 (104m) feet wide and three to four feet (1m) deeper

*Figure 7.5 (Continued)* 1977: All has been buried in floodplain. The person is 6 ft (2 m) tall. (Continued)

than the main channel of the Mississippi." This would be an unusual configuration for this minor stream, and one wonders whether some sort of turning basin had been dug. The presence of so many miners gives the idea more credence, but no evidence of such an excavation has been found. A painting of the town vista in 1856 does show a wide river with several large steamboats docked (Hobbs, 1939). Based on the painting, a width of 300–400 ft (91–122 m) does not seem unlikely. However, a painting of Galena in 1843 in Bale (1939) suggests a much narrower stream. In any case, the river at Galena was being dredged by 1867 (WPA, 1937), presumably to improve navigation. But by this time, railroads were outcompeting the steamboats (WPA, 1937), and any decrease of steamboat traffic was probably general to the region. Nevertheless, navigation locks were constructed between the town of Galena and the Mississippi River in 1891 (Adams, 1942). Although these locks were removed about 1911, they probably would have increased the rate of sediment infilling at Galena. A detailed 1937 map (WPA, 1937) shows the river to still be as wide as 250 ft (76 m) at Galena, but photographs of that period show a much narrower river. Superimposition of a modern USGS topographic map (Figure 7.11) over a 1828 map shows the narrowing of the river from about 350 ft (107 m) to about 80 ft (24 m). Photographs showing the narrowing are given in Adams (1940, pp. 24–25). Presently, Galena is separated from the river by a huge flood dike.

A few miles north of Galena lay the mining village of Potosi, Wisconsin. About 3 mi (5 km) from the Mississippi River, it became a depot and enjoyed heavy river traffic until the late 19th century. But later sediment deposition made the port unusable.

*Figure 7.5 (Continued)* 1977. Rectangular pier to hold flashboards (seen in the 1946 photo), excavated 1977. An elevation placed on the dam in 1932 tied into earlier surveys by the WI Dam Commission and allowed reconstruction of village elevations back to 1914 (Figure 7.2).

## Elkport and Garber, Iowa

Elkport, Iowa, at the confluence of Elk Creek and the Turkey River was a thriving community by 1886 (Figure 7.12). By 1905, the Iowa Geological Survey reported that about 6 in. (15 cm) of sediment had been deposited on the floodplain there. Flooding and sedimentation continued, and most of the village eventually had to be abandoned. Of the old village, only the Lutheran church, sitting on higher ground, still exists. The ruins of the rest of the village, including the Catholic church and school, now lie under a large cornfield (Figure 7.13). Local farmers say they sometimes plow up parts of the old buildings. The entire village is now surrounded by a flood dike.

*Figure 7.6* Martiny house, west of Main Street, Chaseburg, refer to Figure 7.2. Top: Excavation of foundation, August 1975. The vertical pipe marks the spot. Bottom: Base of excavated foundation, August 1975. Flag on left marks base of foundation. Flag on right marks contact of old, dark Mollisol and the lighter, highly stratified historical sediment deposited since 1903.

Chapter seven:  The lower main valleys

*Figure 7.7* Early harbingers of a dark future for lower Chaseburg. Looking northwest at the cheese factory on south side of town, early 1900s. Floodplain of tributary from south in foreground with buried fence posts by road (now County Road K). Chaseburg cheese factory to left. For reference, Lutheran cemetery in right background. Bottom, raising the Main Street (now County Road K) bridge over the millpond on south side of village, 1915 presumably, to place it above floods. This channel is now completely occluded.

*Figure 7.8* Top: Gardner's garage (later Vernon County Highway Dept. garage) under construction on top of terrace, 1915. This was directly across Main Street from the Martiny House by the bridge. Viewer stands on road (now WI Highway 162), but eye level is well below the floor of the garage. The house in the background sits on a terrace about 3 ft (1 m) higher than the base of the garage. Bottom: County highway garage, 1977. Façade had been changed, building remains the same. Viewer stands on road, eye level far up on the building. More than 2 ft (60 cm) of building are buried in fill. Approximately 10 ft (3 m) of stream and floodplain aggradation have occurred here, and the road has been raised accordingly (see Figure 7.2). Note that the floodplain had built up the level of the terrace on which the house in background sits. (Continued)

*Figure 7.8 (Continued)* July 2008. Two large floods had passed though in the preceding 10 months and the building was abandoned and derelict.

Directly across the Turkey River from Elkport is the village of Garber, once called East Elkport. Its history appears to be much like that of Elkport (Wilder, 1906).

## Village Creek, Iowa

Village Creek, Iowa (Allamakee County), was founded in 1857 and seemingly destined to become a significant industrial community (Figure 7.14). By 1862, there were two or three mills here (Hancock, 1913). By 1865, a 3 1/2 story stone woolen mill with 15 employees was operating (Alexander, 1882). By c. 1880, three other mills were also operating, and the village population was 167; but later, flooding was a constant problem (Hancock, 1913). By 1913, only one mill still operated (Hancock, 1913). Village Creek is now a ghost town, and there is no sign of the several shoals and mill sites that inspired its location; they are presumably buried. The only old building still standing is the very substantial stone school, built in 1860 (Figure 7.15). Despite being located on the south side of the village on what was apparently higher land, it has been buried about 4 ft (1.2 m) so that the windows are almost at ground level.

## Ion, Iowa

Ion was one of the earliest villages created in the Iowa Hill Country. By 1886, there were several buildings including a church, school, and mill with a population of 55 in 1880 (Alexander, 1882). Photographs from the early 20th century show two mills, several

*Figure 7.9* Beaver, MN, on Beaver Creek, Whitewater River valley, Winona County, looking upstream Beaver Creek flows right to left (eastward). Top, Beaver in 1898. The downstream millpond is visible in the right foreground. Bottom: Repeat of previous shot, 10/15/40. Only two buildings remain. Presently, only a forested valley floor can be seen from this photo site.

Chapter seven: The lower main valleys

*Figure 7.10* A partially buried house in Beaver, 11/28/39. See H. Johnson (1976, pp. 212–213) for another example).

*Figure 7.11* Galena, Ill. Overlay of modern topographic map on 1828 map showing how the Galena River has narrowed from the influx of sediment. As late as 1850, large steamboats could turn around in the river. Early map from the Wisconsin State Library and Archives.

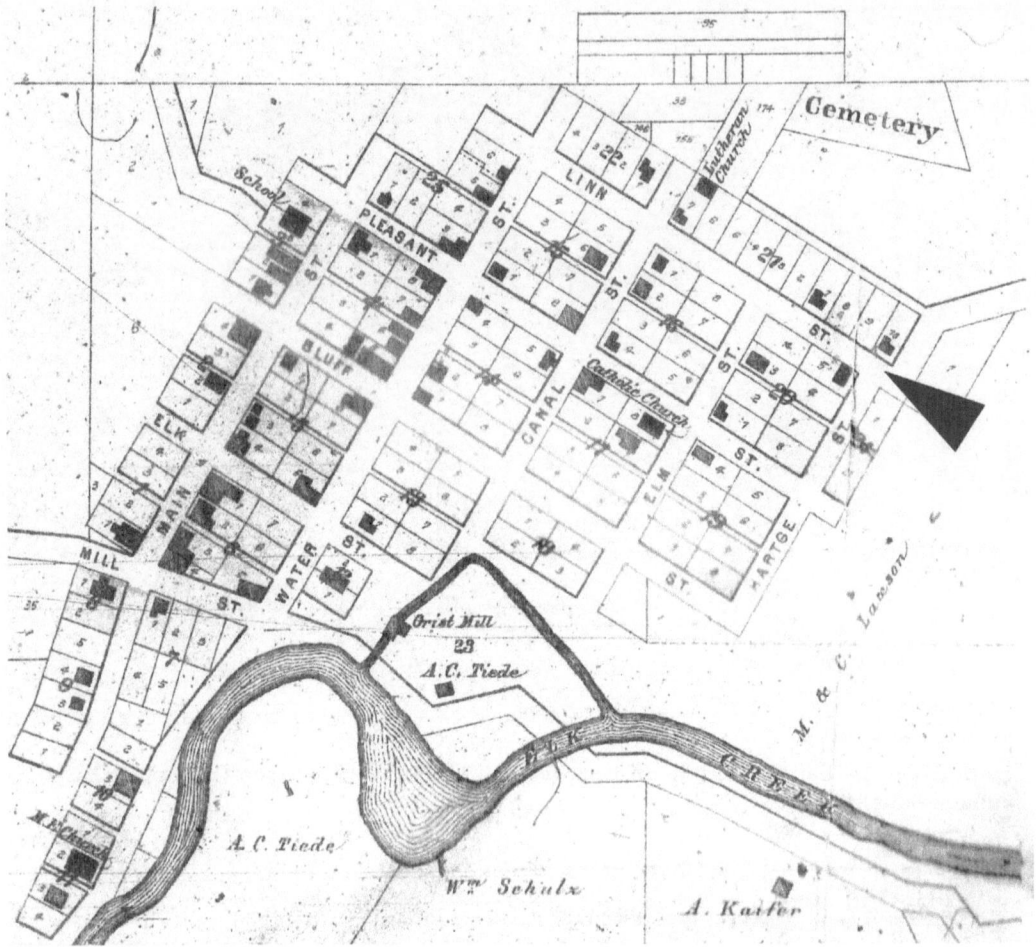

*Figure 7.12* Elkport, Clayton County, IO, 1886. At confluence of Elk Creek and the Turkey River. Arrow marks Linn Street and photo site in Figure 7.13. Note Lutheran church farther down the street on right and Catholic Church and other buildings off to the left. (*Plat Book of Clayton Co, Iowa*, 1886). Stream flows left to right (eastward).

buildings, and a substantial bridge. Flooding became a problem in the early 20th century, one in 1916 so damaging that the town never recovered. There is very little trace of the old village and of the shoals on the Yellow River.

## Rushford, Minnesota

The Root River has been problematic for a long time with flooding and sedimentation. As a result, most towns and villages as well as much farmland along its route now crouch behind flood dikes. Most notable are Peterson, Houston, and Rushford, Minnesota. In addition to having flood dikes, the river and its tributaries near Rushford have been greatly modified, particularly by straightening.

But early on, the Root River was heavily utilized. Indeed, there was a veritable water-powered industrial park south of the Root River in what is now South Rushford (*Illustrated*

Chapter seven: The lower main valleys 161

*Figure 7.13* View down Linn Street in Elkport in 2009 (see arrow in Figure 7.12). The Lutheran church on higher ground to the right is the only building left. The remains of the other buildings lie beneath the sediment on the field to the left now growing corn. The whole village is now protected by a flood dike from which this photo was made.

*Historical Atlas of the State of Minnesota*, 1874). Now, there is only a featureless floodplain to be seen there, most everything having been removed or buried (Figure 7.16). However, if one looks about 500 ft (150 m) east (downstream of the south end of the causeway) across the Root River floodplain, one will find a derelict early 20th century Art Deco–style hydro-electric power station (Figure 7.16), the only vestige of the water power development at this location. There is no longer a canal leading into and away from it, indicating the massive sediment accretion that has occurred there.

## Arcadia, Wisconsin

The migration of sediment continues, and the lower portions of the LMV are still aggrading. In Arcadia, Wisconsin, at the confluence of Turton Creek and the Trempeleau River (Trempeleau County), basements have become increasingly wet during the past three decades, an indication of rising groundwater. My investigation indicated that migrating sand has aggraded the stream channels of both Turton Creek and the Trempeleau River, thereby raising local groundwater. In some places, stream levels are almost level with the streets and yards of the city (Figure 7.17). Thus, the physical model given in Chapter 3 is not only explanatory, it is also predictive; the downstream portions of the LMV zone can

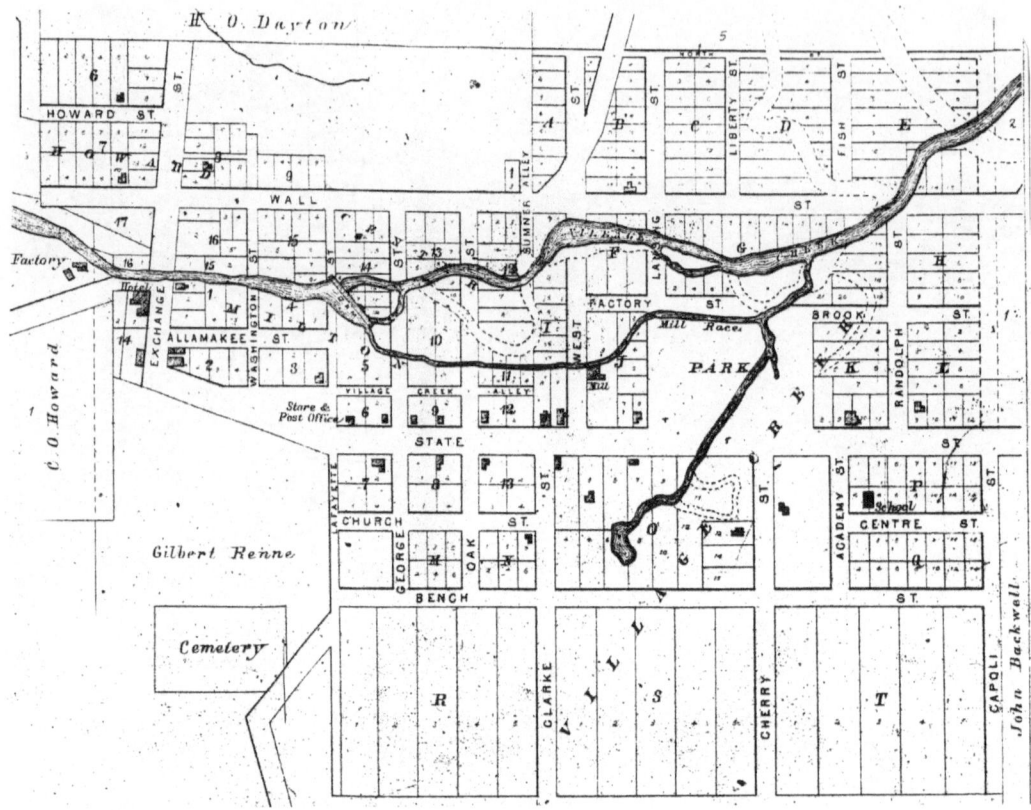

*Figure 7.14* Village Creek, Iowa, on Village Creek about 10 mi (16 km) southwest of Lansing, Iowa, 1886. Note the number of water-powered developments there (*Plat Book of Alamakee County, Iowa,* 1886). There is presently no trace of any of these dams, factories, and canals, all presumably buried by sediment. Stream flows eastward (left to right).

expect continued aggradation of stream channels and floodplains with the attendant problems for decades to come.

Arcadia is a perfect example of a place "trapped in its regional history" as Church and Slaymaker (1989) eloquently put it. Continuing aggradation here from legacy sediment means rising groundwater table, flooded basements, and more danger of flooding. Dikes have been built around the city, and large pumps continuously remove ground and surface water from the city. The only real remediation attempted so far has been to pump sand from the Trempeleau River downstream of Arcadia, thus lowering the local base level, but that cannot be sustained.

## Farms and farmsteads of the LMV

Intermediate terrace surfaces were often good locations for farmsteads as well as for villages. Most were high enough to escape burial, but there are scores of farmsteads that disappeared or just barely survived. As at Coon Valley, many farmsteads throughout the region may be presently observed where the sediment has built up to within a few feet or even inches of the terrace surface. And as with villages, those farmsteads now in the UMV are generally safe for the present, while those in the LMV face further problems.

Chapter seven: The lower main valleys 163

*Figure 7.15* Village Creek. The schoolhouse, built in 1860, is the only old building left. Although in a higher part of the village, it has been buried 4 ft (1.2 m) so that downstairs windows are at ground level. (Photo credit: Nadine Kleinhenz, 2006.)

An excellent example of a farmstead impinged upon by an aggrading floodplain is the very handsome Jacob Roth house on Crooked Creek, Minnesota. A beautiful drawing of the farm was included in an 1874 atlas (Figure 7.18). It shows the house sitting high above the floodplain as evidenced by a high stone wall in front of the house and by a tributary at an even lower level with a waterfall. Scaling from drawings is always problematic, but it suggests that the house sat 14–15 ft (4.5 m) above the lowest level of water shown in the drawing, which would approximate the level of the floodplain. The same drawing shows a train going up the valley on the new railroad (Caledonia branch of the Chicago, Milwaukee, and St. Paul), which was completed in 1871 (Hancock, 1913). Crooked Creek has since aggraded deeply, burying most of the railroad embankment and almost destroying the village of Freeburg 5 mi (8 km) upstream (Chapter 6). A photo of the Roth house made in 2009 looking across the floodplain from the public road suggests the floodplain has aggraded to near the level of the house since 1874, suggesting 14–15 feet (4.5 m) of accretion here (Figure 7.18).

Farmsteads in the lower Whitewater Valley of Minnesota were often located on a low terrace about 5 ft (1.5 m) above the floodplain that occupied much of the valley (SCS, 1940). Such terrace farmland was especially valuable because it was fertile, easily worked, level, and close to the farmstead. As the sediment accumulated on the floodplain to the level of the terrace, the terrace surfaces became part of the floodplain so that flooding was not only higher but more widespread. Thus, farmsteads became more susceptible to the sediment and hydrologic problems already described. By 1941, the homes of 90 farm inhabitants were subject to inundation, and the farmsteads of 50 farm

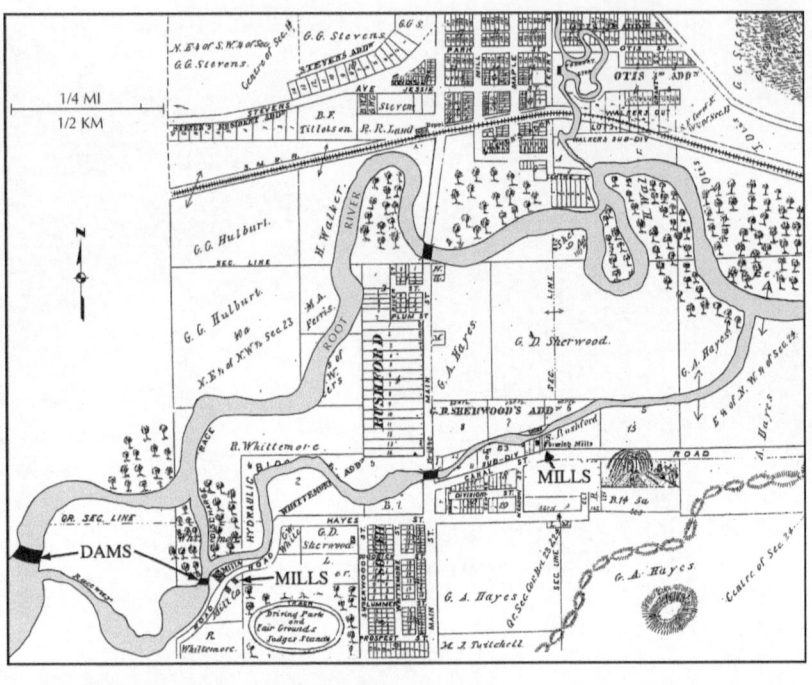

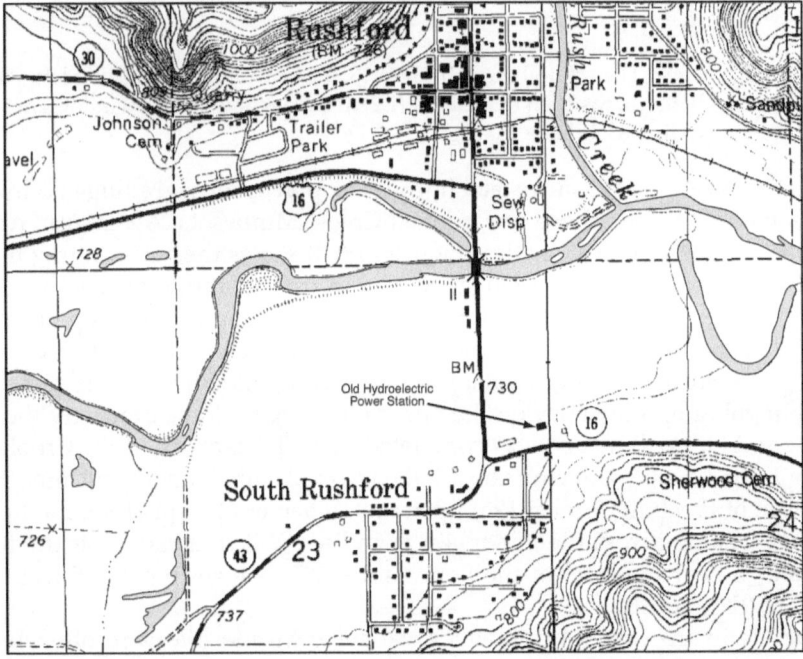

*Figure 7.16* Rushford, MN, Root River. Top: 1874. Stream flows eastward (left to right). Note the large "industrial park" of canals, dams, mills, and houses south of the Root River (*Illustrated Historical Atlas of the State of Minnesota*, 1874). 1980s (USGS 1:24,000 topographic maps): Note the straighter configuration of both the Root River and Rush Creek coming from the north and running through North Rushford. There is now little trace of the old industrial area, it being buried, but there are the remains of a hydroelectric power station about 500 ft (150 m) east of the south end of the present causeway. Styled in Art Deco, the power station was probably built in the early 1920s.

*Figure 7.17* Arcadia, WI. Turton Creek at an old mill site, looking downstream towards the Trempeleau River, 2011. The stream flows along the north side of the town. Note that the water surface of the creek at normal flow is presently almost as high as the land on the other side of the flood dike. The dike to the left is about 4 ft (1.2 m) high. The fact that most of the houses in town had a basement suggests that the original water level was at least 10 ft (3 m) lower. Even the level of the millpond here would have been considerably lower than the present level of the stream. Otherwise, it would have flooded the areas to either side and would have raised local groundwater levels. The higher present stream levels, including the Trempealeau River, just 2 mi (3 km) downstream from this site, have caused groundwater to rise and flood basements in the town.

families were scheduled to be acquired as public land (SCS 1942b). H. Johnson (1976) reported that 95 farmsteads eventually went to public acquisition, but it is not clear whether all of these purchases were due entirely to the hydrologic problems described here. However, there is no doubt that many farmers in the Whitewater were driven from their homes. An eyewitness to the Whitewater exodus says that 75 families were forced to leave and described it thus: "The people were actually driven out the same as an invading army would cause people to evacuate their homes. They were driven out by the farmers who were living on the uplands and they were helpless to do anything about it" (Siebenaler, 1955). Although he lived farther upstream in the UMV, Siebenaler himself was a victim and his family home is shown in Figure 6.1. The excellent quality of the homes in the Whitewater may be judged by the Siebenaler home and by the Appleby house (Chapter 4). Stafford C. Happ, who grew up on a New York farm and first visited the Whitewater Valley c. 1935, told me in 1975 that in his view, the farms in the Whitewater Valley were the most beautiful in the entire Hill Country. But all over the Hill Country in the LMV, house sites were buried or so affected by sedimentation and flooding that they were demolished.

*Figure 7.18* Drawing of the residence of Jacob Roth, Crooked Creek (NW ¼, Sec 27, R4W, T13N, Houston County, MN, from *Illustrated Historical Atlas of the State of Minnesota*, 1874). Note the stone wall in front of the house which, in comparison to the house and the figures, was about 9–10 ft (3 m) high. Note further that the stream in the foreground is at least 5 ft (1.5 m) lower than the base of the wall. For this water to flow out, the floodplain (to the left or south) would have had to be even lower. Thus the house sat at least 14–15 ft (4.5 m) above the floodplain. Also note the train in the distance going up the valley. The tracks to Caledonia were completed in 1871. (Continued)

Besides damaging and burying buildings, another problem in all valley zones was that such floods also destroyed crops and covered pasture and good soils with often less fertile modern alluvium (Trimble and Lund, 1982; Figure 7.19). In a region so dependent on dairying, the loss of pasture meant an immediate decrease in farmer income from the local creamery or cheese factory. In some cases, farmers could replant crops, but the wet soils delayed replanting and made it more difficult. If a flood came too late in the relatively short growing season of the region, the crop was simply lost. In some years, there was more than one large flood during the growing season, and second crops were wiped out also. Areas near streams were sometimes covered by deposits of sandy material with little agricultural value. Such land was usually removed from production.

*Figure 7.18 (Continued)* View of the Roth house from the public road across the floodplain, 2009. The level of the floodplain appears to be close to the base of the house, suggesting aggradation of more than 10 ft (3 m) since 1874.

Livestock were also vulnerable to floods. Large animals such as cows and horses could often seek higher ground or swim to safety, but more tightly penned animals such as chickens and pigs were more at risk (Figure 6.1).

Additionally, much floodplain area was swamped as streams aggraded both their channels and banks, often leaving the more distal floodplain relatively low and with poor drainage: such areas became swamps and even ponds (for example, see the Appleby house, Figure 4.3). The floodplain shown in the top photo is now a lake). In Winona County, Minnesota more than 27% of alluvial soils was idle by 1940, mostly because of sedimentation and attendant problems (Brown and Nygard, 1941). Some farmers dug drainage ditches to counter swamping, but continued deposition of sediment often blocked these ditches, creating more work and expense.

Another major problem was destruction and burial of fences. The frequent floods and debris often destroyed long stretches of fence (Figure 7.19). With the high rates of sediment accretion prevailing in the 1920s and 1930s, it was not uncommon to find three generations of fence posts along the same fence line (Figure 7.19). The continual replacement of fences was a major problem and expense for many farmers, especially in the LMV. A similar problem from floods was the loss of farm bridges. These bridges were usually the only access a farmer had to his usable bottom land fields, so the loss of access meant the loss of the crop.

Note that the foregoing discussion of damage to agriculture on LMV floodplains is primarily related to flooding and deposition by streams. Many areas on the distal floodplains near upland slopes were also subject to gully fans (see Chapter 4, especially the

*Figure 7.19* Floodplain covered with sandy sediment with buried and damaged fences, Whitewater River Valley, 1940. Note three generations of fences in lower photo. (From Trimble, S. and Lund, S., *U.S. Geol.Surv. Prof. Paper* 1234, 1982.

Zink farm and Figures 4.7. 4.15, and 4.18). The fans generally ranged in texture from sand to boulders. Sandy fans were the largest, with one fan along the Whitewater River covering 68 acres or 27 ha (Happ, 1940). But what both types of fans had in common was that they greatly restricted or even precluded the use of land for agriculture. Land seriously damaged or destroyed by coarse sediment deposition was by c. 1940 estimated to be about 400 acres (160 ha) in the Whitewater watershed and about 1300 acres (520 ha) along the Zumbro River, just to the north of the Whitewater.

The burial or swamping of floodplain cropland and pasture put new pressure on uplands. For example, the story of Appleby in the Whitewater Valley is told in Chapter 4: by the 1930s he could no longer use his floodplain pasture because of flooding and sediment. So he turned his animals loose on the slopes above his house, destroying the hillside by gullies and partially burying his house.

## Roads, bridges, and communication lines

Roads and railroads were frequently covered with sediment or damaged by floods. In the upstream zones, as shown in Chapters 4, 5, and 6, erosion of roads was the main problem. Reaches were sometimes covered by overbank sediment and fans, but it was usually local, with at most a few hundred feet of road affected. But in the LMV, long reaches of road and even highways were buried, in some cases up to 10 ft (3 m) deep, as seen in Chaseburg and elsewhere. And of course, bridges have been buried to similar depths.

The LaCrosse and Southeastern Railroad, opened in 1904, ran from Stoddard to Westby, Wisconsin, along the Coon Creek Valley. Because the hydrologic problems were just getting under way, it was not a good time to build a railroad down a stream valley, and problems were evident within a decade (Figure 7.20). After only 30 years, the line was abandoned in 1934, in part from escalating costs of maintaining the right of way caused by sediment and frequent floods. A reading of the records of the railroad in the archives of the Wisconsin Railroad Commission showed constant bridge repairs, but whether the culprit was floods and sediment was not always clear.

But perfectly clear was the sedimentary history: much of the railroad bed, normally elevated 5–15 ft (1.5–4.5m) above the floodplain, was vulnerable to small floods by the 1930s and buried by the 1970s. A valley profile about halfway between Chaseburg and Stoddard (Figure 7.20) shows that the railroad embankment was originally built on the top edge of a terrace escarpment, the surface of which had been about 10 ft (3 m) above the floodplain at the time of settlement. Evidence at two other sites, 2–5 mi (3–5 km) downstream, respectively, indicates the floodplain had already accumulated about 2 ft (0.6 m) of historical sediment accretion in 1904 when the railroad was built (Figure 3.15, "1904" on profiles). Thus, with the added height of the embankment (3 ft or 0.9 m), the railroad, when built in 1904, would still have been over 10 ft (3 m) above the active floodplain (Figure 7.20). By 1938, the floodplain had aggraded a total of about 11–12 ft (3.5 m), bringing it to within a few inches of the embankment top. This suggests a precarious existence for the railroad during its twilight time of the early 1930s. By 1974 when I resurveyed the profile, the embankment was completely buried and there was no visible sign at this site that the railroad had ever existed (Figure 7.20). A similar situation existed in the Crooked Creek Valley across the river in Minnesota. The Caledonia branch of the Chicago, Milwaukee, and St. Paul Railroad was completed in 1871, but there are few signs of the railroad embankment or bridges to be seen in the valley today.

Not only were valley railroads greatly affected by sedimentation, they sometimes enhanced the process. In order to maintain a constant grade, railroads were often routed

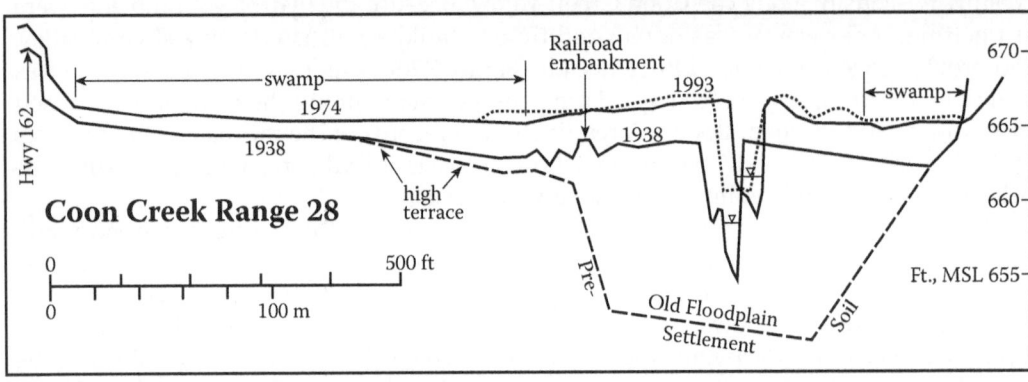

*Figure 7.20* La Crosse and Southeastern Railroad, Coon Creek Valley, built 1904. Top: A railroad bridge seen near Chaseburg in the early 20th century, probably before 1915 based on the apparel. Note that the top of the rails is about 7 ft (2.2 m) above the channel, giving a small bridge opening. Note also woody debris from flooding partially occluding the bridge opening. Bottom: Surveyed cross section between Chaseburg and Stoddard. Railroad embankment was built here on edge of terrace about 10 ft (3 m) above the old floodplain. By 1938, the embankment was only 1–2 ft (30–60 cm) high. By 1974, it was completely buried. Note also the highway (left or north side of the profile), which had to be raised to stay well above the aggrading floodplain of Coon Creek (NW1/4, Sec. 29, T14N, R6W).

from one side of the valley to the other, necessitating many bridge crossings of the stream. To some degree, each bridge slowed the flow of water and thus increased the deposition of sediment. And as shown in Figure 7.20, woody debris carried by floods collected along bridge openings, turning bridges into dams, further enhancing aggradation of the stream and floodplain. To make matters even worse, some railroad companies installed bridges with small stream openings that restricted streamflow and caused deposition of sediment (Gray et al., 1929). And, of course, aggradation of the stream made the bridge openings smaller over time.

In a few cases, heavily used railroads and bridges were later raised above the aggrading and frequently flooding floodplain. An example was the Chicago, Milwaukee, St. Paul, and Pacific railroad right of way along the Root River in Minnesota, but the line was later abandoned. In another effort to prolong use of rights of way, railroad companies also modified stream channels by diverting and straightening, often to little avail.

Railroad and road embankments built along stream valleys often had an effect, unanticipated by the builders but fully expected by fluvial geomorphologists, in that they acted as dikes. Because until they were covered by the aggrading sediment they were at a higher elevation than the floodplain, they often tended to hold off sediment-laden floodwater from the distal floodplain. There are several such locations on the north side of the Coon Creek Valley, but the one most easily seen is on the Hunder Coulee Road (Hilltop Road) where it crosses the old railroad embankment (SW1/4, SW1/4, Sec 25, T14N, R7W, Vernon County). On the stream (south) side, the Coon Creek floodplain has accreted to the height of the railroad embankment, but there is a deep pond on the other (north, or distal) side where floods and sediment could not easily reach. Another such pond is located alongside State Highway 162 just a half mile east of Hunder Coulee Road. The pond was created by the railroad embankment as described before. Later, the Wisconsin State Highway 162 here was placed on the old railroad embankment, creating an even higher dike. Such ponds have deep soft mud on their bottoms, but below that at most places lies the more dense floodplain extant when the dike was built. Such locations are like a view into the past landscape.

There are many buried highways in the LMV zone of the Hill Country. One may be seen in a profile surveyed across Coon Creek (left or north side of Figure 7.20, bottom). Between 1938 and 1974, Wisconsin State Highway 162 had to be raised about 2 ft to stay well above the aggrading floodplain of Coon Creek. A particularly striking example is Minnesota State Highway 74, which ran along the west side of the main valley of the Whitewater River (Figure 7.21). The narrow concrete roadway was built in 1918 on a constructed embankment about 6 ft (1.9 m) high, the latter being necessary because extreme flooding was already evident and a route across high terraces was not available. By 1939, the floodplain had aggraded to the level of the road and, where alluvial fans debouched near the road, it was often well below the surface, in some places several feet (Siebenaler, 1955; see Chapter 4). Precise resurveys in the Whitewater Valley in the early 1990s by Robert Bird and William Lorenzen of NRCS showed that the present road has been built up as much as 8 ft (2.4 m) since 1938, but it is barely higher than the floodplain in many places so that any flood (overbank flow) stops traffic.

This highway along the Whitewater River has acted as a dike in some places much like the railroad embankment along Coon Creek described earlier. In surveys done in 1938, 1964 and 1991, the floodplain on the stream side of Highway 74 is in many places higher than the west (upland or distal) side. The wet areas to the west side of the road have been induced in part by this road, beginning in 1918 when it was constructed. An additional problem is that the road acts as a dike preventing water flowing from the distal hillsides

*Figure 7.21* Minnesota State Highway 74 along the north and west side of Whitewater River on the floodplain. In an attempt to escape the increasing flooding and sediment accretion, it was built in 1918 on a 6 ft (2 m) high embankment with a narrow concrete roadway. By the late 1930s, the floodplain had aggraded to the level of the road or even above it. Top, 6/2/38 near Beaver. Bottom: Reach of Highway 74, probably near Beaver. The floodplain had aggraded above the highway, which had to be cleared.

Chapter seven: The lower main valleys 173

*Figure 7.22* Minnesota State Aid Highway 9 on the south side of the Whitewater River, 11/6/37. Bridge in right foreground was covered by 18 in. (45 cm) of sediment by June 1938. This road, the lifeline for people on the south side of the river, was becoming impassable by the late 1930s. By 1989, it was underwater, impassable even on foot. (Photo credit Soil Conservation Service [SCS].)

from reaching the stream. Indeed, the problems at the Appleby farm (Chapter 4) were probably brought on in part by this highway blocking the lateral movement of sediment to the distal parts of the floodplain and also blocking the movement of water from the distal floodplain to the stream. Thus, the Appleby land became wetter as the Whitewater River aggraded. On the south side of the lower Whitewater Valley, State Aid Highway 9 was usually passable with some difficulty as late as the 1930s but had already been rerouted around the mouth of Trout Creek where it had been buried (SE ¼, Sec 31R9W T109N) (SCS, 1940). By the late 1930s, a 400 ft (120 m) stretch of the road had been buried in a fan over 12 ft (3.6m) deep (NW ¼, Sec 6, R9W T108N) (SCS, 1940). The old road is now covered by a swamp with ponded water and is impassable even by foot (Figure 7.22).

Many other state highways have had to be raised to keep them above the floodplain and above flooding. Already seen is State Highway 162 passing through lower Chaseburg, which is at least 10 ft (3 m) higher than in the early 1900s (Figure 7.2). About a mile (1.6 km) upstream from Chaseburg, the highway sits on a terrace that was once about 11–12 ft (3.5 m) above the floodplain. Since 1938, the road has been raised 2 ft (0.6 m), which now places it only about 3 ft (0.9 m) above the floodplain. The road to Halsey Coulee (now Vernon County Road KK) once crossed Coon Creek about 100 yards upstream from the present crossing, but that bridge was buried by the 1930s (McKelvey, 1939) and I excavated part of it in 1976. About a mile (1.6 km) farther east, the Dodson Hollow Road

*Figure 7.23* The south abutment of an old Dodson Hollow Road bridge over Coon Creek (SE1/4, Sec 22, T14N, R6W, Vernon, Co., WI). This bridge was replaced in 1926 as the creek rapidly aggraded, and the abutment was later buried in the floodplain. When built, probably in the early 1900s, this abutment was at least 10 ft (3 m) above the water. A small lateral shift by the stream in 1993 exposed the old abutment. The newest bridge (c. 1990) is seen to the far left.

crosses Coon Creek to the south. The bridge used there from c. 1900 to 1926 was a steel caisson or "pipe" bridge that was probably built about 10–12 ft (3.0–3.6 m) above the water. In 1993, a small shift in the stream exposed the previously buried south abutment for a short time (Figure 7.23). Since there is 14 ft (4.2 m) of modern sediment here covering the old presettlement floodplain, it will be seen that most of the accretion occurred after 1900. Approximately 2 ft (0.6 m) of that occurred after 1938.

Buried bridges and buried bridge sites are almost ubiquitous in the Hill Country, especially in the LMV zone. Coon Creek alone has at least 15 or so that I have been able to identify. Of course, the old roads that crossed the floodplain to the bridges were buried also. In fact, I was able to date floodplain surfaces for various dates by boring to old road surfaces (Figure 3.15). Most of these roads and bridges were public, usually township or county, but there were also many buried private roads and bridges. In many stream valleys of the Hill Country, public roads ran along only one side of the valley, yet some farmsteads were located on the opposite side, sometimes as much as a quarter mile (400 m) across from the public road (Figure 7.24). Such landholders had little choice except to construct and maintain their own road across the floodplain and stream, an extremely difficult and costly endeavor when the floodplains and streams were aggrading so rapidly. In more recent years, townships have extended modern roads across floodplains to accommodate most landowners.

Chapter seven:    The lower main valleys

*Figure 7.24* Looking southward across the lower main valley of Coon Creek along Range 28 as seen in Figure 7.20. In the left middle ground (arrow) is the trace of an old private road across the valley. The tree canopies to the right are willows marking an old house site. Both the private road and the house site are buried in the floodplain, and only a difference in vegetation marks them. In the distance, the old railroad bed runs down the valley but is totally buried (see profile in Figure 7.20) and there is little to mark it at the surface.

## Mills and reservoirs

The fate of mills and reservoirs in the LMV, located directly in the streams, must now be evident from all the preceding discussion. At Chaseburg, for example, the streambed is now at the normal water elevation of the old mill pond, c. 1914. Although the mill is gone, its foundation, including the turbine box (with the turbine intact), is buried in the bank (Figure 7.5). Other similar buried milldams are at Village Creek, Iowa, Freeburg, Minnesota, and Beaver, Minnesota. Most probably, scores of mills in the LMV zone of the Hill Country met fates similar to these.

In Trempeleau County, Wisconsin, Lake Marinuka was built on Beaver Creek as a very large millpond at Galesville in 1867. Although somewhat larger and draining a much larger area, its history parallels Lake Como in some respects (Chapter 5). A local resort area with steam-powered cuise boats until the early part of the 20th century, its recreational value was destroyed when sedimentation rates reached alarming proportions in the 1920s. In 1935, the lower part of the lake was dredged and the spoil was dumped farther uplake on the delta and spread to create a golf course. However, this treated only the symptom, not the problem, and when my team resurveyed the remainder of the lake in 1976–1977, it had almost refilled with sediment. Although the current rates of filling are lower than in the 1930s, the upstream bank erosion described earlier still provides considerable sediment and the current rates are excessive (Trimble, 1983). When adjusted for the sediment-trapping ability of the reservoir (Trimble and Bube, 1990), the sediment inflow rates have hardly changed. Indeed, I used sediment inflow to Lake Marinuka as a surrogate for sediment yield from Coon Creek (Trimble, 1983). The lake was dredged again and rehabilitated in the 1980s, but its eventual fate is to be refilled, largely by migrating legacy sediment.

*chapter eight*

# The great flood of August 2007 and its implications

I began research in the Hill Country in the summer of 1973, and by fall of 1974 had a reasonable idea about what had happened there over historical time (Trimble, 1975a,b; 1976a,c,d). While my ideas then were not as well developed as presented in this book, they were for the most part correct. The overriding point here, however, was that the changes discussed in this book were caused by the human hand. More specifically, I saw the improvement phase since the 1930s, and that it was solely the work of humans. Others were not so sure of those conclusions and thought that some sort of climate amelioration had occurred during that time. Many farmers in the region with whom I spoke were sure that rainstorms of the 1920s and 1930s were greater and more intense than those of the 1970s. But in most cases, they had not measured the rain: rather, they had witnessed the runoff and its effects, and that response had decreased greatly since the 1930s. But even fellow professionals, some of whom had worked in the region, were sometimes skeptical. Some believed that "big storms" could return the region to the conditions of the 1930s. Others thought that the massive amounts of stored sediment were only temporary, and eventually, some extreme event, say, a 500-year or a 1000-year event, would flush out the accumulated historical sediment (e.g., Schumm, 1973, p. 86).

The historical climate (Chapter 1, Figure 1.7) showed that rainfall patterns and geomorphic change were inversely correlated (Trimble and Lund, 1982). That is, the worst erosion and sedimentation of the early 20th century occurred during a period of generally lighter rainfall and smaller storms. Conversely, the profound decreases of erosion and sedimentation since the 1930s have occurred as rainfall and storms increased (Chapter 1, Figure 1.7).

The only way one could ever really know about the role of extreme events was to wait for some big storms and see what happened. As it happened, there was a basin-wide 100-year storm in 1978 and a more local 100-year storm in Timber Coulee of Coon Creek in 1984. While both storms did considerable damage and made some geomorphic changes, most damage was hardly visible by the next year. For example, tributaries did not return to the gravel-road appearance most had in the 1930s. But there remained the question of what would happen in a disastrous storm of, say, the 500-year return-period magnitude. Would the landscape come apart and would it return the region to a period of instability? And would these huge deposits of historical sediment described in this study be swept out into the Mississippi River?

## The storm

The chance to find out came on August 18–19, 2007, when over 12 in. (300 mm) of rain fell on parts of Coon Creek and the Whitewater River (Figure 8.1). If such a storm had to happen, this one was incredibly fortuitous for my research because these are the primary study basins for which I have many baseline data, including thousands of photos. Moreover, I had spent time in both basins in July 2007 making hundreds of photos, and the comparisons with the storm damage in August were used to great advantage. Even

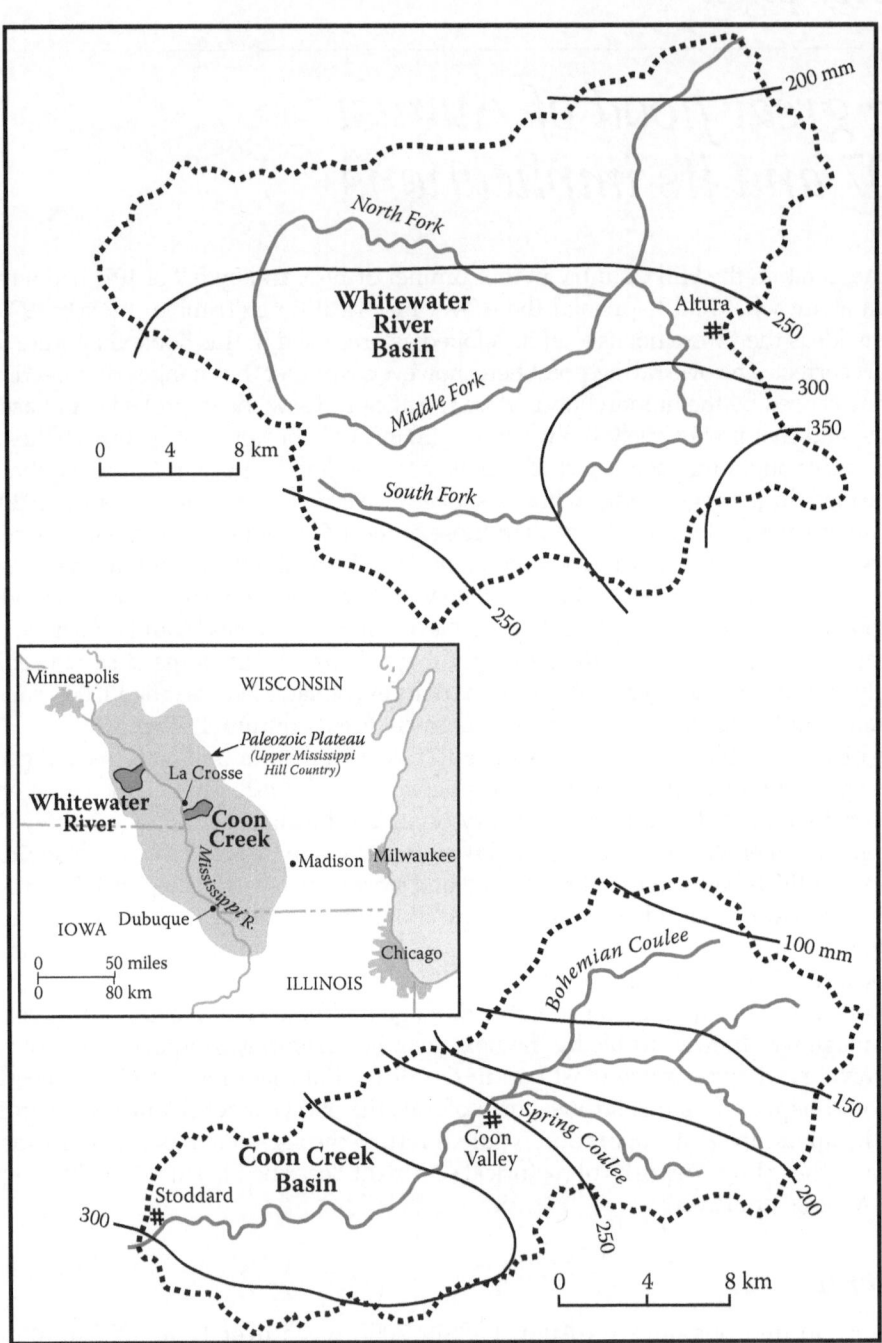

*Figure 8.1* 24-hour precipitation totals for the August 18–19, 2007 storm for Coon Creek and the Whitewater River. 1 in. = 2.5 cm. Data from National Weather Service.

more fortuitous was the fact that there was an array of storm magnitudes ranging from what the USGS later termed a 500-year storm (12 in. or 30 cm/24 h) down to less than a 100-year storm (5 in. or 12.5 cm/24 h; Hershfield, 1961). This range allowed me to ascertain approximate thresholds of geomorphic change as related to storm magnitude. The pattern of rainfall was different in the two basins. In the Whitewater, the heaviest rain was in the headwaters, but in Coon Creek, the heaviest was in the lower valley.

To comprehend the import of such a storm, it might be a good idea to return to the simple quantitative runoff model discussed in the Introduction to this book. Consider, for example, a cultivated field with conservation tillage on a soil of Hydrologic Group A. For a 2 in. (5 cm) rainfall on moderately dry soil, it might be expected to have little or no runoff, whereas a 4–5 in. or 10–12.5 cm (100-year) storm would yield about 1 in. (2.5 cm) of runoff. At the edge of our predictive capability from agricultural experiments, 8 in. (20 cm) of rainfall should increase the runoff to about 3.5 in. (9 cm). The pattern is clear: what we are seeing is that runoff is some power function of rainfall because a fairly constant amount of water will be abstracted by infiltration and storage in the soil mantle (Ward and Trimble, 2004). We have no dependable prediction of runoff for the 12 in. (30 cm) of rain as fell in August 2007, but 6–9 in. (15–23 cm) of runoff would seem reasonable. Such calculations, even if purely estimates, allow some conceptualization of just how much mass and energy is delivered by the runoff from such extreme storms.

One should consider that this event was the acid test for modern soil conservation measures. Two stream basins generally had been fitted with the best soil conservation measures and had been relatively stable for several decades, and then an extreme storm descended on them at the same time. One could not have presciently designed a century-long research plan and have the sequencing occur any better!

As soon as the floodwaters subsided, I was in the region making photos and measurements. Amazingly, another huge spell of rain hit the region in June 2008, and I was there again in early July 2008 to see what had happened. This chapter reports what I found and puts it into context of historical change in the region using a zonal conceptualization similar to that used in the remainder of the book. The major difference is the inclusion of mass movement, a geomorphic process not considered before in this study because it was not manifest.

## Mass movements

Other than slumping of streambanks and severe undercutting of road banks, I don't recall seeing an appreciable mass movement in the Hill Country previous to August 2007. But in steep areas that received 8–12 in. (20–30 cm) of rain, there were literally thousands. Most of these were small slumps along rural roads, but some were very large. On the steep bluffs along the Mississippi River, there were several major slips that took the form of "chutes," or elongated slips of a semicircular or semi-oval cross section. One of these near Goose Island in Vernon County, Wisconsin, was several hundred feet long and closed Wisconsin Highway 35 for some time. In a few cases, there occurred what appeared to be debris flows. In at least one case, part of a hillside above a road cut gave way, and a massive amount of highly liquefied debris collapsed, flowed across a road, and then flowed across the floodplain of a tributary with part of the material actually going into the stream (Figure 8.2). While that event was exceptional, there were many simple mass slips of material along road cuts (Figure 8.2).

While there were thousands of mass movements, small to large, they all had one common characteristic: *they all occurred on undercut slopes*. That is, the base of the slope had been cut away by some sort of construction, usually a roadway. In hundreds of miles of terrain traversed by car, I did not see a single natural slope failure.

*Figure 8.2* Mass movements from the August 2007 storm. Top: What appears to be a debris flow about 2 mi (3 km) north of Chaseburg, WI, at junction of Hohlfeld Road and County K, looking west from north end of bridge (P > 12 in. or 30 cm, NW1/4, Sec 21, T14 N. R 6W). Note that the sediment flow from the undercut road bank moved across the road, across the floodplain, and into Swain Creek. Bottom: Mass movement in Timber Coulee about 3 mi (5 km) north of Westby (P = 6–8 in., or 15–20 cm). (Photos credit: Edyta Zygmunt, 2–9 September 2007.)

Mass movement was clearly related to rainfall amounts. That is, with more than 12 in. (30 cm)/24 h, the mass movements were almost ubiquitous. In the areas having 10–12 in. (25–30 cm), they were quite frequent. There were a few with 6–10 in. (15–25 cm) and almost none with less than 6 in. (15 cm). Time in the field did not permit mapping of mass movements, but such a map would have correlated extremely well with the isohyet map (Figure 8.1).

In mentioning various sites in the two basins, I will give the approximate 24 h rainfall for the storm in parentheses.

## Upland slopes

As discussed in Chapter 2, the soil conservation revolution has continued since its inception in the 1920s and 1930s. Improvements continue, but the greatest leap forward has been no-till agriculture. In Chapter 2, I made the point that the "new" methods of the 1930s (strip contouring, better rotations, etc.) were roughly an order of magnitude better than the old methods. But in my experience, no-till constitutes the same magnitude of improvement over the "new" methods of the 1930s. It requires a large capital outlay, but farmers of the Hill Country have been quietly putting no-till into place over the last two decades. So what are the effects?

Indeed, this is the most startling finding from the 2007 storm. I was able to find no visible erosion (rill or gullies) on no-till land, even under the areas of greatest precipitation with 10% slopes 1000 ft (300 m) long (Figure 8.3)! Not only was I looking but, at my request, NRCS agents were also looking on my behalf as they went about their duties. Two minor exceptions were found. One was where excessive water flowed off a paved road and onto a no-till field (Figure 8.3). The other was in Minnesota, where a small rill formed on a newly planted field in the follow-up storm of June 2008. I note again that the effectiveness of no-till is directly related to productivity: healthier plants grow faster and larger, and produce more residues to leave on the field after harvest.

One might well ask about "sheet" erosion from this storm. While there were effectively no rills or gullies found on no-till fields, how much, if any, soil was eroded by rainfall detachment and so-called "sheet" flow? We have no after-the-fact way to measure this for most fields but, again fortuitously, an opportunity presented itself. About 5 mi (8 km) north of Altura, Minnesota, the outflow from a small watershed in no-till corn and soybeans was caught in a floodwater-retention basin. This location received almost 10 in. (25 cm) of rain in the storm. Considering just the general appearance, and using the root crowns of weeds as a benchmark, I could find no evidence of any deposition in the basin (Figure 8.4).

These findings are in strong contrast to much conventional wisdom of the past three decades or so when many "experts" were crying wolf about US soil erosion rates. There is something ironic about prestigious scientific journals publishing jeremiads about the US soil erosion "crisis" (e.g., Pimentel, et al., 1995) while American farmers, at least those of the Hill Country, were so significantly and quietly improving soil conservation.

Of course, there was highly evident soil erosion from the 2007 storm under less effective forms of conservation management, but these were not common and I had to look for them. Some newly planted fields using conventional agriculture with no evident conservation measures resulted in severe rill erosion (Figure 8.5). Likewise, poor grazing management (by modern standards) allowed rills and gullies to form along animal trails (Figure 8.5). Even some contour strips were overpowered and allowed severe rills (Figure 8.5), but such failures were extremely rare and found only in the areas of greatest rainfall. Indeed, one could drive through Bohemian Coulee of Coon Creek, which received only 4–7 in., or 10–18 cm (Figure 8.1), and hardly know that anything had happened at all. Recalling that 5–6 in., or 12.5–15 cm/24 hr is a 100-year storm, one can readily see the significance.

*Figure 8.3* No-till fields after crops had been removed, April 2008. Note slopes > 10% and several hundred feet long with no evident erosion, P = 12+ in. (30 cm). (Continued)

## Gullies

As seen by the runoff calculations at the beginning of this chapter, the amount of water leaving the slopes had to have been prodigious. Once this much water left upland fields and entered drainage ways with great depth and velocity, it had a much greater ability to erode. Recall that the erosive power of a stream is directly related to velocity, slope, and

Chapter eight: The great flood of August 2007 and its implications      183

*Figure 8.3 (Continued)* No-till fields after crops had been removed, April 2008. Rill caused by runoff from road. All 3 scenes in this figure are on Brinkman Ridge Road, about 4 mi (7 km) north of Chaseburg, WI. Secs 4&9, T14N, R6W.

depth (Figure 3.10). Thus, some old gullies were reactivated, and a few new ones, mostly small, were initiated. For example, both gullies at the Zink Farm (>12 in., or 30 cm) were reactivated, and much of the fan extending westward from the bridge out 500 ft (150 m) onto the Coon Creek floodplain received 0–5 in. (0–13 cm) of material ranging mostly from sand to cobbles (Figure 8.6). Upstream in Timber Coulee (7–8 in. or 19 cm), another rock fan (*cum* rock flume) formed in a farmyard over historical time was reactivated (Figure 8.7).

In Proksch Coulee trench (P > 12 in. or 30 cm), the gully was only mildly and intermittently reactivated, but the 71-year-old dam was overtopped and breached. Huge piles of timber slash and debris left upstream were transported to the dam, where they blocked the drop inlet. Since the dam had no emergency spillway, the huge flow then overtopped the dam and breached it (Figure 8.8). This event is strong evidence that no dam of any size should be built in the eastern United States without an emergency spillway (Trimble, 2011).

Other old gullies fared somewhat better, even when located in the area of maximum rainfall. Certainly, the best documented is a gully about 1 mi (1.6 km) north of Chaseburg, Wisconsin (P > 12 in., or 30 cm; Figure 8.9). In the first photo, made in July 1938, the recentness of the gully is evident. The next photo, made in July 1974, shows a greatly enlarged, but then-stable drainage way, and it has been that way during the entire period that I have been working there. The last photo is after the 2007 flood, and one can hardly tell a flood had passed through (Figure 8.9). Another once-rampant gully, only slightly reactivated, furnished enough debris in the 1930s to continually close Wisconsin State Highway 56 (P = 10–12 in., or 28 cm). To control it, a debris basin was built in 1941, and it filled in four years

*Figure 8.4* A sediment trap after the 2007 storm. Top: A "dry" floodwater retarding dam (arrow) controlling about 8–10 acres (3–4 ha) of mostly no-till crops, 5 mi (8 km) north of Altura, MN, on Beaver-Altura Road looking west. P = 10 in. (25 cm). Corn is to the left and soybeans are to the right. Bottom: Bottom of dry pond. Note lack of any sediment accumulation. NW1/4, Sec. 19, T18N, R9W, Winona County, MN. Photos made Sept. 2–9, 2007.

Chapter eight: The great flood of August 2007 and its implications

*Figure 8.5* Severe soil erosion with lesser levels of conservation management, storm of August 2007. Top: Newly planted field with no apparent conservation measures, Beaver-Altura Road about 4 mi (7 km) north of Altura looking west. P = 10 in. (25 cm). Sec 25, T108N, R10W, Winona Co. Bottom: Gullying on animal trails, heavily grazed area 2 mi (3 km) south of Stoddard, WI, on County O, NW ¼ Sec 3, T13N, R7W. Vernon Co. P = 12 inches (30cm). (All photos made September 2–9, 2007 by Edyta Zygmunt.) (Continued)

*Figure 8.5 (Continued)* Severe soil erosion with lesser levels of conservation management, storm of August 2007. Severe rills in contour strip cropping, about 5 mi (8 km) SE of Stoddard. P = 12 in. (30 cm). (All photos made September 2–9, 2007 by Edyta Zygmunt.)

(Upper Mitchell Debris Dam, Chapter 4, see Trimble and Lund, 1982). Dormant for years and densely vegetated, the gully was reactivated by the 2007 storm but only slightly as evidenced by the 0–4 in. (0–10 cm) deep deposits on the old surface of the basin (Figure 8.10).

## Tributaries

Even in areas having the same rainfall from this storm, there was a great disparity in the amount of change in stream channels. Some channels were ripped up, and others were hardly changed. Some looked as damaged as those of the 1930s, while others changed little. For example, the north Fork of the Whitewater River (P = 9–11 in., or 25 cm), a disaster in the 1930s (see Figure 3.12), was hardly changed (Figure 8.11). While it did not get the maximum amount of precipitation, it is of interest to see what happened at the "benchmark" tributary reach on Bohemian Creek about 4 mi (7 km) northeast of Coon Valley. It will be seen that a 100-year storm (4–7 in., or 10–17 cm) there had little impact (Figure 8.12).

For the region in general, sediment ranging from silt to cobbles was deposited on the "new" lower tributary floodplains, but most visible deposits were sandy and were irregular in extent and depth. A few scattered deposits were found on the high, historical floodplain. In most cases, the irregularity of extent and depth was correlated with particle size: the coarser, the more irregular.

# Chapter eight: The great flood of August 2007 and its implications

*Figure 8.6* Damage to Zink farm from 2007 storm, 1 mi (1.6 km) SW of Stoddard (see Chapter 4, P > 12 in., or 30 cm). Top: View of fan across Coon Creek floodplain from bridge on County O looking west, September 2007. Fresh deposits of mostly coarse material, 1–5 in. (2.5–7.5 cm) thick. Cottonwood tree in right foreground is the one with buried trunk mentioned in Chapter 4. Bottom: between house and barn, looking north up dormant flume formed from rock fan in 1930s to protect house and farmyard. Photo made July 2007. (Photos credit: Edyta Zygmunt.) (Continued)

*Figure 8.6 (Continued)* Damage to Zink farm from 2007 storm, 1 mi (1.6 km) SW of Stoddard (see Chapter 4, P > 12 in., or 30 cm). The same flume September 2007, looking northeast. Accumulated debris from the August 2007 storm has been cleaned out. (Photos credit: Edyta Zygmunt.)

## Upper main valley

Because of the greater rainfall in the headwaters of the Whitewater, there was far more bank erosion in the UMV there than in Coon Creek. Indeed, a bridge was severely damaged about 2 mi (3 km) south of Elba and there was severe bank erosion. In Coon Creek, there was only moderate bank erosion during the storm, but the prolonged bankfull flow saturated the high banks, which are normally dry and extremely stable during the summer. The prolonged saturation and higher pore pressure allowed many similar banks to slump or even collapse after the event (Figure 8.13). The deposition of sediment on floodplains was similar to that of the tributaries.

## Lower main valley

In both basins, the LMV experienced high flows, but the bank and channel erosion again appeared more severe in the Whitewater. In Coon Creek, most LMV bank erosion appeared to be correlated with streams having trees or grazed grass as riparian vegetation (Figure 8.14). Ungrazed grass did not appear to have much erosion (Figure 8.15). This observation also seemed to be correct in the Bad Axe River and accords with published work on the subject (Trimble, 1997, 2004; Lyons et al., 2000). In parts of the Whitewater, however, bank vegetation seemed to make little difference. Indeed, bank erosion observed there from banks having ungrazed grass was just as great as from forested banks (Figure 8.16). This greater bank erosion might result from greater rainfall in headwaters, or it might be from the high sand content of floodplains in the Whitewater. This sand came from the erosion of high terraces such as those seen at Ratz Gully (Chapter 4) and Fairwater (Chapter 5) and would serve to make LMV streambanks more erodible.

*Figure 8.7* Flume formed from rock fan to protect farm, similar to that at Zink farm (see Figure 8.6), photo made from County P looking north, Timber Coulee, NW1/4, Sec 8, T14N, R4W, Vernon Co. P = 7–8 in. (20 cm). Top: July 2007, dormant flume. Bottom: September 2007. Note "gravel road" appearance. (Photo credit: Edyta Zygmunt.)

*Figure 8.8* Proksch Coulee Trench, looking downstream at the dam and drop structure built in 1936 by Soil Conservation Service to stabilize gully, 2 mi (3 km) east of Stoddard WI (P > 12 in., or 30 cm). Viewer stands on reservoir fill. Top: Dam and drop inlet, looking downstream, July 2007. Arrows point to top-right corner of drop structure. Note sediment filled to level of drop inlet (see Figure 4.10). Bottom: Breached dam, September 2007. Note drop inlet to left covered and blocked with woody debris. Gully in foreground extends about 200 ft (60 m) upstream in accumulated sediment. Bottom of gully visible in right background is about 30 ft (9 m) lower than dam. (Photos credit: Edyta Zygmunt.)

*Figure 8.9* Gully 1 mi (1.6 km) north of Chaseburg, WI, looking downstream from bridge on County K, P > 12 in. or 30 cm (SW1/4 Sec21 T14N R6W, Vernon Co.). Top: July 1, 1938. Note about 3 ft (1 m) of light-colored, historical sediment accretion on bank and apparent recentness of gully incision. A storm the next day destroyed the timber and rock revetment. Photo by S.C. Happ. Bottom: 1974. Note that gully more than doubled in size after 1938 but was stable by 1974. (Continued)

*Figure 8.9 (Continued)* Gully 1 mi (1.6 km) north of Chaseburg, WI, looking downstream from bridge on County K, P > 12 in. or 30 cm (SW1/4 Sec21 T14N R6W, Vernon Co.). September 2007 after flood. Note line of rock following thalweg, perhaps rip-rap from plunge pool in foreground, but gully is otherwise stable. (Photo credit: Edyta Zygmunt.)

The only erosion on the floodplain surface itself, as opposed to bank erosion, was where deep, rapid flows occurred as along chutes. A good example is just east of Stoddard, where floodwaters flowed across a meander neck (E1/2, Sec 27, R7W, T14N). But even here, most erosion occurred at points where the flow left the channel and then where it returned. There was little erosion in the interior of the meander neck.

Deposition on the floodplain was very noticeable and of significant volume. Most visible deposition was generally sandy in nature with a highly irregular distribution (Figure 8.17). Most occurred in scattered splays with usual thicknesses of 1 to 5 in. (2.5–7.5 cm), but some small splays almost 2 ft (60 cm) thick were measured (Figure 8.17). It generally appeared that there was more sediment deposition just downstream of severe bank cutting, which suggested that a significant proportion of the deposited sediment was coming from local bank and channel erosion. Indeed, it appeared that material was simply transferred by the turbulent water from steambanks to floodplains immediately downstream.

It is apparently perceived that such extreme floods in the region lay down "horizons of relatively high sand concentration" (Knox, 2001, p. 205) so that "individual large floods deposited 15–25 cm of sediment across the valley floor" (Knox, 2001, p. 212). But as we can clearly see in the photos, such extreme floods tend to lay down sand splays, not "horizons," the latter term implying some sort of uniformity in thickness and composition. Such demonstrated variability of sediment deposition from this large flood brings into doubt the validity of using sediment strata thickness from limited borings to identify major floods of the past. Simply examining the photographs (Figure 8.17) demonstrates the difficulty of identifying this flood from future borings down through the floodplain. For example, one future boring might show 20 in. (50 cm) of vertical accretion from this flood, but another

Chapter eight:   The great flood of August 2007 and its implications    193

*Figure 8.10* Upper Mitchell debris basin, built in 1941 to protect WI Hwy 56 from gully debris, 4 mi (7 km) SE of Genoa (P = 10–12 in. or 30 cm, Sec 36, T13 N R6W). Top: Surveying the basin, 1977. Basin filled in 4 years (1941–1945). Dam to left, sediment basin outlined by dark vegetation, Hwy 56 to top left. Arrow marks edge of fill against dam. Bottom, September 2007. Light material is 1–5 in. (2.5–7.5 cm) of sediment delivered by August 2007 storm. Note that level of fill by dam is hardly higher than in 1977.

*Figure 8.11* North Fork of Whitewater River 1 mi (1.6 km) west of Elba, looking upstream from bridge as seen in Figure 3.12; note barn for reference. Top: 1940. Note disturbed condition of channel. Bottom: 1975, condition was stable, and this has continued to the present. (Continued)

*Figure 8.11 (Continued)* September 2007, P = 9–11 in. (22–27 cm). (Photo credit: Edyta Zygmunt.)

boring only a few feet away might show nothing. For an example of how borings might mislead, even over a longer period, see Trimble, 2009a, Figure 8.16.

## *Sediment yield or efflux*

A major question was the stability of stored legacy sediment in the context of such a major storm. Would a 500-year event "flush" out the vast amounts of sediment stored over the past 150 years? If so, there would have to be a huge delta thrust into Navigation Pool 8 in the Mississippi River.

The first approach was to compare the Coon Creek delta front at successive dates by oblique ground photography (Figure 8.18). I had such photographs for many dates because my research crews often had late afternoon picnics on a bluff overlooking Navigation Pool 8 and the Coon Creek delta. The view is so beautiful that one cannot resist making several photographs at each visit. In Figure 8.18, I compare the extent of the delta in July 2007 to that of September 2007; there is little difference. What we are actually seeing, however, is the plant growth on the delta. The comparison is not perfect because of probable slight differences in pool level and seasonal stage of vegetation growth. July 2008 and summer 2006 were also available (not shown), but all are essentially identical. Indeed, the only way one can tell them apart is by the recorded date. Certainly, we can conclude from these photographs that there was no huge burst of sediment brought to the Mississippi River by the big flood.

The harder line of evidence is a surveyed profile across the delta dating from 1938 (Figure 8.19). The profile is just downstream from the railroad bridge crossing the mouth

*Figure 8.12* Bohemian Creek about 3 mi (5 km) NW of Coon Valley, WI, looking upstream from County G (SW1/4 Sec 34T15NR5W, La Crosse Co.). Top: Photo made 1940 by S.C. Happ as the "typical" tributary in the Hill Country. Bottom: 1974, note stable, grassed floodplain and narrow stream. (Continued)

*Figure 8.12 (Continued)* Bohemian Creek about 3 mi (5 km) NW of Coon Valley, WI, looking upstream from County G (SW1/4 Sec 34T15NR5W, La Crosse Co.). September 2007, completely stable after storm of 4–7 in. (10–18 cm), which is greater than the 100-year storm. Bohemian Creek had only scattered bank erosion from this storm, and I found no mass movements or visible soil erosion of any kind on agricultural fields.

of Coon Creek (Figure 8.15) and is not visible in Figure 8.18, being just out of view to the right (upstream). For clarity, only the profiles for 1975, 1992, and July 2008 are shown. Not only is there no deluge of sediment from the 2007 and 2008 floods, the rate of accretion for 1992–2008 is slightly *less* than that of 1975–1992! As pointed out in Chapter 3, the rate of accretion at the mouth of Coon Creek has been remarkably uniform since the late 1930s. Here, we see that an extreme flood seems to have made little difference.

## Conclusions

In conclusion, perhaps the most significant finding concerning the effects of the 500-yr flood was the lack of visible erosion on no-till fields. Soil conservation measures are normally set up to counter 10- to 25-year events, and I know of no tests of no-till for such extreme conditions as experienced in the Hill Country. Other results of the August 2007 flood appear to challenge conventional wisdom in several ways. First, the extreme storm did not destabilize the landscape and take it back to the conditions of the 1930s. The landscape, by absorbing more rainfall, now allows less force to develop and presents more resistance to the forces deployed. Perhaps if several such events came in rapid succession, with each progressively reducing resistance, it might bring back the conditions of the 1930s, but that appears unlikely.

The inherent stability of stored sediment and the inability of the storm to move it out of the watershed might surprise some. This relates not only to the resistance of the

*Figure 8.13* Collapse of high stream banks 2 weeks after the flood of August 2007, looking downstream Upper Main Valley, Coon Creek, 2 mi (3 km) east of Coon Valley WI, R = 4–10 in. (10–25 cm). For exact location, see photo site in Figure 6.7. Top: July, 2007. Range C from Pleistocene terrace (hump on floodplain in distance) to distant high bank seen in Figure 6.7 is noted by arrows. Bottom: About 10–14 days after the flood. Note logs in background on old floodplain suggesting inundation of the old floodplain (now a terrace). Some erosion of the high bank in the foreground was evident. However, inundation plus the rainfall saturated the normally dry bank, increasing pore pressure and weakening the bank, thus allowing it to collapse and be more susceptible to erosion. (Photos credit: Edyta Zygmunt.)

*Figure 8.14* Erosion of grazed Lower Main Valley streambanks by August 2007 flood, looking upstream from bridge over Coon Creek, 3 mi (5 km) west of Chaseburg, WI (NW1/4 Sec 30 T14N R6W). High bank and terrace in background is shown in Figure 3.17. Top: July 2007; note trampled banks from grazing. Bottom: September 2007. Note severe erosion in left foreground, but far bank in middle ground appears to have accretion. (Photos credit: Edyta Zygmunt.)

*Figure 8.15* Stable ungrazed grass banks in the Lower Main Valley after 2007 flood. Top: Lower Coon Creek looking downstream from Hwy 35 bridge 1 mi (1.6 km) south of Stoddard WI. Bottom: North Fork, Bad Axe River looking downstream from Hwy 56 bridge at Romance WI (SE1/4 Sec 36 T13N R&W. Vernon Co.). (Photos credit: Edyta Zygmunt.)

*Figure 8.16* Extremely severe erosion of ungrazed grass streambanks, looking downstream from bridge over Whitewater River at Beaver MN, 6 mi (10 km) north of Elba, R = 10–14 in. (25–35 cm). Top: July 2007. Bottom: September 2007. (Photos credit: Edyta Zygmunt.)

*Figure 8.17* Sediment deposition on Lower Main Valley floodplains from August 2007 flood, highly variable in distribution and thickness, Top: Splays of sand ranging up to almost 2 ft (60 cm) thick just downstream of bridge in Figure 8.14. Bottom: Splays of sediment on Coon Creek floodplain looking southwest from Cedar Valley Drive, 2 mi (3 km) east of Stoddard, WI (E1/2 Sec 26 T14N R7W, Vernon Co.). (Photos credit: Edyta Zygmunt.)

Chapter eight: The great flood of August 2007 and its implications 203

*Figure 8.18* Repeat views of Coon Creek delta front in Pool 8 of Mississippi River, looking northeast from scenic overlook on bluff, off Hwy 35, 3 mi (5 km) north of Genoa, WI. Top: July 2007, arrow indicates westernmost growth of vegetation on delta front. Bottom: Sept 2007. There was no discernible change. A photo in July 2008 also showed no change, and there was no change from July 2006. (Photos credit: Edyta Zygmunt.)

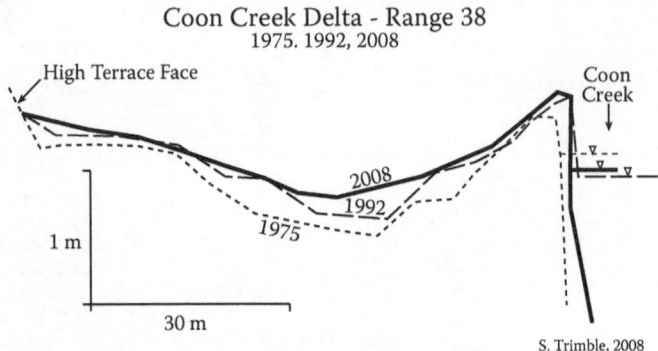

*Figure 8.19* Instrumented cross-sectional survey of north bank of the Coon Creek delta extending into navigation Pool 8 of the Mississippi River, 1975, 1992, and July 2008. The profile is located 500 ft (160 m) downstream of the railroad bridge seen in Figure 8.15 and off to the right and out of sight in Figure 8.18. The period 1975–1992 included one basin-wide, 100-year storm, and the period 1992–2008 included the August 2007 flood and another large storm in June 2008. The average rate of accretion for the earlier period was *greater* than for the later period including the extreme storms of August 2007 and June 2008. These data do not support the concept that extreme floods necessarily scour out legacy sediment, at least in humid area streams with broad floodplains.

landscape vis-à-vis the forces arrayed against it but also to the distribution of forces from such a storm. Recall that during a flood, the flow by definition exceeds the main or low flow channel and uses the flood channel or floodplain. With a wide floodplain such as the larger streams possess in the Hill Country, the flow spreads out and loses much of its tractive force to erode. Not only that, but the floodplain, unlike the channel, is protected against erosion by vegetation. The same vegetation, by its hydraulic resistance, slows the flow and induces deposition of sediment transported by the water. Thus, the huge flood may erode the channel and banks, but storage on the floodplain subsumes much or most of this. The result is that the process of removing stored sediment from a watershed in humid, vegetated regions is a slow one, especially with the presence of a wide floodplain. The movement of sediment down through a stream system has been described as a "jerky conveyor belt" (Ferguson, 1981). If so, a wide, vegetated floodplain may be considered to induce significant "friction" in the conveyor belt.

One might remember that much of the historical sediment was stored in the stream basins in just a few decades. But what we see here suggests that it will take centuries or even millennia for the sediment to be removed. Thus, with such highly disturbed basins as found in the Hill Country, there is an apparent "fast in, slow out" principle operating (Trimble, 2010b).

Indeed, the relatively minor transport of sediment to the mouth of Coon Creek brings into question the concept of sediment from similar tributaries filling the backwaters of the navigation pools on the upper Mississippi River. This was suggested by studies such as McHenry et al. (1984) and Ritchie et al. (1986), which presumed to show alarming rates of aggradation in these pools. However, an extensive but precise resurvey in the backwaters of Pool 8 near Stoddard, Wisconsin, showed as much erosion as accretion since the area was precisely mapped in 1933 by the US Army Corps of Engineers before Pool 8 was created (Carson, 2003). Further research would seem warranted on the sediment contribution of direct tributaries to the Mississippi River.

I was not able to quantify the total sediment fluxes of the 2007 flood as I did for earlier time periods. The first reason for this was that profiles had not been surveyed immediately prior to the event, and even had I known the flood was coming, a highly unlikely scenario, it takes several years to complete all the surveys. The second problem is that changes were often local and isolated rather than being general so that the profiles, normally spaced about a mile (1.6 km) apart, might not be a good sample of a highly diverse set of conditions. Thus, the photographic record was of major importance. This was an instance when modern remote sensing techniques such as LIDAR (Light Detection and Ranging) could have proved useful.

However, it is imperative that we be able to make measures of changes from such extreme events so that we can compare them quantitatively with the many more modest events that continually occur. But based on what I've seen so far, it would seem that the latter are much more important than the former. That is, the many moderate events make far more significant changes in the long run than do the infrequent extreme events. Of course, this really is not news; Wolman and Miller laid out the principle over 50 years ago when they suggested that it was moderate size events that collectively do most of the geomorphological work, not the extreme events (Wolman and Miller, 1960).

It is generally thought that a huge storm like that of August 2007 destabilizes a stream basin so that it "bleeds" for some number of years with higher sediment yields coming from eroded streams and that may be true in some cases. But what I have observed since the 2007 storm is that stream channels, highly eroded by the 2007 storm, are now using transported sediment to rebuild channels to their normal size, that of conveying the 1-year or 2-year storm flow. In a sense, then, the channel erosion from this great storm created morphological "sinks" such as channel and bank "blow-outs" that have subsumed much of the sediment transported by the stream since the storm. In summary, the extreme streamflows eroded channels and banks and appeared to deposit most of the material on the floodplain directly downstream from the erosion. The process since then seems to be the filling of these morphological voids. Under these conditions, it may be argued that an extreme storm actually reduces sediment yielded over a decadal period.

# *Conclusions*

Seemingly haphazard and random, there is actually a systematic time and space pattern to the profound physical changes shown in this book and to the resulting effects on human activity. And there are explanations for most of it.

The second conclusion is that while geography may not be a determinant, it is a powerful context within which people act. Most settlers of the Hill Country were not wanton exploiters of the land. Rather, they tended to be careful farmers who did the best they could with the tools and technology at hand. So why did such landscape destruction take place within a few decades? The primary answer is that these people were farming in an environment for which soil erosion control was simply beyond their ken, or perhaps better said, beyond the knowledge and technology of that time. Indeed, given the steep slopes, intense rainfall, and crops grown, there was no methodology available that would have allowed them to economically carry on their agriculture without the severe environmental damage that actually occurred. Whereas the environmental problem at the overall scale was the severe climate, the problem at the more local scale was clearly slope. While the soil erosion described would not have happened without agriculture, the correlation between land use and erosion was only general, but the correlation with slope was much more evident.

The third conclusion is that the destruction here, as well as most of the worst soil erosion of the world historical record, was accomplished with very simple technology. In this case, it was simple, animal-drawn implements. Indeed, this is true for most of the world. Some critics of American agriculture are quick to condemn tractors and other large farm machinery as promoting erosion (e.g., Worster, 2004; see Trimble 2010a for a counterargument). Tractors were coming on the scene in the 1920s and did not become common until after World War II when soil erosion was rapidly decreasing in the Hill Country. Indeed, mechanization greatly helped soil conservation by allowing farmers to break up plow pans and to create proper terraces. Further, tractors could not be used on very steep slopes where horse-drawn plows had allowed tilling. Thus, such slopes were often retired with the advent of tractors and often succeeded into forest.

The fourth conclusion is that climate change did not cause the landscape changes described in this study. While indeed there were climate changes, they were negatively or inversely correlated with the landscape changes. In particular, the strong wet trend with greater storms since the late 1930s (Figure 1.7) is negatively correlated with the great decrease of flooding and erosion since then. Thus, we can discount climate amelioration as the cause of the improvement. In geomorphological terms, while the *forces* on the landscape have become progressively greater since the 1930s, the *resistance* to those forces has become even greater so that less geomorphic work (erosion) is being done.

The fifth and perhaps most obvious conclusion of this study is that while humans were responsible for the environmental degradation of the Hill Country, they were also responsible for the almost miraculous recovery of the region. This most certainly is not the wanton destruction of landscape and persistent if not worsening soil erosion crisis sometimes reported over the past three decades or so, often even in scholarly journals. So is the Hill Country merely an exception? My own experience says "no," that other regions of the country have also seen great improvements, (Trimble and Crosson, 2000). As just one example, much is made in this study of improved land management having increased base flow of streams. But this is not only true for the Hill Country but is also true for the entire state of Iowa (Schilling and Libra, 2003). The Southern Piedmont, mentioned in the preface, is a large region showing similar hydrologic improvement as the Hill Country (Price, 1998). But there is an important difference between the Southeast and the Corn Belt. The great decrease of soil erosion in the Piedmont was largely the result of cropland reverting to forest. That is, the incredible productive capacity of American agriculture means that marginal-quality farmland in the eastern United States, such as the Piedmont, is no longer needed. Expressed another way, why grow corn in Georgia with yields of 50–75 bushels per acre when corn yields in Wisconsin, Iowa, Minnesota, and Illinois may be more than 200 bushels per acre (Trimble, 2009b). Not many realize that American agricultural productivity increased almost 2% per year over the 20th century (Conkin, 2008). During the period c. 1945–2005, corn yields per acre increased fourfold, soybeans and wheat yields tripled, hay yields doubled, and potato yields increased sixfold (Karlen et al., 2010)! Expressed another way, while land area used for agriculture decreased by 10% during the period 1947–1994, productivity increased 150% (Helms, 2003). Hence, millions of formerly cultivated acres between Texas and Maine are now in forest because they are not needed. Just between 1940 and 2007, 37 million acres of cropland reverted (Hart, 2010). Most of that is now forest. One might extend this idea further and suggest that agricultural exports from US plentitude allow other agricultural areas of the world, some very fragile, to be used less intensely and perhaps remain in forest. Thus, we see here a case of soil conservation at a distance, of "virtual soil conservation," a concept analogous to "virtual water" (Allen and Wichelns, 2008).

These "new" forests in the United States and elsewhere are great carbon sinks and have greatly enhanced wildlife habitat as well as water quality. Once-scarce animals such as deer, turkey, wildcats, and beaver have returned (McKibben, 1995). The point is that applied US agricultural technology and practice have not only reduced erosion in the Hill Country but they have also greatly increased productivity there and elsewhere, which in turn allows a return to a more natural environment in other areas of the United States and perhaps in the world. Indeed, McKibben (1995) calls this forest regrowth and attendant environmental improvements in the eastern United States the "environmental story of the century."

Ironically, the implications of the dramatic reversion of cropland to forest in the United States have received little attention from scientists who have seemingly failed to grasp its environmental significance. Rather, many studies continue to emphasize *conversion* of forest to cropland (e.g., de la Cretz and Barten, 2007). However, this process had mostly ended by c. 1940, whereas the main process since then has been reversion.

The sixth conclusion is that well-designed governmental programs, along with an enlightened and informed populace, can help ameliorate environmental problems. While perhaps some mistakes were made, many of the programs were inspired. Driven by necessity the United States has led the world in soil conservation technology, and administration, and this continues, as epitomized by the Hill Country.

A seventh conclusion is that good policies and actions sometimes have some bad results. As also occurred in the Piedmont, soil conservation measures reduced the sediment load in tributaries, which made them "hungry" for more sediment, thus eroding their banks and channels and sending more sediment downstream, much of it historic or "legacy" sediment. Hence, we have the irony of soil conservation helping to sustain downstream sedimentation problems.

A corollary of the above is that the stream processes demonstrated in this study are much more complex in time and space than generally perceived, even by many environmental professionals. An example is the Great River Environmental Action Team (GREAT) study (1980), which suggests that merely using good soil conservation techniques on uplands will diminish the amount of sediment moving into the Mississippi River and its adjacent wetlands. But as this study shows, matters are not so simple (e.g., Chapter 3, Figure 3.8). As yet, there is no evidence that 75 years of soil conservation in the Hill Country has reduced sediment inflow to the Mississippi River.

Stemming from the above, the eighth conclusion is that many downstream Lower Main Valley (LMV) reaches will continue to have channel and floodplain aggradation from legacy sediment. This means that flooding will become worse and that more areas may have to be permanently evacuated. Already, the village of Soldiers Grove on the Kickapoo River has been moved uphill, and Gays Mill, also on the Kickapoo, may meet the same fate. In September 2010, hundreds in Arcadia, Wisconsin, were evacuated to escape a flood. In Chapter 7, I describe how migrating sediment has set up Arcadia, a town of about 4000 people, for a disaster of this magnitude. The remains of lower Chaseburg, Wisconsin were demolished in 2009, effectively and permanently moving the village uphill to the high terrace. We must now watch villages and towns on streams such as Root River in Minnesota, the Yellow and Turkey Rivers in Iowa, the Trempealeau and Kickapoo in Wisconsin, and the Galena in Illinois. While most of the localities are protected by flood dikes, this is no guarantee of future safety, especially with the increasing trend of rainfall as shown in Chapter 1. Even in the Upper Main Valley (UMV) reaches, eventually there may be problems at such places as Coon Valley. The eighth conclusion might therefore be restated thus: parts of the Hill Country may be a victim of its history.

Although I do not presume to be a theorist, the ninth conclusion is that the study clearly shows the utility of much hydrologic and geomorphic theory. Indeed, this study is in some ways a long set of textbook-quality examples. Such principles as stream equilibrium, channel size as related to flood frequencies, stream sediment loading, and stream behavior as related to land use have been clearly demonstrated. But not only have they been demonstrated, their *utility* has been demonstrated. A particularly interesting example was in Coon Valley, where the height of the floodplain was cut down from the natural 6 ft (1.9 m) to 3 ft (0.9 m) in order to accommodate handicapped fishermen. But the stream simply began building the floodplain back up to its normal height, which reflected the present flood regime (Chapter 6, Figure 6.4). A more esoteric example is demonstrating the lag of stream behavior to changes of rural land use and management largely as the result of changes in the soil (Chapter 3, Figure 3.7). The principle had been articulated but never before demonstrated. Yet, another example is demonstrating that improving land use can increase the flow of springs and the base flow of streams on a large regional basis.

The tenth conclusion is that the effect of sedimentation on milldams and reservoirs was much greater than the effect of milldams and reservoirs on sedimentation. Much has recently been made in geomorphic circles about the effect of milldams and small reservoirs on the distribution and amount of accumulated historical sediment but the effect of such small dams in the Hill Country was only local. Many dams were completely buried. Expressed another way, everything else being equal, valleys without dams accumulated about as much sediment as valleys with dams. To be sure, if dams and reservoirs were head to toe along a stream, there would be significantly enhanced accumulation but historical records will show that very few areas of the US had such a heavy distribution of mills and reservoirs.

The eleventh conclusion is that significant landscape changes, even those that threaten life and property, may soon be forgotten unless well documented and made known. In traversing the Hill Country, one would hardly discern the environmental violence and human frustrations that once existed here. Indeed, it is presently difficult to imagine a more pastoral, bucolic, and attractive landscape. But the violence and frustration were real. It is a tribute to the citizens of that region, and to our society, that in something more than half a century, the Hill Country has been transformed from an environmental disaster to what is a near model for an agricultural landscape. But at the same time, fluvial processes continue to bring problems. I think it remarkable that more has not been written on the history of the Hill Country and that there is no museum to keep people aware of that history and its implications. Perhaps that should become a regional goal. There is already a small but excellent museum at the Whitewater River State Park Visitor Center in Elba, Minnesota, which shows some of the changes discussed in this book. Perhaps it could be expanded as the first step.

# References

Adams, C. 1940. Modern sedimentation in the Galena River Valley, Illinois and Wisconsin: M.S. thesis, University of Iowa, Iowa City.
\_\_\_\_ 1942. Accelerated sedimentation in the Galena River Valley, Illinois and Wisconsin: Ph.D. dissertation, University of Iowa, Iowa City.
\_\_\_\_ 1944. Mine waste as a source of Galena River bed sediment. *Journal of Geology* 52: 275–287.
Alexander, W. 1882. *History of Winneshiek and Allamakee Counties, Iowa*. Sioux City, IA: Western Publishing Company.
Allen, A. and D. Wichelns. 2008. Virtual water. In S. W. Trimble (ed.), *Encyclopedia of Water Science*, 2nd Ed. Vol. II. Boca Raton, FL: CRC Press, 1285–1288.
Anfinson, J. 2003. *The River We Have Wrought: A History of the Upper Mississippi*. Minneapolis: University of Minnesota Press.
Argabright, M., R. Cronshey, D. Helms, G. Pavelis, and R. Sinclair, 1996. *Historical Changes in Soil Erosion, 1930–1992: The Northern Mississippi Valley Loess Hills*. Natural Resources Conservation Service, U.S. Dept. Agr. Historical Note No. 5.
Bale, F. 1939. *Historic Galena: Yesterday and Today*. Waukegan, IL: Skokie Press.
Bates, C.G. 1931. Chaining the father of waters. *American Forests* 36: 66–69, 106, 127.
\_\_\_\_ 1936. Forest influence on streamflow under divergent conditions. *Journal of Forestry* 34: 961–969.
\_\_\_\_ 1937. A study of public aspects of soil erosion in southwestern Wisconsin: A progress report for 1937. Unpublished report.
Bates, C.G. and O. Zeasman. 1930. Soil erosion—a local and national problem. *Wisconsin Research Bulletin* 99. Madison: University of Wisconsin Agricultural Experiment Station.
Bates, R. 1939. Geomorphic history of the Kickapoo region. *Bulletin of the Geological Society of America* 50: 819–880.
Beach, T. 1994. The fate of eroded soil: Sediment sinks and sediment budgets of agrarian landscapes in southern Minnesota. *Annals of the Association of American Geographers*. 84: 5–28.
Beltrami, J. 1828. *A voyage in Europe and America Leading to the Sources of the Mississippi River*, Vol. 2. London: Hunt and Clarke
Bennett, H.H. 1939. *Soil Conservation*. New York: McGraw-Hill.
Bennett, H.H. and W. Chapline, 1928. Soil erosion, a national menace. USDA *Agricultural Circular* 33.
Benton, T. and N. Russell. 1927. *Soil Survey of Winneshiek County, Iowa*. USDA Bureau of Soils. Washington, DC: Government Printing Office.
Bissen, B. 1975. *Greetings from Hokah, Minnesota*. Mimeographed. No place, No publisher.

Blanchard, W.O. 1924. The geography of southwestern Wisconsin. Wisconsin Geological and Natural History Survey, *Bulletin* No. 65, Educational Series No. 8. Madison: The State of Wisconsin.

Bogue, A. 1963. *From Prairie to Corn Belt*. Chicago: University of Chicago Press.

Bogue, M. 1990. Exploring Wisconsin's waterways. Reprinted from the 1989–1990 *Wisconsin Blue Book*. Madison: Wisconsin Legislative Reference Bureau.

Bray, E.C. and M. Bray, 1976. *Joseph N. Nicolet on the Plains and Prairies: the Expeditions of 1838–39 with Journals, Letters, and Notes on the Dakota Indians*. St Paul, MN: Historical Society.

Brick, E. 2010. Repairing driftless area fields and streams. *Big River*. July–August 28–29, 43.

Brown, C.B. 1941. Dynamics of entrainment of erosional debris and sedimentation. *Transactions American Geophysical Union* 22: 305–309.

Brown, M.H. and I. Nygard, 1941. Erosion and related land use conditions in Winona County, Minnesota: U.S. Department of Agriculture, Soil Conservation Service *Erosion Survey* 17.

Brown, R.H. 1948. *Historical Geography of the United States*. New York: Harcourt: Brace & Co.

Brunson, A. 1884. *History of Vernon County, Wisconsin*. Springfield, IL: Union Publishing Company.

Bunnel, L.H. 1897. *Winona and Its Environs on the Mississippi in Ancient and Modern Days*. Winona, MN: Jones and Krueger.

Burke, W. 2000. *The Upper Mississippi Valley*. Waukon, IA: Mississippi Valley Press.

Butterfield, C.W. 1881. *History of Grant County, Wisconsin*. Chicago: Western Historical Company.

Butzer, K. 1974. Soil erosion, a problem in man-land relationships. In I. Manners and M. Mikesell (eds.), *Perspectives on Environment*. Washington, DC: Association of American Geographers.

Carson, J. 2003. Dynamic fluvial processes in the upper Mississippi River backwaters M.A. thesis, University of California, Los Angeles.

Chamberlain, T.C. and Salisbury, R.D. 1885. The Driftless Area of the Upper Mississippi Valley. 6th Annual Report of the U.S. Geological Survey. Washington, DC: U.S. Government Printing Office, 205–308.

Church, M. and Slaymaker, O. 1989. Disequilibrium of Holocene sediment yield in glaciated British Columbia. *Nature* 337: 452–454.

*City of Winona and Southern Minnesota*. 1858. Winona, MN: D. Sinclair & Co.

Cohee, M.H. 1934. Erosion and land utilization in the Driftless Area of Wisconsin. *Journal of Land and Public Utility Economics* 10: 243–253.

Colby, C.C. 1924. Agricultural adjustments of the natural environment in southeastern Minnesota during the period of bonanza wheat farming. *Transactions of the Illinois State Academy of Science* 17: 213–225.

Commoner, B. 1971. *The Closing Circle: Nature, Man and Technology*. New York: Alfred Knopf.

Conkin, P. 2008. *A Revolution down on the Farm: The Transformation of American Agriculture since 1929*. Lexington: University of Kentucky Press.

Conzen, M. 1997. The European settling and transformation of the Upper Mississippi Valley lead region. In R. Ostergren and T. Vale (eds.), *Wisconsin Land and Life*. Madison: University of Wisconsin Press.

Conzen, M. (ed.). 2010. *The Making of the American Landscape*, 2nd ed. New York: Routledge.

Cooke, R. and R. Reeves. 1976. *Arroyos and environmental change in the American Southwest*. Oxford: Clarendon Press.

Coues, E. 1895. *The Expeditions of Zebulon Montgomery Pike*, Vol. 1. New York: Harper.

Costa, J. and V. Baker. 1981. *Surficial Geology*. New York: Wiley.

Cross, J.M. and M.C Davis. 1971. Coon Valley proves the claim. *Soil Conservation*.

Curtis, E. 1959. *The Vegetation of Wisconsin: An Ordination of Plant Communities*. Madison: University of Wisconsin Press.

Curtis, W. 1966. Forest zone helps minimize flooding in the Driftless Area. *Journal of Soil and Water Conservation* 21: 101–102.

Curtiss-Wedge, F. and E.D. Pierce 1917. *History of Trempealeau County, Wisconsin*. Winona: H.C. Cooper, Jr., & Co.

Daniels, E. 1854. *First Annual Report of the Geological Survey of the State of Wisconsin*. Madison: Wisconsin State Geological Survey.

de la Cretez, A. and P. Barten, 2007. Land use effects on streamflow and water quality in the northeastern United States. Boca Raton: CRC Press.

# References

Derleth, A. 1940. *Bright Journey.* New York: Scribner's.
Dole, R.B., and F. Wesbrook. 1907. The quality of surface waters in Minnesota, *USGS Water-Supply and Irrigation* Paper 198. Washington: DC Government Printing Office.
Dinsmore, J.J. 1994. *A Country So Full of Game: The Story of Wildlife in Iowa.* Iowa City: University of Iowa Press.
Edwards, M., A. Anderson, A. Meyer, J. Chucka, and D. Wilcox. 1928. *Soil Survey of Vernon County, Wisconsin.* USDA Bureau of Chemistry and Soils. Washington, DC: Government Printing Office.
Faulkner, D. 1998. Spatially variable historical alleviation and channel incision in west-central Wisconsin. *Annals of the Association of American Geographers* 88: 666–685.
FEMA (Federal Emergency Management Agency). 2008. Gays Mill long term community recovery plan. http://www.gaysmills.org/pdfs/GayMills_Longterm_Recovery_PlanWEB.pdf.
Ferguson, R.I. 1981. Channel forms and channel changes. In J. Lewin (ed.), *British Rivers.* London: Allen and Unwin. pp. 90–125.
Fraczek, W. 1987. Assessment of the effects of changes in agricultural practices on the magnitude of floods in Coon Creek watershed using hydrograph analysis and air photo interpretation: MS thesis, University of Wisconsin, Madison, WI.
Fremling, C. 2005. *Immortal River: The Upper Mississippi River in Ancient and Modern Times.* Madison: University of Wisconsin Press.
Friis, H. 1969. The David Dale Owen map of southwestern Wisconsin. *Prologue, The Journal of the national Archives* 1: 9–28.
Garland, H. 1917. *A Son of the Middle Border.* New York: Macmillan.
_____ 1926. *Trail Makers of the Middle Border.* New York: Grosset and Dunlap.
Gatwood, E. 1989. Applied field, lab and computer exercise in fluvial geomorphology: bankfull discharge capacity and drainage area in Coon Creek WS. Unpublished paper in partial fulfillment of MA degree in Geography, University of California, Los Angeles.
Gebert, W. and W.R Krug. 1996. Streamflow trends in Wisconsin's driftless area. *Journal of the American Water Resources Association* 32: 733–744.
Gilbert, G.K. 1917. Hydraulic mining debris in the Sierra Nevada. *U.S. Geological Survey Professional Paper*, No. 105.
Glanz, J. 1999. Sharp drop seen in soil erosion rates. *Science* 285: 1187–1189.
Gottschalk, L. 1945. Effects of soil erosion on navigation in upper Chesapeake Bay. *Geographical Review* 35: 219–238.
Goudie, A. (ed.). 1997. *The Human Impact Reader.* Oxford, UK: Blackwell.
_____ 2006. *The Human Impact on the Natural Environment.* Malden, MA: Blackwell.
Gray, A., W. Moran, A. Hasty, S. Hill, C. Mattson, H. Newman, and H. Petraborg. 1929. *Soil Survey of Houston County, Minn.* Washington, DC: USDA Bureau of Soils.
Great River Environmental Action Team (GREAT). 1980. *GREAT I, Study of the Upper Mississippi River.* St. Paul (sponsored by several state and Federal government agencies. Volume 4, Sediment and Erosion.
Gross, E. 1973. Buried soils of the drainageways in the Driftless Area of the Upper Mississippi River Valley: Ph.D. dissertation, University of Minnesota, Minneapolis.
Hancock, E.M. 1913. *Past and Present of Allamakee County, Iowa.* Chicago: S.J. Clarke. 2 vols.
Happ, S.C. 1940. Sedimentation related to flood problems in the Whitewater River valley. Unpublished office report in files of the Whitewater River project,
_____ 1944. Effect of sedimentation on floods in the Kickapoo valley, Wisconsin. *Journal of Geology* 52: 53–68.
_____ 1975. Valley sedimentation as a factor in sediment yield determinations. In *Present and Prospective Technology for Predicting Sediment Yields and Sources*, USDA-ARS Publication 5-40.
_____ 1977. Interim report, Coon Creek sedimentation study. Typed report to Roger Wolfe, USGS Hydrologist, Menlo Park, CA. June 1, 1977.
_____ 1985. Letter to Olin C. Fimreite, District Conservationist, Soil Conservation Service, Whitehall, WS, 19 Feb. 1985.
Happ, S.C., G. Rittenhouse, and G.C. Dobson. 1940. Some principles of accelerated stream and valley sedimentation. U.S. Department of Agriculture *Technical Bulletin 695.*
*Harper's New Monthly Magazine.* June 1853. No. 37. Vol. 7: 176–191.

Hart, J. 2010. From bolls to boles. *Southeastern Geographer* 50: 189–199.
Hays, O.E., A.G. McCall, and F.G. Bell. 1949. Investigations in erosion control and the reclamation of eroded land at the Upper Mississippi Valley Conservation. Experiment Station near LaCrosse, Wisconsin, 1933–43: U.S. Department of Agriculture *Technical Bulletin* 973.
Helms, D. 2003. The evolution of conservation payments to farmers. In N. DeCuir, A. Sokolow, and J. Woled (eds.), *Compensating Landowners for Conserving Agricultural Land*, 123–132. Davis: University of California Community Services Extension.
Hendrickson, J. 1990. Whitewater River project hydraulic/sediment analysis. Unpublished summary report, U.S. Army Corps of Engineers, St. Paul, Minnesota
Hershfield, D. 1961. Rainfall frequency atlas of the United States. U.S. Department of Commerce, Weather Bureau *Technical Paper* 40.
*Hoard's Dairyman*. 15 November, 1895. (erosion history card file, Record Group 118, National Archives, Washington, D.C.)
Hobbs, G. 1939. *Glamorous Galena*. Galena, IL: Galena Gazette Press.
Hood, E. 1896. *Plat Book of Vernon County, Wisconsin*. Minneapolis: Foote.
Horn, B. 1972. Beaver: Death of a village, lesson in conservation. *Winona* (Minnesota) *Sunday News*. 16 July 1972 (In "Beaver" folder, Winona County Historical Society).
Hudson, J. 1994. *Making the Corn Belt: A Geographical History of the Middle-Western Agriculture*. Bloomington: University of Indiana Press.
\_\_\_\_ 2010. Remaking the prairies. In M. Conzen, *The Making of the American Landscape*, 2nd Ed. New York: Routledge.
Huff, F. and J. Angel. 1992 Rainfall frequency atlas of the Midwest. Illinois State Water Survey. Champaign: *Bulletin* 71.
*Illustrated Historical Atlas of the State of Minnesota*. 1874. Chicago: A. T. Andreas.
Iowa Geological Survey, 1895 Third Annual Report. Iowa City.
Jackson, G.A., C.E. Korschgen, P.A. Thiel, J.M. Besser, D.W. Steffeck, and P. Bockenhauer. 1981. *A Long-Term Resource Monitoring Plan for the Upper Mississippi River System*. Vol. 2. Bloomington, MN: Upper Mississippi River Basin Commission.
Jarchow, M.E. 1949. *The Earth Brought Forth: A History of Minnesota Agriculture to 1885*. St. Paul: Minnesota Historical Society.
Johansen, H.E. 1969. Spatial diffusion of contour strip cropping in Wisconsin: MA thesis, Dept. of Geography, University of Wisconsin, Madison.
Johnson, H.B. 1957. King wheat in southeastern Minnesota: A case study of pioneer agriculture. *Annals of the Association of American Geographers*. 47: 350–362.
\_\_\_\_ 1976. *Order upon the Land*. New York: Oxford University Press.
\_\_\_\_ 2010. Gridding a national landscape. In Michael Conzen (ed.), *The Making of the American Landscape*. New York: Rutledge. pp. 142–161.
Johnson, L. 1991. *Soil Conservation in Wisconsin, Birth to Rebirth*. Madison University of Wisconsin Department of Soil Science.
Johnson, W.C. 1976. The impact of environmental change of fluvial systems: Kickapoo River, Wisconsin: Ph.D. dissertation. Madison: University of Wisconsin.
Juckem, P., H. Randall, and M. Anderson. 2006. Scale effects of hydrostratigraphy and recharge zonation on base flow. *Groundwater* 44: 362–370.
Juckem, P., R. Hunt, M. Anderson, and D. Roberson. 2008. Effects of climate and land management on streamflow in the driftless area of Wisconsin. *Journal of Hydrology* 355: 123–130.
Judson, S. 1963. Erosion and deposition of Italian streams during historic time. *Science* 140: 898–899.
Kane, L., J. Holmquist, and C. Gilman (eds.). 1978. *The Northern Expeditions of Stephen H. Long: The Journals of 1817 and 1823 and Related Documents*. St. Paul: Minnesota Historical Press.
Karlen, D., D. Dinnes, and J. Singer. 2010. Midwest soil and water conservation: Past, present and future. In T. Zobeck and W. Schillinger (eds.)., *Soil and Water Conservation Advances in the United States*. Soil Science Society of America Special Publication 60: 131–162.
Kent, C. 1999. The influence of changes in land cover and agricultural land management practices on baseflow in southwest Wisconsin: Ph.D. dissertation. University of Wisconsin, Madison.
Kessinger, M. 1888. *History of Buffalo County*. Alma, WI: No publisher.

Knox, J.C. 1972. Valley alluviation in southwestern Wisconsin. *Annals of the Association of American Geographers* 62: 401–410.

____ 1977. Human impacts on Wisconsin stream channels *Annals, Association of American Geographers* 67: 323–342.

____ 1985. Geologic history of valley incision in the driftless area. In *Pleistocene Geology and Evolution of the Upper Mississippi Valley: A Working Conference*. Winona State University, Minnesota, pp. 5–8.

____ 1987. Historical valley floor sedimentation in the upper Mississippi Valley. *Annals of the Association of American Geographers* 77: 224–244.

____ 1993. Letter to Western District, Wisconsin Department of Natural Resources, Eau Claire, dated 22 November, 1993.

____ 2001. Agricultural influence on landscape sensitivity in the upper Mississippi River Valley. *Catena* 42: 193–224.

____ 2006. Floodplain sedimentation in the upper Mississippi River Valley: natural versus human accelerated. *Geomorphology.* 79: 286–310.

Knox, J.C., P.J. Bartlein, K.K. Hirschboek, and R.J. Muckenhirn. 1975. *The Response of Floods and Sediment Yields to Climatic Variation and Land Use in the Upper Mississippi Valley*. University of Wisconsin-Madison, Institute for Environmental Studies.

Knox, J.C. and D. Faulkner, 1994. Post-settlement erosion and sedimentation in the lower Buffalo River Watershed. Report to Western District, Department of Natural Resources. Eau Claire, Wisconsin.

Kondolf, M. 1997. Hungry water: effects of gravel mining and dams on river channels. *Environmental Management* 21: 533–551.

Krug, W.R. 1996. Simultation of temporal changes in rainfall-runoff characteristics, Coon Creek Basin, Wisconsin. *Journal of American Water Resources Association* 32: 745–752.

Kuechler, A.W. 1975. Potential Natural Vegetation of the Coterminous United States. *American Geographical Society Special Publication* No. 36.

Kunsman, H.S. 1944. Stream and valley sedimentation in Beaver Creek Valley, Wisconsin: M.Ph. thesis. University of Wisconsin, Madison.

Lapham, I.A. 1846. *A Geographical and Topographical Description of Wisconsin*. 2nd Ed. Milwaukee, Wisconsin: I.A. Hopkins.

Laudermilk, W. 1953. Conquest of the land through 7,000 years. *USDA Agricultural Information Bulletin* 99.

Leopold, A. 1935. Coon Valley: an adventure in cooperative conservation. *American Forests* 41: 205–208.

Leverett, F. and F.W. Sardeson. 1919. Surface formations and agricultural conditions of the south half of Minnesota. Minnesota Geological Survey. *Bulletin* No. 14. Minneapolis: University of Minnesota.

Loehr, R.C. (ed.). 1939. *Minnesota Farmers' Diaries: William R. Brown, 1845–46, Mitchell Y. Jackson, 1852–63*. St. Paul: Minnesota Historical Society.

Lyons, J., S.W. Trimble, and L.K. Paine. 2000. Grass versus trees: managing riparian areas to benefit streams of central North America. *Journal of the American Water Resources Association* 36: 919–930.

Magilligan, F.J. 1985. Historical floodplain sedimentation in the Galena River valley, Wisconsin and Illinois. *Annals of the Association of Amerian Geographers* 75: 583–594.

Marks, J.B. 1942. Land use and plant succession in Coon Valley, Wisconsin. *Ecological Monographs.* 12: 115–133.

Martin, L. 1932. The physical geography of Wisconsin. Wisconsin Geological and Natural History Survey, *Bulletin* 36.

____ 1965. *The Physical Geography of Wisconsin*. Madison: University of Wisconsin Press.

McHenry, J.R., J.C Ritchie, C.M. Cooper, and J. Verdon, 1984. Recent rates of sedimentation in the Mississippi River. In J.G. Wiener et al. (eds.), *Contaminants in the Upper Mississippi River*. Stoneham, MA: Butterworth. pp. 47–62.

McKelvey, V.E. 1939. Stream and valley sedimentation in the Coon Creek drainage basin. Wisconsin: M.A. thesis, University of Wisconsin–Madison.

McKibben, B. 1995. An explosion of green. *Atlantic Monthly*, 27: 61–83, April.

McLeod, R. 1984. Soil conservation work began in 1910. *LaCrosse Tribune*, January 22, 1984.

McMaster, S. 1893. *60 Years on the Upper Mississippi*. Rock Island, IL: No publisher.

Meyer, L. and W. Moldenhauer, 1985. Soil erosion by water: The research experience. In D. Helms and S. Flader (eds.), *The History of Soil and Water Conservation*. Washington, DC: The Agricultural Society. pp. 90–102.

Mirk, W. 1997. *An Introduction to the Tallgrass Prairie of the Upper Midwest: Its History, Ecology, Preservation and Reconstruction*. Boscobel, WS: The Prairie Enthusiasts.

Mississippi River Commission. 1900. Survey of the Upper Mississippi River.

National Research Council, 2008. *Mississippi River Water Quality and the Clean Water Act*. Washington: National Academies Press.

Nesbit, R.C. 1973. *Wisconsin: A History*. Madison: University of Wisconsin Press.

No author. 1883 *History of Winona County*. Chicago: H.H. Hill and Co.

No author. 1960. They remember village killed by soil erosion. *Winona* (Minnesota) *Daily News*. June 28, 1960. In "Beaver" file at the Winona County Historical Society.

Ostergren, R.C. 1997. The Euro-American settlement of Wisconsin, 1830–1920. In R. Ostegren and T. Vale (eds), *Wisconsin Land and Life*. Madison: University of Wisconsin Press.

Owen, D.D. 1844. Report of a geological exploration of part of Iowa, Wisconsin, and Illinois. U.S. 28th Congress, 1st session, *U.S. Senate Document* 407.

_____ 1847. Report of geological reconnaissance of the Chippewa Land. District of Wisconsin: U.S. 30th Congress, 1 st. session, U.S. Senate *Executive Document* 57.

_____ 1852. *Report of a Geological Survey of Wisconsin, Iowa and Minnesota*. Philadelphia: Lippincott, Grambo and Co.

Paullin, C. and J. Wright. 1932. *Atlas of the Historical Geography of the United States*. Washington, DC: Carnegie Institution of Washington and American Geographical Society of New York.

Perfect, D.E. and D.A. Sheetz. 1942. Physical land conditions on the Farmersburg McGregor Project. Clayton County, Iowa. USDA Soil Conservation Service *Physical Land Survey* No. 28.

Pimentel, D., C. Harvey, P. Resosudarmo, K. Sinclair, D. Kruz, M. McNair, S. Crist, L. Shpritz, R. Saffour, and R. Blair, 1995. Environmental and economic costs of soil erosion and conservation benefits. *Science*. 267: 1117–1123.

*Plat Book of Allamakee County, Iowa*. 1886. Minneapolis: Warner & Foote.

Potter, K.W. 1991. Hydrological impacts of changing land management practices in a moderate-sized agricultural catchment. *Water Resources Research* 27: 845–855.

Price, P. 1998. The effect of climate and land use on the hydrology of the Upper Oconee River Basin, Georgia. Unpublished Ph.D. dissertation. University of California, Los Angeles.

Price, R. 1916. *History of Clayton County, Iowa*. Chicago, Robert Law.

Pyne, S. 1997. *Fire in America: A Cultural History of Wildlife and Rural Fire*. Seattle: University of Washington Press.

Read, M.J. 1941. A population study of the driftless hill land during the pioneer period, 1832–1860: Ph.D. dissertation, Department of Geography, University of Wisconsin-Madison.

Renwick, W., S. Smith, J. Bartley, and R. Buddemeier. 2005. The role of impoundments in the sediment budget of the conterminous United States. *Geomorphology* 71: 99–111.

Ritchie, J.C., C.M. Cooper, and J.R. McHenry. 1986. Sediment accumulation rates in lakes and reservoirs in the Mississippi River Valley in S.Y. Wang, H.W. Shen, and L.Z. Ding (eds.), *Proceedings of the Third International Symposium on River Sedimentation*. University of Mississippi. pp 122–137.

Robinson, E. 1915. Early economic conditions and the development of agriculture in Minnesota, *University of Minnesota Studies in the Social Sciences Bulletin* 3.

Rohe, R. 1997. Lumbering, Wisconsin's northern urban frontier. In R. Ostergren and T. Vale (eds),. *Wisconsin Land and Life*. Madison: The University of Wisconsin Press.

Saari, M. 1956. The town that vanished. *Minneapolis Sunday Tribune*. April 14, 1956.

Sartz, R.S. 1959. The trees are coming back in the Coulee Region. *The Badger Sportsman*. November.

_____. 1961. *Analysis of Watershed Management Problems on Forest and Other Critical Lands of the Driftless Area*. St. Paul: Lake States Forest Experiment Station, U.S. Forest Service.

_____. 1975. Soil erosion in the Lakes States Driftless Area—an historical perspective. Unpublished report in files of North Central Forest Experiment Station, St. Paul, Minnesota.

____. 1976. Sediment yield from steep lands in the driftless area. In *Proceedings of the Third Federal Inter-Agency Sedimentation Conference*. Washington, D.C.: Water Resources Council, Section 1, 123–131.

Scarpino, P. 1985. *Great River: An Environmental History of the Upper Mississippi River*. Columbia: University of Missouri Press.

Schilling, K. and R. Libra. 2003. Increased baseflow in Iowa over the second half of the 20th century. *American Water Resources Association* 39: 851–860.

Schumm, S.A. 1973. Geomorphic thresholds and complex response of drainage systems. In M. Morisawa (ed.), *Fluvial Geomorphology*. pp. 299–310. Binghamton State University of New York. Publications in Geomorphology.

____ 1977. *The Fluvial System*. New York: Wiley.

Schafer, J. 1922. The yankee and the teuton in Wisconsin. *The Wisconsin Magazine of History* 6: 125–145.

Schafer, J. 1932. *The Wisconsin lead region*. Madison: State Historical Society of Wisconsin.

Shaw, E. and A. Trowbridge, 1916. Geologic Atlas of the United States, Galena-Elizabeth Folio. Washington, D.C.: US Geological Survey

Siebenaler, A.M. 1955. 75 farm families driven from their homes: while villages wiped out. *Lewiston (Minn.) Journal*. April 1, 1955.

Sillman, W. 1976. *The Whitewater Valley Report*. Winona, MN: Winona County Historical Society.

Smith, L.S. 1908. Water powers of Wisconsin. *Wisconsin Geological and Natural History Survey Bulletin*. No. 20.

Smith, W. R. 1975 (reprint of 1838 edition). *Observations on Wisconsin Territory*. State Historical Society of Wisconsin Library. Reprinted by Arno Press, New York.

Soil Conservation Service (SCS) 1940. Field maps of gullies and fans in the Whitewater River basin. Various scales. Unpublished studies in the files of the Whitewater River.

____ 1942a. Hydrologic studies at the Coon Creek demonstration project, Soil Conservation Service-Wisconsin-1, Coon Valley, Wisconsin: *Technical Publication* 46.

____ 1942b. Whitewater River watershed survey report (mimeographed). Milwaukee Wisconsin.

____ undated-a, Farming against erosion: the experiences and recommendations resulting from five years of efforts to check erosion in the Coon Creek area at Cook Valley, Wisconsin: Madison, Soil Conservation Office, unpublished manuscript.

____ undated-b Unpublished data collected for the Whitewater and Zumbro Rivers and other regional streams of SE Minnesota.

*Standard Atlas of Vernon County, Wisconsin*. 1930. Rockford, IL: Hixson Map and Atlas Co.

Stanley, D.G., L. Stanley, and J.R. Sellars. 1985. Final report of supplementary phase II investigations at the Motor Townsite. Clayton County, Iowa. Prepared for Clayton County Iowa. Decorah, IA: Bear Creek Archeology, Inc.

Strahler, A. 2010. *Introducing Physical Geography*. New York: Wiley.

Stoeckeler, J. 1959. Trampling of livestock reduces infiltration rates of soil. USDA Forest Service, Lakes States Experiment Station *Technical Note* 556.

Stroessner, W. and J. Habeck. 1966. The presettlement vegetation of Iowa County, Wisconsin. *Transactions of the Wisconsin Academy of Sciences, Arts and Letters* 55: 167–180.

Thorn, W., C. Anderson, W. Lorenzen, D. Henrickson, and J. Wagner, 1997. A review of trout management in southeast Minnesota streams. *North American Journal of Fisheries Management* 17: 860–872.

Thrower, N.J.W. 1960. *Original Survey and Land Subdivision: A Comparative Study of the Form and Effect of Contrasting Cadastral Surveys*. Chicago: Rand McNally.

Trewartha, G. 1940a. The vegetal cover of the driftless cuestaform hill land: Pre-settlement record and postglacial evolution. *Transactions of the Wisconsin Academy of Sciences, Arts, and Letters* 3: 361–382.

____ 1940b. A second epoch of destructive occupance in the driftless hill land. *Annals of the Association of American Geographers*. 30: 109–142.

Trewartha, G. and G.-H. Smith. 1941. Surface configuration of the driftless cuestaform hill land. *Annals of the Association of American Geographers* 31: 25–45.

Troeh, F., J. Hobbs, and R. Donahue, 2004. *Soil and Water Conservation for Productivity and Environmental Protection*. Upper Saddle, NJ: Prentice Hall.

Trowbridge, A.C. 1921. The *Erosional History of the Driftless Area*. University of Iowa Studies in Natural History: Iowa City.

Trowbridge, A.L. undated. Conservation survey report of Coon Creek Watershed, Coon Valley, Wisconsin: Section of Conservation Surveys, Soil Conservation Service, unpublished study on file at office of State Conservationist, Soil Conservation Service, Madison, Wisconsin.

Trimble, S.W. 1969. Culturally accelerated sedimentation on the middle Georgia piedmont, MA Thesis, University of Georgia, Athens. Reprinted and distributed in 1970 by USDA Soil Conservation Service.

―――― 1970. The Alcovy River Swamps: The result of culturally accelerated erosion. *Georgia Academy of Science Bulletin*, 28, 131–141.

―――― 1974. *Man-Induced Soil Erosion on the Southern Piedmont, 1700–1790*. Ankeny, IA: Soil Conservation Society of America. Enhanced 2nd Ed., 2008.

―――― 1975a. Response of Coon Creek, Wisconsin to soil conservation measures. In B. Zakrewska-Borowiecki (ed.), *Landscapes of Wisconsin: A Field Guide*. Washington: Association of American Geographers. pp. 24–29.

―――― 1975b. Field trip notes, Coon Creek Watershed Wisconsin. Mimeographed report given to participants in Geomorphology Field Trip, Annual Meeting. Association of American Geographers. Milwaukee, Wisconsin.

―――― 1975c. A volumetric estimate of man-induced erosion on the Southern Piedmont. In Present and Prospective technology for predicting sediment yields and sources. USDA Agricultural Research Service *Publication S-40*, pp. 142–154.

―――― 1975d. Denudation studies: Can we assume steady state? *Science* 188: 1207–1208.

―――― 1976a. Sedimentation in Coon Creek Valley, Wisconsin, in *Proceedings of the Third Federal Inter-Agency Sedimentation Conference*, Denver, Water Resources Council, Section 5. pp. 100–112.

―――― 1976b. Unsteady state denudation: *Science* 19: 871.

―――― 1976c. Sedimentation rates in Coon Creek Valley, Wisconsin: oral presentation at the Inter-Agency Sedimentation Conference, Denver, Colorado, March 1976.

―――― 1976d. Modern stream and *valley* sedimentation in the driftless area, Wisconsin, USA. In L.P. Gerasimov (ed.), *Geomorphology and Paleogeography*. pp. 228–231. Proceedings of the 23rd International Congress, Moscow.

―――― 1981. Changes in sediment storage in the Coon Creek basin, driftless area, Wisconsin, 1853 to 1975. *Science* 214: 181–183.

―――― 1983. A sediment budget for Coon Creek basin in the Driftless Area, Wisconsin, 1853–1977. *American Journal of Science* 283: 454–474.

―――― 1985. Perspectives on the history of soil erosion control in the eastern United States. *Agricultural History* 59: 162–180.

―――― 1993. The distributed sediment budget model and watershed management in the Paleozoic Plateau of the upper Midwestern United States. *Physical Geography* 14: 285–303.

―――― 1997a. Stream channel erosion and change resulting from riparian forests. *Geology* 25: 467–469.

―――― 1997b. Streambank fish-shelter structures help stabilize tributary streams in Wisconsin. *Environmental Geology* 32: 230–234.

―――― 1998. The use of historical data in fluvial geomorphology. *Catena* 32: 283–304.

―――― 1999. Decreased rates of alluvial sediment storage in the Coon Creek Basin, Wisconsin, 1975–1993. *Science* 285: 1244–1246.

―――― 2004. Effects of riparian vegetation on stream channel stability and sediment budgets. In Sean Bennett and Andrew Simon (eds.), *Riparian Vegetation and Fluvial Geomorphology: Hydraulic, Hydrologic, and Geotechnical Interaction*. Washington, DC American Geophysical Union, 2004. pp. 153–169.

―――― 2008. Some principles of accelerated stream and valley sedimentation by S.C. Happ, G. Rittenhouse, and G. Dobson. *USDA Technical Bulletin* 695, 1940. Invited paper for the series Classics in Physical Geography Revisited. *Progress in Physical Geography* 32: 337–345.

―――― 2009a. Fluvial processes, morphology, and sediment budgets in the Coon Creek Basin, Wisconsin, 1975–1993. *Geomorphology* 108: 8–23

―――― 2009b. Americans should be proud of their accomplishments in soil conservation: A tale of two regions. *Journal of Soil and Water Conservation* 63: 42.

_____ 2010a. Donald Worster's "Dust Bowl." *Aeolian Research* 2: 1–4
_____ 2010b. Streams, valleys, and floodplains in the sediment cascade. In T. Burt and R. Allison (eds), *Sediment Cascades: An Integrated Approach*. Chichester, UK: Wiley. pp. 307–343.
_____ 2011. The Historical decrease of soil erosion in the eastern United States—the role of geography and engineering. Chapter 77 in S. Brunn, *Engineering Earth*. Berlin: Springer. pp. 1383–1393.
Trimble, S.W. and S.W. Lund. 1982. Soil conservation and the reduction of erosion and sedimentation In the Coon Creek basin, Wisconsin. *U.S. Geological Survey Professional Paper* 1234. Washington, DC: Government Printing Office.
Trimble, S.W. and Bube, K.P. 1990. Improved reservoir trap efficiency prediction. *The Environmental Professional*. 12: 255–272.
Trimble, S.W. and Mendel, A. 1995. The cow as geomorphic agent—a critical review. *Geomorphology*, 13: 233–253.
Trimble, S.W. and Crosson, P. 2000. U.S. soil erosion rates—myth or reality. *Science* 289: 248–250.
Veatch, O., and C. Orben. 1923. *Soil Survey of Dubuque County, Iowa*. USDA Bureau of Soils. Washington, DC: Government Printing Office.
Vita-Finzi, C. 1967. *The Mediterranean Valleys*. Cambridge: Cambridge University Press.
Vondracek, B., K. Blann, C. Cox, K. Mumford, B. Nerbonne, and J. Nerbonne. 2005. Land use, spatial scale, and stream systems: Lessons from an agricultural region. *Environmental Management* 36: 775–791.
Ward, A.D. and S.W. Trimble. 2004. *Environmental Hydrology*. Boca Raton, FL.
Ward, R.C. 1975. *Principles of Hydrology*, 2nd Ed., McGraw-Hill,:Maidenhead, UK.
Warren, G. 1867. Survey of Upper Mississippi River: *House Executive Document 58*, 39th Congress, 2nd Session.
Whitson, A., J. Geib, J. Dunnewald, and C. Lounsbury. 1914. Soil survey of La Crosse County. Wisconsin. Madison: Wisconsin Geological and Natural History.
Whitson, A. and T. Dunnewald. 1916. Keep our hillsides from washing. Agricultural Experiment Station of the University of Wisconsin, *Bulletin* 272.
Whitson, A., J. Geib, T. Dunnewald, O. Noer, C. Lounsbury, and L. Cantrell. 1917. Soil Survey of Buffalo County, Wisconsin. Madison: Wisconsin Geological and Natural History Survey, *Bulletin* 54-A.
Whitson, A., J. Geib, T. Dunnewald, A. Goodman, G. Musgrave, and C. Clevenger. 1923. Soil survey of Jackson County, Wisconsin. Madison: Wisconsin Geological and Natural History Survey, *Bulletin* 54B.
Wilder, F. and T. Savage. 1906. *Annual Report of the Iowa Geological Survey, 1905*, Vol. 16. Des Moines: Iowa Geological Survey.
Williams, M. 2010. Clearing the forests, In M. Conzen (ed.), *The Making of the American Landscape*, 2nd Ed. New York, Routledge. pp. 162–187.
Winchell, N.H. 1884. *The Geological and Natural History Survey of Minnesota, 1872–1882*. Vol. 1. Minneapolis: Johnson, Smith, and Harrison.
*Winona [MN] Republican-Herald*, Mud buries two cars on highway 61, June 8, 1940.
Wischmeier, W. and D. Smith, 1978. Predicting rainfall losses: A guide to conservation planning. *Agricultural Handbook* 537, USDA Science and Education Administration.
Wisconsin Board of Commissioners of Public Lands, 1848. Land survey field notes, http://libtext.library.wisc.edu/cgi-bin/SurveyNotes.
Wolman, M. and J. Miller. 1960. Magnitude and frequency of forces in geomorphic processes. *Journal of Geology* 68: 54–74.
Woltemade, C.J. 1994. Form and process: Fluvial geomorphology and flood-flow interaction, Grant River, Wisconsin. *Annals, Association of American Geographers* 84: 462–479.
Works Progress Administration (WPA). 1937. Federal Writers Project. American Guide Series. Galena, Illinois. No place, no publisher.
Worster, D. 2004. *Dust Bowl: The Southern Plains in the 1930s* (Twenty-fifth anniversary edition). New York, Oxford University Press.

Zakrzewska, B. 1971. Valleys of driftless areas. *Annals of the Association of American Geographers.* 61: 441–459.

Zeasman, O.R. and I.O. Hembre. 1963. *A Brief History of Soil Erosion Control in Wisconsin.* Madison: Wisconsin Soil and Water Conservation Committee Special Publication.

Zon, R. 1929. Relation of forestry to the control of floods in the Mississippi Valley. *House Document* 513, 70th Congress, 2nd Session.

# *Glossary*

Many of the definitions are based on *ASAE Standard: ASAE 5256 Soil and Water Engineering Terminology*, American Society of Agricultural Engineers, St. Joseph, MI.

Terms not contained in this glossary can possibly be found at these glossary Web sites:

- EPA Terms of Environment: http://www.epa. gov/docs/OCEPAterms/
- USGS Unofficial Glossary: http://wwwga.usgs. gov/edu/dictionary.html
- Water Quality Association Glossary of Terms: http://www.wqa.org/

**Accelerated erosion** Erosion much more rapid than normal, natural, or geological erosion, primarily as a result of the influence of the activities of humans or, in some cases, of animals.

**Accretion (of sediment)** Deposition of sediment on some surface when the stream velocity becomes too low to transport the material. Vertical accretion normally occurs on floodplains and tends to to be finer. Lateral accretion normally occurs on point bars and tends to be coarser than vertical accretion.

**Accuracy** The proximity of the measured value to the actual value.

**Acre** A measure of area covering 43,560 ft2 or 0.4047 hectare. While it might be perceived as a square 209 ft on a side, it can take any shape.

**Acre-foot** A measure of volume. It is one acre, 1 ft deep (43, 560 ft3) or 1234 m3 ( see table of conversions).

**Adsorption** Attachment of molecules of gas or molecules in solution to the surface of solid materials with which they come into contact.

**Aggradation (of stream bed)** An increase of bed elevation created by accretion (accumulation) of sediment

**Alluvial** Relating to streams

**Alluvial fan** A deposit of sediment caused by a sharp decrease of gradient in a sediment-laden stream. It is fan-shaped in plan and conical in cross section, normally being thickest in the middle.

**Alluvium** Sediment deposited by a stream, usually in the channel or on the flood plain. Such sediment is susceptible to being reentrained by the stream at some future date ranging from hours to millennia with that along the channel being much more vulnerable than distal deposits.

**Anaerobic decomposition** The decay of organic matter by microorganisms in the absence of oxygen.

**Aquifer** A geologic formation that holds and yields usable amounts of water. Aquifers can be classified as confined or unconfined.

**Aquitard** Underground geologic formation that is slightly permeable, but yields inappreciable amounts of water compared to an aquifer.

**Arroyo** A gully or trenched ephemeral stream. The term is normally used for deep, steep-walled gullies found in the American Southwest (see Cooke and Reeves, 1976).

**Alfisol** Moderately fertile soils generally formed under forest in mid-latitudes with grey to brown surface horizons, medium to high supply of bases, and B horizons of illuvial clay accumulation. Formerly called grey-brown forest podzols

**Available plant water** The portion of water in a soil that can be readily absorbed by plant roots. It is the amount of water released between in situ field capacity and the permanent wilting point.

**Average rate of precipitation** Derived by dividing the amount of precipitation that occurs during a given time period by the length of that period. Common units of intensity are inches, millimeters, or centimeters per hour. Average rate is also known as intensity.

**Bank storage** Water leaving a stream channel and flowing into the bank during rising stages of streamflow. It normally flows back out during falling stages and leaves banks weakened from the saturation. See Floodplain storage.

**Bankfull discharge** The streamflow that fills the main channel and begins to spill onto the active floodplain (see Wolman and Miller, 1960). It is a range of flows that is most effective in forming a channel, benches (floodplains), banks, and bars). The bankfull discharge is considered to be the channel-forming or effective discharge.

**Bankfull indicators** Physical characteristics of a channel that denote bankfull.

**Bankfull width, depth, or cross-sectional area** The channel width, depth, or cross-sectional area, respectively, when the streamflow is at bankfull discharge.

**Base flow** Sustained low flow of a stream often due to groundwater inflow to the stream channel. Often written as a single word. See Groundwater flow.

**Bed features** The sequence of bed forms found in streams, such as riffle–pools, step–pools, cascades, and convergence/divergence. The particular form a stream achieves is dependent on channel plan form and gradient.

**Bed load** Coarse sediment or material moving on or near the bottom of a flowing channel by rolling, sliding, or bouncing.

**Best management practice (BMP)** Structural, nonstructural, and managerial techniques recognized as the most effective and practical means to reduce surface and groundwater contamination while still allowing the productive use of resources.

**Braided river** A river channel characterized by an abundance of islands that continually divide and subdivide the flow of the river.

**Bulk density (Soil)** The mass of dry soil per unit bulk volume. The bulk volume is determined before drying to constant weight at 105°C (220°F). It is one indicator of porosity, lower bulk density indicating more porosity. Most soils fall between 1.2-1.5 kg/l or 75-95 lbs/cubic foot.

**Bushel** A volumetric measure of agricultural products= 9 liquid gallons=4 pecks=2150.2 cu. in.=35.24 liters. In practice, the measure is by weight, specific to the crop and moisture condition. Corn (maize) for example is 56 lbs. or 27.2 kg.

**Canopy** Vegetative cover over the land surface of a catchment area.
**Capacity (of streams)** Streams have a finite sediment transport rate.
**Catchment** See Watershed.
**Channel** The bed and banks of a stream or river.
**Channel capacity** Flow rate in a channel when flowing full.
**Channel evolution** Channels will change shape over time to seek equilibrium of the multiple factors, such as sediment supply, valley geology, water surface slope, vegetation, and the like, which influence their form. The stages of the evolution process are typically predictable.
**Channel improvement** Increasing the cross section, straightening, or clearing vegetation from a channel to change its hydraulic characteristics, increase its flow capacity, and reduce flooding. Often used perjoratively.
**Channel stabilization** Erosion prevention and stabilization of a channel using vegetation, jetties, drops, revetments, or other measures.
**Chute** A linear depression, or minor channel, usually discontinuous, created by flood waters flowing over a floodplain. A common location is the across the neck of a meander bend where floodwater sometimes takes a "shortcut" during a flood.
**Clay** Soil or sediment particles less than 2 µm in equivalent diameter.
**Climatology** The study of the long term movement of air masses from their sources (continents, lakes, oceans, transpiration from land areas) and their associated effects on precipitation and temperature over time.
**Coefficient of variation** The standard deviation expressed as a percentage of the mean.
**Conventional agriculture** The traditional form of agriculture with tillage of the bare soil and planting of crops. See conventional tillage.
**Colluvium** Sediment deposited by slope wash, usually not susceptible to being reentrained
**Competence** The ability of a stream to carry large sediment particles.
**Connectivity** The propensity of water and sediment to rapidly move from uplands downslope to stream valleys.
**Conservation tillage** A tillage practice that leaves plant residues on the soil surface for erosion control and water conservation.
**Contour plowing** Furrows that are placed across a slope at constant elevation so that the depression of the furrow can hold rainwater that might otherwise flow downslope.
**Contour strips** Bands of crops placed across slopes at constant elevation, designed to slow and detain the flow of water downslope. Any plowing is done on the contour also.
**Convective** Air that expands when heated by solar energy and becomes lighter than the air around it. The lighter air rises by convection, potentially causing convective precipitation.
**Conventional tillage** The traditional tillage practice that involves inverting the tillage layer, burying most of the plant residues, and leaving the soil bare.
**Conversion (to agricultural land)** The converting of land, often forest or prairie, to agricultural land.
**Correlation** The intensity or level of association between the two variables.
**Cover crop** Close-growing crop that provides soil protection, seeding protection, and soil improvement between periods of normal crop production or between trees in orchards and vines in vineyards. When plowed under and incorporated into the soil, cover crops may be referred to as green manure crops.
**Creek** A small stream (smaller than a river), often a shallow or intermittent tributary to a river. Also called branch, brook, or run.
**Crop residue** Portion of a plant or crop left in the field after harvest.

**Crop rotation** A system of farming in which a succession of different crops is planted on the same land area, as opposed to growing the same crop time after time (monoculture).

**Crusting** The dry condition when water will not enter the coarse-textured subsurface layer until it becomes wet; it affects many soils, especially those with low organic matter.

**Cut bank** The stream bank on the outside of a stream meander that is being eroded away. This is a major mechanism of stream channel erosion. It often presents a bare and vertical appearance.

**Debris flow** An unusual and sometimes dangerous form of mass movement. These are fast moving, liquefied landslides of mixed and unconsolidated water and sediment that look like flowing concrete.

**Deep percolation water** Water that moves downward through the soil profile below the root zone and cannot be used by plants.

**Degradation (of stream bed)** The lowering of stream bed elevation by loss of sediment, sometimes accompanied by erosion or mass wastage (slumping) of stream banks.

**Dendritic** A treelike system of channels.

**Dendrochronology** The study of estimating annual precipitation rates using tree growth records. Rings are thin for dry years and thick for wet years, and the record can be extended back for centuries.

**Discharge** Rate of water movement.

**Discharge curve** Rating curve that shows the relation between stage and flow rate of a stream, channel, or conduit.

**Dominant channel materials (D50)** The median size particle determined by a channel material size distribution analysis, typically a pebble count.

**Drainage** Process of removing surface or subsurface water from a soil or area.

**Drainage basin** See Watershed.

**Edaphic** Soil conditions as they relate to biological life.

**Effective discharge** The streamflow that transports the most sediment over the long term (Wolman and Miller, 1960).

**Ephemeral gully** Small channels eroded by runoff that can be easily filled and removed by normal tillage, only to re-form again in the same location.

**Ephemeral stream** A stream that is dry most of the year and only contains water during and immediately after a rainfall event.

**Equilibrium (watershed)** A condition when input equals output or storage is zero. For a stream system or watershed, it means that sediment yield over some period of time equals erosion from the basin so that there is no gain or loss of sediment storage. For a stream reach, it means that the accumulation of sediment is offset by erosion, so that there is neither aggradation or degradation. Also termed Steady State.

**Erosion** The wearing away of the land surface by running water, wind, ice, or other geological agents, including such processes as gravitational creep.

**Erosivity** The ability of rainfall to detach and transport soil particles.

**Evaporation** The physical process by which a liquid is transformed to a gaseous state.

**Evapotranspiration** The combination of water transpired from vegetation and evaporated from the soil and plant surfaces.

**Exceedence probability** The probability that an event with a specified magnitude and duration will be equaled or exceeded in some time period.

**Field capacity** Amount of water remaining in a soil when the downward water flow due to gravity becomes negligible.

**Flood series** An array, or a probability analysis of historic high stream flows for a location

**Flood spillway** An auxiliary channel to carry a flood flow that exceeds a given design rate to the channel downstream (the preferred term is emergency spillway).

**Floodplain** The relatively level area on either side of the bankfull channel that carries the flow greater than the bankfull flow, that is, all discharges greater than the 1- to 2-year return frequency. It is created by a combination of overbank flow and lateral migration of meander bends.

**Floodplain storage** Volume of water that spreads out and is temporarily stored on a floodplain.

**Flow rate** Rate of water movement. Often written as a single word and expressed in cubic feet per second or cubic meters per second.

**Flume** Open conduit for conveying water across obstructions.

**Fluvial** Pertaining to streams.

**Frequency–magnitude (of storm events)** The relationship between frequency of occurrence and the amount of rainfall or runoff.

**Friable** Poorly cemented or highly weathered material, usually rock, that crumbles easily.

**Gaging station** Section in a stream channel equipped with a gage or facilities for obtaining streamflow data.

**Gaining stream** Stream or part of a stream that has an increase in flow because of inflow from groundwater, often termed an effluent stream.

**Geological erosion** The normal or natural erosion caused by geological processes acting over long geological periods (synonymous with natural erosion).

**Geomorphology** The study of landforms and the processes which create them

**Glacial outwash** Rock debris carried by a glacier which is washed downslope and downstream as a glacier melts.

**Grade** Degree of slope to a road, channel, or ground surface usually expressed as a dimensionless fraction.

**Graded stream** A stream channel adjusted to flow and sediment conditions. Often used interchangeably with steady-state and equilibrium.

**Grassed waterway** Natural or constructed channel covered with an erosion-resistant grass that transports surface runoff to a suitable discharge point at a nonerosive rate.

**Groundwater** Water occurring in the zone of saturation in an aquifer.

**Groundwater flow** Flow of water in an aquifer. The portion of the discharge of a stream derived from groundwater, often termed baseflow.

**Growing season** The period, often the frost-free period, during which the weather is such that crops can be produced.

**Gully** Eroded channel where runoff concentrates, usually so large that it cannot be obliterated by normal tillage operations.

**Gully erosion** The erosion process by which water accumulates in narrow channels and, over short periods, removes the soil from this narrow area to considerable depths, ranging from 0.5 m (1.6 ft) to as much as 30 m (97 ft.)

**Gully head advance** Upstream migration of the upper end of a gully. Sometimes referred to as a head cut.

**Head** The height to which water can raise itself above a known datum (commonly sea level), exerting a pressure on a given area, at a given point. It is synonymous with hydraulic head.

**Humid climate** Climate characterized by high rainfall and low evaporation potential. A region is usually considered humid when precipitation averages more than 500 mm (20 in.) per year.
**Hydraulic length** Longest flow path on a watershed.
**Hydraulic resistance** Friction along the wetted boundary of a channel or conduit that causes a loss in head.
**Hydrograph** Graphical or tabular representation of the flow rate of a stream with respect to time.
**Hydrologic cycle** Term used to describe the movement of water in and on the earth and atmosphere. Numerous processes, such as precipitation, evaporation, condensation, and runoff, comprise the hydrologic cycle.
**Hydrologic groups** Classification of soils based on infiltration capacities after a prolonged period of wetting. See Figure 1.6.
**Hydrologist** A person who considers such phenomena as precipitation, temperature, and land use in relation to water supplies and movement to, on, and beneath the earth's surface.
**Hysteresis** (1) A differential relationship between the dependent and independent variables, the difference generally being between increasing and decreasing values of the independent variable ( see Figure 3.7). (2) The condition that is caused during wetting when the small pores fill first, while during drainage and drying, the large pores empty first.
**Impermeable layer** (Soil) Layer of soil resistant to penetration by water, air, or roots. See plow pan.
**Infiltration** The downward entry of water through the soil surface into the soil.
**Infiltration capacity** The point at which additional precipitation is not infiltrated.
**Infiltration rate** The quantity of water that enters the soil surface in a specified time interval. Often expressed in depth of water per unit of time.
**Influent stream** Stream or portion of stream that contributes water to the groundwater supply. See Losing stream.
**Interception** The portion of precipitation caught by vegetation and prevented from reaching the soil surface.
**Interflow** Water that infiltrates into the soil and moves laterally through the upper soil horizons until it returns to the surface, often in a stream channel. Also called throughflow.
**Interfluve** The boundary between streams.
**Intermittent stream** Natural channel in which water does not flow continuously.
**Interrill erosion** The removal of a fairly uniform layer of soil on a multitude of relatively small areas due to raindrop impact and by shallow surface flow.
**Lacustrine** Relating to lakes.
**Lag time** (Hydrology) The interval between the time when one half of the equivalent uniform excess rain (runoff) has fallen and the time when the peak of the runoff hydrograph occurs.
**Lateral point bar accretion** The process by which point bars are created on the inside of the bend where the flow and velocity are lower, allowing deposition, thereby forming a point bar.
**Legacy sediment** Sediment from historical soil erosion, usually stored as alluvium, which is subject to reentrainment by stream action and further transportation.
**Levee** A ridge or berm along a stream. A natural levee is created by overbank flows which selectively deposit the coarser part of the sediment load as velocity slows along

**Losing stream** A channel that loses water into the bed or banks. See Influent stream.

**Overland flow** That portion of rainfall that cannot infiltrate the soil and thus cascades downslope on the surface.

**Magnetic declination** The difference at a location, between true north and the direction shown by a compass (magnetic azimuth).

**Mass movement** The downward movement of material under the influence of gravity on a slope (e.g., landslips, mudflow, landslides, etc).

**Meandering** The propensity of natural channels to vary from a straight line and follow a winding and turning course. In the process, they erode the outsides of bends (see Cut bank) and deposit sediment on the inside of bends (see point bars).

**Mollisol** Very fertile soils formed under grass cover with very dark, organic-rich surface horizons and high supply of bases. Formerly termed prairie in the Hill Country.

**Natural Resources Conservation Service** An agency of the U.S. Department of Agriculture (USDA) charged with assisting farmers, builders and everyone in conserving soil, water, and all associated natural resources. Formerly known as the Soil Conservation Service (SCS).

**Natural erosion** Wearing away of the earth's surface by water, ice, or other natural agents under natural environmental conditions of climate, vegetation, and the like undisturbed by humans. See Geological erosion.

**No tillage or no till** A tillage system in which the soil is not tilled except during planting, when a small slit is made in the soil for seed and agrichemical placement. Pest control is achieved using pesticides, crop rotation, and biological control rather than tillage. Sometimes called zero tillage.

**Nonpoint source pollution (NPS)** Pollution originating from diffuse areas (land surface or atmosphere) with no well-defined source.

**Normal erosion** The gradual erosion of land used by society; does not greatly exceed natural erosion. See Natural erosion.

**Occluded channel** A condition when the channel has been completely filled or blocked, usually by sediment but also by other materials such trees or debris

**Overbank vertical accretion** The process by which materials that are finer than those deposited during lateral point bar accretion are deposited on the banks and floodplain when flows exceed the effective discharge and overtop the banks (except in deeply incised channels).

**Overland flow** Surface runoff occurring at relatively shallow depths across the land surface when precipitation rates exceed infiltration rated. May cause sheet and rill erosion.

**Particle size analysis** Determination of the various amounts of gain sizes in a sample, usually by sedimentation, sieving, or micrometry.

**Pattern, profile, and dimension** These terms are used by geomorphologists to describe completely the features of a stream channel. Pattern refers to the plan view of a stream. The meander is the pattern of concern. Profile refers to a longitudinal cross section of the channel. Slope is observed in the profile. Dimension refers to a cross section. The flow cross section is observed here.

**Percolating water** Subsurface water that flows through the soil or rocks. See Seepage.

**Percolation** Downward movement of water through porous media such as soil.

**Perennial stream** A stream that flows throughout the year.

**Permeability** (1)(Qualitative) The ease with which gases, liquids, or plant roots penetrate or pass through a layer of soil or porous media. (2) (Quantitative) The specific soil property designating the rate at which gases and liquids can flow through the soil or porous media.

**Plow pan** A hard and often less permeable layer of soil at plow depth caused by the downward pressure of a plow passing through the soil.

**Point bar** Accumulation of sand and other sediments on the inside of meander bends in stream channels.

**Porosity** The volume of the void spaces through which water or other fluids can be stored or travel in a rock or sediment divided by the total volume of the rock or sediment.

**Precipitation depth** The total amount of precipitation, usually in inches, centimeters, or millimeters.

**Precipitation duration** The time from the beginning of the storm until the end of the storm.

**Precision** The scatter resulting between repeated measurements.

**Raindrop erosion** Soil detachment resulting from the impact of raindrops on the soil. See Erosion.

**Rainfall erosivity** A measure of rainfall's ability to detach and transport soil particles.

**Rainfall frequency** Frequency of occurrence of a rainfall event with an intensity and duration that can be expected to be equaled or exceeded (the preferred term is return period).

**Rainfall intensity** Rate of rainfall for any given time interval, usually expressed in units of depth per time.

**Range** Equal to the difference between the smallest and largest measurement in a data set.

**Reach** A length of a stream or channel with relatively constant characteristics.

**Recession curve** Descending portion of a streamflow or hydrograph.

**Recharge** Process by which water is added to the zone of saturation to replenish an aquifer.

**Recharge area** Land area over which water infiltrates and percolates downward to replenish an aquifer. For unconfined aquifers, the area is essentially the entire land surface overlaying the aquifer; for confined aquifers, the recharge area may be a part of or unrelated to the overlaying area.

**Recurrence interval** See Return period.

**Regime** Condition of a stream with respect to short and long term variance in its rate of flow.

**Relative humidity** Ratio of the amount of water present in the air to the amount required for saturation of the air at the same dry bulb temperature and barometric pressure, expressed as a percentage.

**Reservoir** Body of water, such as a natural or constructed lake, in which water is collected and stored for use.

**Return period** The frequency of occurrence of a hydrologic event with an intensity and duration that can be expected to be equaled or exceeded, usually expressed in years.

**Reversion (of agricultural land)** By intention or default, allowing agricultural land to return to a natural condition such as forest or prairie.

**Riffle** The shallow area of a stream where water accelerates, the water surface becomes rippled, and in more turbulent waters, contains a hydraulic jump.

**Riffle–pool sequence** The pattern of consecutive riffle and pools that naturally form in many streams and rivers. In steeper valleys, a step–pool pattern will form.

**Rill** Small channel eroded into the soil surface by runoff that can be filled easily and removed by normal tillage.

**Rill erosion** An erosion process by concentrated overland flow in which numerous small channels only several centimeters deep are formed; occurs mainly on recently cultivated soils. See Rill.

**Riparian** Pertaining to the banks of a body of water, as in riparian vegetation.

**River** A large natural stream of water emptying into an ocean, lake, or other body of water and usually fed along its course by converging tributaries.

**Rip-rap** Heavy rock, typically limestone, broken into sub-angular pieces to make them interlock and used to protect streambanks and other critical areas from erosion.

**River flow** Flow of water in a channel.

**Root zone** Depth of soil that plant roots readily penetrate and in which the predominant root activity occurs.

**Row crop** Crops such as corn (maize), soybeans, potatoes, and cabbage which are planted in rows separated by a space of normally bare soil. Such crops are considered to be more erosive. Also termed 'clean-tilled crops' because the spaces between the rows are kept clean of weeds and other vegetation.

**Row grade** The slope in the direction of crop rows.

**Runoff** The portion of precipitation, snowmelt, or irrigation that flows over and through the soil, eventually making its way to surface water supplies, and leaving the watershed. Not synonymous with overland flow.

**Sample** A subset of data collected from the population or the collection of observations actually available for analysis.

**Sand** Soil or sediment particles ranging from 50 to 200 µm in diameter. Soil material containing 85% or more particles in this size range.

**Sediment basin** Pond at the upper end of a conveyance or reservoir for detaining particle-laden water for a sufficient length of time for deposition to occur. Often use downstream of construction sites as BMP.

**Sediment budget** An accounting for all the flux or movement and storage of sediment for a stream basin or stream reach. It normally considers upland erosion, colluvial deposition on slopes, alluvial deposition in streams and on floodplains, stream channel erosion, and sediment yield.

**Sediment load** Amount of sediment carried by running water or wind.

**Sediment rating curve** The relationship between sediment load and stream discharge.

**Sediment transport** Amount of sediment transported by a stream; calculated from the sediment rating curve and the flow duration curve, usually given in tons/year or, more commonly, tons/unit area/year.

**Sediment yield** The amount of sediment leaving a watershed, sometimes termed efflux in sediment budget studies.

**Sedimentation** Deposition of waterborne or wind-borne particles resulting from a decrease in transport capacity.

**Sheet erosion** The removal of soil from the land surface by rainfall and surface runoff. Often interpreted to include rill and interrill erosion.

**Sheet flow** Water, usually storm runoff, flowing in a thin layer over the soil or other smooth surface.

**Silt** (1) Soil or sediment particles between 2 and 50 µm in diameter. (2) (Colloquial) Deposits of sediment that may contain soil particles of all sizes.

**Silt bar** A deposition of sediment in a channel.

**Sinuosity** A bending or curving shape or movement. Defined as ratio of the length of the channel to the valley length .

**Small grains** Wheat, oats, barley and rye

**Soil** The unconsolidated minerals and material on the immediate surface of the earth that serves as a natural medium for the growth of plants. Soils have properties due to the integrated effect of climate and living matter acting upon the parent material, as influenced by landforms, over periods of time.

**Soil aeration** Process by which air and other gases enter the soil or are exchanged.

**Soil compaction** Consolidation, reduction in porosity, and collapse of the structure of soil when subjected to surface loads.

**Soil conservation** Protection of soil against physical loss by erosion and chemical deterioration by the application of management and land use methods that safeguard the soil against all natural and human-induced factors.

**Soil erodibility** A measure of the soil's susceptibility to erosional processes.

**Soil erosion** Detachment and movement of soil from the land surface by wind or water. See Erosion.

**Soil map unit** A delineation on a map of an area dominated by one major kind of soil or several soils with similar properties.

**Soil series** A group of soils that have similar characteristics and a similar arrangement of soil layers (horizons).

**Soil structure** The combination or arrangement of primary soil particles into secondary particles, units, or peds that make up the soil mass. These secondary units may be, but usually are not, arranged in the profile in such a manner as to give a distinctive characteristic pattern. The principal types of soil structure are platy, prismatic, columnar, blocky, and granular.

**Soil texture** Classification of soil by the relative proportions of sand, silt, and clay present.

**Soil water** All water stored in the soil.

**Soil and water conservation district (SWCD)** A local governmental entity within a defined water or soil protection area that provides assistance to residents in conserving natural resources, especially soil and water.

**Spatial** Area or space.

**Spillway** Conduit through or around a dam for the passage of excess water. May have controls.

**Splash erosion** The detachment and airborne movement, caused by the impact of raindrops on soils, of small soil particles.

**Stage** Elevation of a water surface above or below an established datum; gage height.

**Steady state (streams)** See Equilibrium and graded stream.

**Stratified soils** Soils that are composed of layers, usually varying in permeability and texture.

**Stream** A current or flow of water running along a channel; specifically, a creek or small river .

**Stream bank stabilization** Vegetative or mechanical control of erodible stream banks, including measures to prevent stream banks from caving or sloughing, such as lining banks with riprap or matting and constructing jetties or revetments, as necessary, for permanent protection.

**Stream instability** A stream that is not self-sustaining, changes geometry, and is not balanced between import and export of sediment.

**Stream stability** Requires a stream to be self-sustaining, retain the same general geometry over time (decades), and be balanced between the import and export of sediment.

**Stream channel erosion** Scouring of soil and the cutting of channel banks or beds by running water. Sometimes called streambed erosion or stream bank erosion.
**Streamflow** The rate of water movement in a stream. Often written as two words.
**Subsurface drain** Subsurface conduits used primarily to remove subsurface water from soil. Classifications of subsurface drains include pipe drains, tile drains, and blind drains.
**Succession** The sequence of changes or stages in a plant community as it develops over time and eventually leading to a climax stage.
**Surface drainage** The diversion or orderly removal of excess water from the surface of land by improved natural or constructed channels, supplemented when necessary by shaping and grading of land surfaces to such channels.
**Surface runoff** Precipitation, snowmelt, or irrigation in excess of what can infiltrate or be stored in small surface depressions. See overland flow.
**Surface sealing** Reorienting and packing of dispersed soil particles in the immediate surface layer of soil and clogging of surface pores, resulting in reduced infiltration.
**Surface water** Water flowing or stored on the earth's surface.
**Suspended load** Fine materials such as clay, silt, and fine sand that remain suspended in the water column, but can settle in locations where the travel velocity is low or the settling depth is small. The suspended load can be more than 90% of the transported material.
**Suspended sediment** Material moving in suspension in a fluid; caused by the upward components of the turbulent currents or by colloidal suspension. Sometimes called suspended load.
**Terrace** (1) A broad channel, bench, or embankment constructed across the slope to intercept runoff and detain or channel it to protected outlets. (2) A level plain, usually with a steep front, bordering a river, lake, or sea. (3) A relict floodplain or lake bed.
**Thalweg** The line of flow in a stream channel, normally the deepest part.
**Throughflow** See interflow.
**Transpiration** The process by which water in plants is transferred to the atmosphere as water vapor.
**Trend line or regression line (of precipitation)** A statistical technique to determine the annual average of precipitation for a given period of time.
**Turbidity** The degree of opaqueness in a liquid caused by the presence of suspended material.
**Underflow** Stream flow which passes through porous streambed deposits, often undetected.
**Wash load** That part of the sediment load of a stream composed of suspended clay and silt particles.
**Water table** The upper surface of saturated zone below the soil surface where the water is at atmospheric pressure.
**Watershed** Land area that contributes runoff (drains) to a given point in a stream or river. Synonymous with catchment and drainage or river basin.
**Watershed gradient** The average slope in a watershed; measured along a path of water flow from a given point in the stream channel to the most remote point in the watershed.
**Wetlands** Area of wet soil that is inundated or saturated under normal circumstances and would support a prevalence of hydrophytic plants.
**Xeric** Dry

# Appendix: Unit Conversion Factors

| Multiply the U.S. Customary Unit | | By | To Obtain the SI Unit | |
|---|---|---|---|---|
| Name | Symbol | | Symbol | Name |
| **Acceleration** | | | | |
| feet per second squared | ft/sec$^2$ | 0.3048 | m/sec$^2$ | meter per second squared |
| inch per second squared | in./sec$^2$ | 0.0254 | m/sec$^2$ | meter per second squared |
| **Area** | | | | |
| acre | acre | 0.4047 | ha | hectare |
| acre | acre | 4.0469 × 10–3 | km$^2$ | square kilometer |
| square foot | ft$^2$ | 9.2903 × 10$^{-2}$ | m$^2$ | square meter |
| square inch | in.$^2$ | 6.4516 | cm$^2$ | square centimeter |
| square mile | mi$^2$ | 2.5900 | km$^2$ | square kilometer |
| square yard | yd$^2$ | 0.8361 | m$^2$ | square meter |
| **Energy** | | | | |
| British thermal unit | Btu | 1.0551 | kJ | joule |
| foot pound-force | ft·lbf | 1.3558 | J | joule |
| horsepower-hour | hp·h | 2.6845 | MJ | megajoule |
| kilowatt-hour | kWh | 3600 | kJ | kilojoule |
| kilowatt-hour | kWh | 3.600 × 106 | J | joule |
| watthour | W·h | 3.600 | kJ | kilojoule |
| watt-second | W·sec | 1.000 | J | joule |
| **Force** | | | | |
| pound-force | lb$_f$ | 4.4482 | N | newton |

233

# Appendix: Unit Conversion Factors

| Multiply the U.S. Customary Unit | | By | To Obtain the SI Unit | |
|---|---|---|---|---|
| Name | Symbol | | Symbol | Name |
| **Flow Rate** | | | | |
| cubic foot per second | ft³/sec | $2.8317 \times 10^{-2}$ | m³/sec | cubic meter per second |
| gallon per day | gal/d | $4.3813 \times 10^{-2}$ | L/sec | liter per second |
| gallon per day | gal/d | $3.7854 \times 10^{-3}$ | m³/d | cubic meter per day |
| gallon per minute | gal/min | $6.3090 \times 10^{-5}$ | m³/sec | cubic meter per second |
| gallon per minute | gal/min | $6.3090 \times 10^{-2}$ | L/sec | liter per second |
| million gallon per day | Mgal/d | 43.8126 | L/sec | liter per second |
| million gallon per day | Mgal/d | $3.7854 \times 10^{3}$ | m³/d | cubic meter per day |
| million gallon per day | Mgal/d | $4.3813 \times 10^{-2}$ | m³/sec | cubic meter per second |
| **Length** | | | | |
| foot | ft | 0.3048 | m | meter |
| inch | in. | 2.54 | cm | centimeter |
| inch | in. | 0.0254 | m | meter |
| inch | in. | 25.4 | mm | millimeter |
| mile | mi | 1.6093 | km | kilometer |
| yard | yd | 0.9144 | m | meter |
| **Mass** | | | | |
| ounce | oz | 28.3495 | g | gram |
| pound | lb | $4.5359 \times 10^{2}$ | g | gram |
| pound | lb | 0.4536 | kg | kilogram |
| ton (short: 2000 lb) | ton | 0.9072 | Mg (metric ton) | megagram (10³ kilogram) |
| ton (long: 2240 lb) | ton | 1.0160 | Mg (metric ton) | megagram (10³ kilogram) |
| **Power** | | | | |
| British thermal units per second | Btu/sec | 1.0551 | kW | kilowatt |
| foot-pound (force) per second | ft·lb$_f$/sec | 1.3558 | W | watt |
| horsepower | hp | 0.7457 | kW | kilowatt |
| **Pressure (force/area)** | | | | |
| atmosphere (standard) | atm | $1.0133 \times 10^{2}$ | kPa (kN/m²) | kilopascal (kilonewton per square meter) |
| inches of mercury (60°F) | in. Hg (60°F) | $3.3768 \times 10^{3}$ | Pa (N/m²) | pascal (newton per square meter) |
| inches of water (60°F) | in. H$_2$O (60°F) | $2.4884 \times 10^{2}$ | Pa (N/m²) | pascal (newton per square meter) |
| pound-force per square foot | lb$_f$·ft² | 47.8803 | Pa (N/m2) | pascal (newton per square meter) |
| pound-force per square inch | lb$_f$·in.² | $6.8948 \times 10^{3}$ | Pa (N/m²) | pascal (newton per square meter) |

# Appendix: Unit Conversion Factors

| Multiply the U.S. Customary Unit | | By | To Obtain the SI Unit | |
|---|---|---|---|---|
| Name | Symbol | | Symbol | Name |
| pound-force per square inch | lbf·in.$^2$ | 6.8948 | kPa (kN/m$^2$) | kilopascal (kilonewton per square meter) |
| **Temperature** | | | | |
| degree Fahrenheit | °F | 0.555(°F − 32) | °C | degree Celsius (centigrade) |
| degree Fahrenheit | °F | 0.555 (°F + 459.67) | °K | degree Kelvin |
| **Velocity** | | | | |
| feet per second | ft/s | 0.3048 | m/sec | meters per second |
| mile per hour | mi/h | 4.4704 × 10$^{-1}$ | km/sec | kilometer per second |
| **Volume** | | | | |
| acre-foot | acre-ft | 1.2335 × 10$^3$ | m$^3$ | cubic meter |
| cubic foot | ft$^3$ | 28.3168 | L | liter |
| cubic foot | ft$^3$ | 2.8317 × 10$^{-2}$ | m$^3$ | cubic meter |
| cubic inch | in.$^3$ | 16.3871 | cm$^3$ | cubic centimeter |
| cubic yard | yd$^3$ | 0.7646 | m$^3$ | cubic meter |
| gallon | gal | 3.7854 × 10$^{-3}$ | m$^3$ | cubic meter |
| gallon | gal | 3.7854 | L | liter |
| ounce (U.S. fluid) | oz (U.S. fluid) | 2.9573 × 10$^{-2}$ | L | liter |

# Index

## A

AAA. *See* Agricultural Adjustment Agency
*Acer saccharum*, 16
Agricultural Adjustment Agency, 47, 51
Agricultural Stabilization and Conservation Service, 47
Agriculture, historical, 26–33
    conservation effects, 65–68
    crops, 33–36
    European settlement, 26–33
    high terraces, banks, stream erosion, 80–84
    hydrologic change, Hill Country, 57–64
    landscape effects, 55–84
    lower main valley, 75–80
    modified hydrology, 55–57
    sediment budgets, 68–70
    soil erosion, Hill Country, 57–64
    tributaries, 70–74
    upper main valley, 74–75
    zones of physical processes, 70–80
Alfisol, 222
Alluvial fan, 90–91, 104, 221
Alluvium, 5, 20, 60, 144, 166, 221, 226
Anaerobic decomposition, 222
Animal husbandry, settlers, 37–41
Appleby Farm, 86, 89
Aquifers, 5, 222, 225, 228
Aquitard, 222
Arcadia, Wisconsin, 161–162
Arroyo, 100, 222
ArtDeco-style hydroelectric power station, 161
Artifacts, cultural, 18
ASCS. *See* Agricultural Stabilization and Conservation Service

Available plant water, 222

## B

Bank storage, 222
Bankfull discharge, 72–73, 222
Base flow, 222
Basswood, 16–17
Beaver Falls, Minnesota, 147–149
Bed load, 222
Beech growth, 16
Best management practice, 222, 229
Black River, 28, 62, 98
BMP. *See* Best management practice
Braided river, 222
Bridges, 118–122, 169–175
Brook trout, 20–21
Brown forest, 17
Buffalo grazing, 15, 32–33, 59–60, 62, 73, 85, 98
Buffalo River terrace, gully erosion, 98
Bulk density, 222
Bur oak, 16
Bushhogging, 16

## C

Cambrian strata, 5
Capacity of streams, 223
Catchment, 58, 223, 231
CCC. *See* Civilian Conservation Corps
Changes of land use, 23–54
    agriculture land management revolution, 44–47
    animal husbandry, 37–41
    conservation agencies, 47–50
    crops, historical, 33–36

237

early settlement, 24–26
erosive land use, 52–54
grazing, 37–41
historical agriculture, 26–33
land use management, 41–44
soil conservation, 51–52
Channels
capacity, 223
evolution, 223
improvement, 223
stabilization, 223
Chaseburg, Wisconsin, 142–147
Chippewa River, 6, 19, 81
Civilian Conservation Corps, 49
Clay, 18, 21, 222–223, 230–231
Climate changes, 10, 13, 208
negatively correlated with changes in landscape, 13–14
Coefficient of variation, 223
Colluvium, 4–5, 38, 55–56, 66, 68, 223
Communication lines, 118–122, 169–175
Connectivity, 39, 65, 67, 223
Conservation
agencies, 47–50
tillage, 51, 179, 223
Conservation Reserve Program, 16
Construction, 8, 25, 29, 145, 151, 156, 179, 229
Contour plowing, 23, 45, 223
Contour strips, 23, 45–47, 51, 86, 181, 186, 223
Conventional agriculture, 181, 223
Conventional tillage, 223
Cooke, Sir Ron, 100, 222
Coon Creek, 17, 43–47, 49, 51, 57–59, 61, 65–73, 75–76, 79–84, 90–92, 94, 98–100, 102, 104, 118, 120, 124–125, 131, 133–136, 138–139, 141–144, 148–149, 169–171, 173–179, 181, 183, 187–188, 195, 197–200, 202–204
Coon Valley, 131–134
Corn, 27, 31–32, 34–36, 39, 41–43, 45, 51–52, 136, 153, 161, 181, 184, 208, 222, 229
Cover crops, 223
Crop residue, 223
Crop rotation, 45–46, 224, 227
Crop set-aside programs, 16
Crops, historical, 33–36
CRP. *See* Conservation Reserve Program
Crusting, 224
Cultivation decline, wheat, 34
Cultivation of wheat, decline in, 34
Cultural artifacts, 18
Curve, discharge, 224
Cut bank, 72, 81, 116, 136, 139, 224, 227

D

Dairy Belt, 37
Debris flow, 224
Declination, magnetic, 227
Deep percolation water, 224

Degradation of stream bed, 224
Dendrochronology, 3, 224
Density of grazing animals, 40
Discharge curve, 224
Dolomite, 4–5
Dominant channel materials, 224
Drainage, 3, 21, 28, 40, 57–59, 73, 94, 100, 124, 167, 182–183, 224, 226, 231
Drainage basin, 124, 224

E

Early grazing practices, 37–41
Early settlement, 24–26
Effective discharge, 222, 224, 227
Elba, Minnesota, 129–131
Elk grazing, 15, 153, 160
Elkport, Iowa, 153–157
Enlightenment concepts, 23
Ephemeral gully, 224
Ephemeral stream, 222, 224
European settlement, 19–21, 23–54
agriculture land management revolution, 44–47
animal husbandry, 37–41
conservation agencies, 47–50
crops, historical, 33–36
early settlement, 24–26
erosive land use, 52–54
grazing, 37–41
historical agriculture, 26–33
land use management, 41–44
soil conservation, 51–52
Evaporation, 224, 226
Evapotranspiration, 224
Exceedence probability, 224

F

*Fagus grandifolia. See* Beech
Fairwater, Minnesota, 109–112
Fan, alluvial, 90–91, 104, 221
Farming contour-strips, 23, 45–47, 51, 86, 181, 186, 223
Federal Resettlement Agency, 49
Field capacity, 225
Fish habitat, 125–126
Flood series, 225
Flood spillway, 225
Flooding, 177–205
efflux, 195–197
gullies, 182–186
lower main valley, 188–195
mass movements, 179–181
sediment yield, 195–197
storms, 177–179
tributaries, 186–188
upland slopes, 181–182
upper main valley, 188
Floodplain, 56, 131, 139, 155, 168, 170, 222, 225
storage, 222, 225

# Index

Flow rates, 223, 225, 234
Flume, 92, 189, 225
Forest
    grazed, 39, 41, 52
    ungrazed, 52
Forestry, 211
Franconia shale, 5
Freeburg, Minnesota, 134
Fremling, Calvin, 8, 15, 19
Frequency-magnitude relationships, precipitation, 9
Frequency of storm events, 225
Friability, 225

## G

Gaging stations, 58, 225
Gaining stream, 225
Galena, Illinois, 151–153
Garber, Iowa, 153–157
Garland, Hamlin, 6, 14, 27–28, 33, 41, 116
Gays Mills, Wisconsin, 149–151
Geological erosion, 225
Geomorphologists, 8, 171, 227
Geomorphology, 9, 57, 59, 68, 71, 225
Glacial outwash, 6, 81, 225
Glaciation, 5
Goat prairies, 16
Golf course development, 175
Goodhue County, 31–34
Gorges, 8
Graded streams, 225, 230
Grains, 31–32, 35, 42–43, 45, 52, 230
Grassed waterway, 225
Grazed forest, 39, 41, 52
Groundwater, 39, 44, 56, 73, 129, 141, 161–162, 165, 222, 225–226
Groundwater flow, 225
Growing seasons, 51, 166, 225
Gulf of Mexico, 8
Gully erosion, 39, 59, 64, 85, 87, 89, 91, 93, 95, 97, 99, 101, 103, 105, 107, 225
    Black River terrace, 98
    Buffalo River terrace, 98
    high terrace, 98
    Hillside gullies, fans, 86
    Peterson event, 107
    Proksch Coulee, 98–101
    Ratz gully, 101
    upland, 85–108
    Zink farm, 86–98
Gully head advance, 225
Guttenberg, 25, 33

## H

Happ, Stafford C., 18, 56, 64–68, 71–72, 76, 78, 80, 85–86, 94, 99, 102, 104, 129, 149, 151, 165, 169, 191, 196
Haugen, John, 48

Hershfield, David, 11
Hillside gullies, fans, erosion, 86
Historical agriculture, 26–33
    conservation effects, 65–68
    crops, 33–36
    European settlement, 26–33
    high terraces, banks, stream erosion, 80–84
    hydrologic change, Hill Country, 57–64
    landscape effects, 55–84
    lower main valley, 75–80
    modified hydrology, 55–57
    sediment budgets, 68–70
    soil erosion, Hill Country, 57–64
    tributaries, 70–74
    upper main valley, 74–75
    zones of physical processes, 70–80
*Historical Atlas of the State of Minnesota,* 161, 164, 166
Hogging off, 41
Humid climates, 226
Humidity, 228
Husbandry, animal, of settlers, 37–41
Hydraulic length, 226
Hydraulic resistance, 204, 226
Hydrologic cycles, 226, 228
Hydrologic groups, 10, 179, 226
Hydrologists, 226
Hysteresis, 67, 226

## I

Infiltration capacity, soils, 226
    *vs.* average rainfall rates, 11
Influent stream, 226
Interflow, 226
Interfluve, 226
Intermittent stream, 226
Interrill erosion, 226, 229
Ion, Iowa, 157–160

## J

Jefferson, Thomas, 23, 45
Juniper growth, 16
*Juniperus virginiana.* See Juniper growth

## K

Keokuk, Iowa, 10, 29
Kickapoo River, 2, 67, 149, 209
Kinetic energy, 1, 10

## L

La Crosse, 6, 18, 25, 28, 33, 45, 47, 69, 120, 170, 178, 196–197
Lacustrine, 18, 226
Lag time, 226
Land use management, 41–44
Lansing, 18, 25, 162

Lateral point bar accretion, 226–227
Legacy sediment, 162, 176, 195, 204, 209, 226
Levees, 80, 116, 226
Lightning, 14
LMV. *See* Lower main valley
Local relief, Hill Country, 6
Long-term grass cover, prairie soils, 17
Lorentz, Pare, 40, 50
Lower main valleys, 141–176
    Arcadia, Wisconsin, 161–162
    Beaver Falls, Minnesota, 147–149
    bridges, 169–175
    Chaseburg, Wisconsin, 142–147
    communication lines, 169–175
    Elkport, Iowa, 153–157
    farms, 162–169
    farmsteads, 162–169
    Galena, Illinois, 151–153
    Garber, Iowa, 153–157
    Gays Mills, Wisconsin, 149–151
    Ion, Iowa, 157–160
    Potosi, Wisconsin, 151–153
    reservoirs, 175–176
    roads, 169–175
    Rushford, Minnesota, 160–161
    Soldiers Grove, Wisconsin, 149–151
    towns, 141–162
    Village Creek, Iowa, 157
    villages, 141–162
    Whitewater Falls, Minnesota, 147–149
    zone of perennial sedimentation, 141

## M

Magnetic declination, 227
Magnetude of storm events, 225
Management-weighting factors, historic erosive land use estimation, 52
Marshes, 8, 18
Mass movement, 86, 179–181, 224, 227
Mills, 122–125
Mining, 24–25, 57, 59, 152
Mollisols, 17–18, 51, 106, 154, 227

## N

Native Americans, 2, 14, 16, 18–19
Natural Resources Conservation Service, 47, 227
Nonpoint source pollution, 227
NPS. *See* Nonpoint source pollution
NRCS. *See* Natural Resources Conservation Service

## O

Occluded channels, 227
Older alluvium, radiocarbon dates, 20
Overbank vertical accretion, 227
Overland flow, 227

## P

Particle size analysis, 227
Pedological principles, 18
Percolating water, 227
Percolation, 224, 227
Perennial stream, 227
Permeability, 228, 230
Peterson event, gully erosion, 107
Physiographic diagram, Hill Country, 7
Physiography, 5–8
Pleistocene loess, 5
Pleistocene terraces, untrenched drainageways, 99
Plow pan, 226, 228
*The Plow That Broke Plains*, 49–50
Plowing, contour, 23, 45, 223
Point bar, 131, 226–228
Porosity, 222, 228, 230
Potosi, Wisconsin, 151–153
Poverty, 57
Power, water, 122–125
Prairie du Chien, 5, 8, 18–19, 24–25, 33, 47, 113
Precipitation, 8–9, 11–13, 20, 66, 116, 178, 181, 186, 222–224, 226–229, 231
    depth, 228
    duration, 228
    frequency-magnitude relationships, 9
    rate, 222
    regression line, 231
    trends, Hill Country, 12–13
Primeval landscape, 1–22
    brook trout, 20–21
    climate, 8–13
    floodplain soils, 19–20
    physiography, 5–8
    radiocarbon dates, older alluvium, 20
    soils, 14–19
    streams, 19–21
    vegetation, 14–19
Proksch Coulee, gully erosion, 98–101
Proksch Trench, alluvial fan from, 104
Public roads, 50, 61, 101, 111, 163, 167, 174
Puget Sound, 9–10

## Q

*Quercus macrocarpa*. *See* Bur oak

## R

Radiocarbon dates, older alluvium, 20
Railroads, 31–33, 50, 61, 70, 85, 94–95, 97, 109, 118–122, 125, 141, 152, 159, 163, 169–171, 175, 195, 204
Raindrop erosion, 228
Rainfall, 9–11, 20, 35, 52, 55, 57, 65, 83, 92, 177, 179, 181, 183, 186, 188, 197–198, 207, 209, 224–229
    erosivity, 228
    frequency, 228
    intensity, 228

# Index

Rainfall rates, *vs.* infiltration capacities, soil, 11
Ratz gully erosion, 101
Recession curve, 228
Recharge areas, 38–39, 56, 73, 228
Recurrence interval, 228
Red Wing, 21, 25, 33
Regionalization, Hill Country, 26
Regression line, precipitation, 231
Relative humidity, 228
Reservoirs, 65, 100, 122–125, 175–176, 190, 228–229
Residue, crop, 223
Resistance to forces, 13–14
Return periods, 11, 228
Revolution in agriculture land management, 44–47
Rhine River, 8
Riffle-pool sequence, 228
Rill erosion, 56–57, 59, 65–70, 181, 183, 227, 229
Rip-rap, 74, 192, 229
Riparian waterways, 9, 17, 74, 76, 80–81, 83, 115, 188, 229
Roads, 118–122, 169–175
    public, 50, 61, 101, 111, 163, 167, 174
Rolling uplands, 3, 6, 14, 38
Root River floodplain, 161
Root zone, 224, 229
Rotation, crops, 45–46, 224, 227
Row crops, 31–32, 34–35, 42, 46, 52, 229
Row grade, 229
Runoff, 10, 37, 39, 43–44, 50, 56–60, 66, 70, 72, 75, 85, 100, 117, 177, 179, 182–183, 224–227, 229, 231
Rushford, Minnesota, 160–161

## S

Sediment basin, 193, 229
Sediment budge, 229
Sediment load, 132, 136, 209, 226, 229, 231
Sediment rating curve, 229
Sediment transport, 19, 223, 229
Sediment yield, 65–66, 68–70, 75, 109, 176, 195, 224, 229
Set-aside programs, crop, 16
Settlement, early, 24–26
Sheet erosion, 60, 181, 229
Sheet flow, 229
Sheet metal flume, 63
Silt, 3, 5, 18, 72, 78, 136, 186, 229–231
Silt bar, 229
Sinuosity, 230
Size analysis, 227
Small grains, 31–32, 35, 42–43, 45, 52, 230
Sod busting, 17
Soil aeration, 230
Soil and water conservation district, 230
Soil compaction, 230
Soil Conservation Service, 47
Soil map unit, 230
Soil series, 17, 230
Soil structure, 34–35, 55, 230
Soil texture, 230
Soil water, 38, 56, 227, 230
Soldiers Grove, Wisconsin, 149–151
Source pollution, 227
Soybeans, 27, 31–32, 35–36, 136, 181, 184, 208, 229
Spillway, 183, 225, 230
Splash erosion, 230
St. Peter sandstone, 4–5
Steady state streams, 230
Steamboat, 25, 33, 151–152
Storms, 9, 11–13, 20, 44–45, 57, 59–60, 73, 75, 85–86, 90, 98, 113, 125, 147–148, 177–181, 184–188, 191, 193, 195, 197, 204–205, 208, 225, 228–229
Stratified soils, 230
Stream bank stabilization, 230
Stream channel erosion, 65, 70, 224, 229, 231
Stream instability, 230
Stream stability, 230
Structure, soil, 34–35, 55, 230
Subsurface drain, 231
Sugar maple, 16
Surface drainage, 231
Surface runoff, 70, 225, 227, 229, 231
Surface sealing, 231
Surface water, 94, 162, 229, 231
Suspended load, 231
Suspended sediment, 20, 231
Swamps, 8, 18
SWCD. *See* Soil and water conservation district

## T

Thalweg, 192, 231
Thomson, Virgil, 50
Throughflow, 231
Thunderstorms, 9
*Tilia Americana. See* Basswood
Transpiration, 223, 231
Trempealeau River, 5, 165
Trend lines of precipitation, 13, 231
Tributaries, 109–128
    bridges, 118–122
    changes, 109–128
    communication lines, 118–122
    farms, 113–118
    farmsteads, 113–118
    fish habitat, 125–126
    mills, 122–125
    railroads, 118–122
    reservoirs, 122–125
    roads, 118–122
    villages, 109–113, 129–135
    water power, 122–125
Trout, presence of, 20–21
Turbidity, 21, 231

## U

UMV. *See* Upper main valley
Underflow, 231
Ungrazed forest, 52
Untrenched drainageways, Pleistocene terraces, 99
Upland gully erosion, 85–108
    Black River terrace, 98
    Buffalo River terrace, 98
    high terrace gullies, fans, 98
    Hillside gullies, fans, 86
    Peterson event, 107
    Proksch Coulee, 98–101
    Ratz gully, 101
    Zink farm, 86–98
Upper main valleys, 129–140
    changes, 129–140
    Coon Valley, Wisconsin, 131–134
    Elba, Minnesota, 129–131
    farms, 135–140
    farmsteads, 135–140
    Freeburg, Minnesota, 134
U.S. Rectangular Survey System, 24

## V

Variation coefficients, 223
Vertical accretion, overbank, 227
Village Creek, Iowa, 157
Violent storms, 9

## W

Wabasha, 18–19, 25, 33, 88, 96–97, 124
Wash load, 231
Water power, 122–125
Water table, 148, 231
Watershed, 20, 68, 84–85, 94, 116, 120, 136, 169, 181, 197, 204, 223–224, 226, 229, 231
Watershed gradient, 231
Weighting factors, management, historic erosive land use estimation, 52
Western Europe, 9–10
Wetlands, 6, 209, 231
Wheat, 14, 20, 26–27, 31–35, 208, 230
White-tailed deer, 15
Whitewater Falls, Minnesota, 147–149
Winona, 18–20, 24–25, 33, 47, 59, 61, 68, 74–77, 86, 89, 94, 105–106, 131, 148–149, 158, 167, 184–185

## X

Xeric edaphic conditions, 16, 111
Xeric surface, 231

## Z

*Zea mays. See* Corn
Zebulon Pike, 15
Zink, Gene, 86, 90–94, 118, 169, 183, 187–189
Zink farm, 86–98, 118, 169, 183, 187–189
Zumbro River, 17, 21, 68, 120, 124, 169

A partially buried house in Beaver, Minnesota, 28 Nov. 1939. See Figure 7.10 on page 159.